XVI Meeting of Physics 2017

Journal of Physics: Conference Series
Volume 1143

Lima, Peru
2 - 4 August 2017

Editors:

Rosendo Ochoa
Ener Salinas
Harold Blas

ISBN: 978-1-5108-8024-5
ISSN: 1742-6588

Printed from e-media with permission by:

Curran Associates, Inc.
57 Morehouse Lane
Red Hook, NY 12571

Some format issues inherent in the e-media version may also appear in this print version.

Printed by Curran Associates, Inc. (2019)

For permission requests, please contact the Institute of Physics
at the address below.

Institute of Physics
Dirac House, Temple Back
Bristol BS1 6BE UK

Phone: 44 1 17 929 7481
Fax: 44 1 17 920 0979

techtracking@iop.org

Additional copies of this publication are available from:

Curran Associates, Inc.
57 Morehouse Lane
Red Hook, NY 12571 USA
Phone: 845-758-0400
Fax: 845-758-2633
Email: curran@proceedings.com
Web: www.proceedings.com

Table of contents

Volume 1143

XVI Meeting of Physics

2–4 August 2017, Lima, Peru

Accepted papers received: 12 November 2018
Published online: 21 December 2018

Preface

Papers

IOP Conf. Series: Journal of Physics: Conf. Series **1143** (2018) 011001 doi:10.1088/1742-6596/1143/1/011001

UNIVERSIDAD NACIONAL DE INGENIERÍA (UNI)
NATIONAL UNIVERSITY OF ENGINEERING, FACULTY OF SCIENCES, LIMA, PERU

XVI MEETING OF PHYSICS

2nd to 4th August, 2017

In memoriam of Prof. Holger Valqui

Editors:

R. Ochoa, E. Salinas and H. Blas

Preface

It is a great pleasure for the editorial committee to introduce the Proceedings of the XVI Meeting of Physics. This event is an initiative of a team of Peruvian scientists –local and living abroad– in conjunction with various renowned international guests that, since 1999, year after year gather to present and share the results of their latest investigations in physics and related areas. A major objective of these meetings is to provide students with the superb opportunity to closely interact with experts. Moreover, students are also invited and encouraged by their supervisors to present their work in poster sessions and then, within a reasonable time, write an improved version of their results in the form of a scientific article, in such a way that those papers can pass a peer review and become published in proceedings such as this, along with the works of keynote speakers and senior researchers. For some students this is the first time to publish. For this reason, the editorial committee and their supervisors had especial patience and interacted several times with them until their articles reached a publishable level.

IOP Conf. Series: Journal of Physics: Conf. Series **1143** (2018) 011001 doi:10.1088/1742-6596/1143/1/011001

This XVI edition of the Meeting of Physics is an especial one as we honor the memory of our mentor, teacher and friend, Prof. Holger Valqui, who passed away in Nov 2016. Prof. Valqui was not only a rigorous mathematical physicist but also a highly intuitive scientist, who contributed to popularize classical and new scientific ideas among students and general public. He also thought us the value of skepticism as key part of scientific validation and discovery. Another quality of Prof. Valqui was his sense of humor, which was one more motivation for us to deal with science.

Prof. Valqui got two first degrees, one in Electrical Engineering and another in Mathematics from UNI (1954) and UNMSM (1956) respectively. He continued his studies in the Heidelberg University, Germany, graduating as "Diplom-Physiker" in 1962, having as supervisor Prof. J. Hans D. Jensen who in 1963 shared the Nobel Prize with Maria G. Mayer "*For their discoveries concerning nuclear shell structure*". Since 1962 when he joined the Faculty of Science of the "Universidad Nacional de Ingeniería" (UNI) and for 54 years until he passed away Prof. Valqui was an extraordinary teacher and researcher, participating in every possible activity related to science. Conferences, seminars, science fairs (e.g. EXPOCIENCIA), students thesis supervision, lectures, etc. In fact, he was the initiator of these Meetings of Physics, which started as a set of seminars in theoretical physics then it was expanded to become a local/international event.

At the moment, several former students of Prof. Valqui occupy relevant positions in various research organizations in Peru and around the world. Some of us would like to dedicate our time and efforts to invite our international colleagues and friends to this event and in that way continue his legacy.

We would like to thank the Dean of the Faculty of Sciences of UNI, Prof. Orlando Pereyra, for the procedures that lead to the organization of this event, to our colleagues in Peru and abroad that contributed with its realization, to the IOP Publishing members for their patience while waiting for the final versions of the papers and to every author who has contributed with an article in these proceedings.

Rosendo Ochoa
Ener Salinas
Harold Blas

Lima, November 13, 2018

Prof. Holger Valqui
1931-2016

IOP Conf. Series: Journal of Physics: Conf. Series **1143** (2018) 011002 doi:10.1088/1742-6596/1143/1/011002

Pictures of the XVI Meeting of Physics

IOP Conf. Series: Journal of Physics: Conf. Series **1143** (2018) 011002 doi:10.1088/1742-6596/1143/1/011002

IOP Conf. Series: Journal of Physics: Conf. Series **1143** (2018) 011003 doi:10.1088/1742-6596/1143/1/011003

Poster of the event

IOP Conf. Series: Journal of Physics: Conf. Series **1143** (2018) 011004 doi:10.1088/1742-6596/1143/1/011004

Peer Review Statement

All papers published in this volume of *Journal of Physics: Conference series* have been peer reviewed through processes administered by the proceedings Editors. Reviews were conducted by expert referees to the professional and scientific standards expected of a proceedings journal by IOP Publishing.

IOP Conf. Series: Journal of Physics: Conf. Series **1143** (2018) 012001 doi:10.1088/1742-6596/1143/1/012001

The Holographic Correspondence : Origins

Sumit R. Das

Department of Physics and Astronomy, University of Kentucky, Lexington, KY 40506, U.S.A.

E-mail: das@pa.uky.edu

Abstract. Over the past several decades, developments in String Theory has led to a radically different description of gravity. In this description, gravity emerges as an approximate description of a quantum field theory in lower number of dimensions. The latter quantum theory is similar to those used to describe the other fundamental interactions, defined on a non-dynamical space-time. This talk traces the conceptual basis of such a description at a fairly non-technical level.

1. Introduction

It is a remarkable fact that we are now able to trace most forces of nature to several fundamental interactions and we understand how these interactions work to a very large extent. Three of these forces are very similar. Electromagetism, Weak Interactions and Strong Interactions are described very well by a kind of quantum field theory called "gauge theory". They can be thought of being mediated by particles which are the "force carriers" : photons for electromagnetism, W and Z bosons for weak interactions and gluons for strong interactions. These particles propagate in space-time which is *fixed* - the space-time is non-dynamical.

The fourth force is gravity. The nature of this force appears to be quite different from the others. In fact as Einstein showed - in a sense gravity is not a force of the usual type. Rather it is a manifestation of a curvature of space-time itself. Objects move on such a curved space-time as usual along the path of least action - but this path (a geodesic) appears to be curved, so we can alternatively describe this due to the action of a force. Gravity is then the dynamics of space-time itself - ripples of space-time are gravitational waves. This key difference could be at the root of our troubles in trying to make gravity consistent with quantum mechanics.

Finally, there are two huge elephants in the room : Dark Matter and Dark Energy - these account for 96 % of the energy of the known universe, and we know almost nothing about these.

This talk is about a modern point of view in which gravity is not really a fundamental force. Rather it is a kind of approximate description of something more fundamental. This fundamental theory is a quantum field theory defined on a fixed space-time background and lives in a *lower* number of dimensions - as if it were a hologram of the bulk. This is called the "Holographic Correspondence": at this time there are several concrete realizations of this idea. It has often been conjectured that at very short distances the notions of space and time break down and need to be replaced by some other mathematical structures. The holographic correspondence provides a way to find what this structure could be.

IOP Conf. Series: Journal of Physics: Conf. Series **1143** (2018) 012001 doi:10.1088/1742-6596/1143/1/012001

2. Strings in QCD

The origins of this remarkable correspondence can be traced to a rather different area of physics : the theory of strong interactions. Before the correct theory of strong interactions - quantum chromodynamics (QCD) - was found, it was realized that hadrons like mesons and baryons can be considered as quarks held together with *strings*. With the advent of QCD it became clear that these strings are really chromoelectric flux tubes. However, there was a puzzle. Consider for example a usual quantum theory of point particles. These particles are not really point-like : rather vacuum polarization effects render them a finite size. Nevertheless in theories like quantum electrodynamics we can consider e.g. an electron as a point particle since these vacuum polarization effects are proportional to powers of the dimensionless coupling, the fine structure constant $\alpha = e^2/(\hbar c)$. In QED the fine structure constant is a small number - so these effects are small. Indeed in the limit of an infinitesimally small α an electron can be genuinely considered as a point particle. Thus, if QCD can be described in terms of strings, there should be a small number in QCD which controls the interaction between strings. In the limit where this number vanishes, these strings would be infinitely thin and non-interacting. The puzzle was that there is no dimensionless number in QCD : if we for the moment ignore the quarks, QCD becomes pure Yang Mills theory with a gauge group $SU(3)$. This theory has no free parameter - it only has a dynamically generated scale. So it appears that there should be no useful way to consider this as a theory of strings !

2.1. The Large N limit

In a landmark work, 't Hooft solved this puzzle by considering a generalization of QCD in which the gauge group is $SU(N)$, and showed that in this theory the efffective strings interact with a strength $1/N$. Therefore in the $N = \infty$ these strings are free and thin.

More precisely, starting with the usual Yang-Mills action

$$S = \frac{1}{4g_{YM}^2} \int d^4x \mathrm{Tr} F_{\mu\nu} F^{\mu\nu} \tag{1}$$

where $F_{\mu\nu} = \partial_\mu A_\nu - \partial_\nu A_\mu + i[A_\mu, A_\nu]$ is the field strength of the $SU(N)$ gauge field $A_\mu = (A_\mu)_{ij}, i, j = 1 \cdots N$ and g_{YM} is the coupling. The large N limit is defined as

$$N \to \infty \qquad g_{YM} \to 0 \qquad \lambda \equiv g_{YM}^2 N = \text{fixed} \tag{2}$$

To see how $1/N$ can appear as a coupling constant of the effective theory of invariants, let us consider the problem of quantum mechanics of a particle moving in a central potential in N space dimensions. In cartesian coordinates $x_i, i = 1 \cdots N$ the hamiltonian is

$$H = \frac{1}{2} \sum_i -\frac{\partial^2}{\partial x_i^2} + V(|x|) \tag{3}$$

This problem has a $O(N)$ symmetry of rotations among the x^i. Suppose we are interested only in the *singlet* wave functions. It is clearly convenient to go to polar coordinates $(r, \theta_1, \cdots \theta_{N-2}, \phi)$ and rescale r to have

$$r^2 = \frac{1}{N} \sum_i (x^i)^2 \tag{4}$$

The singlet wave function $\Psi(r)$ is only a function of r. Generalizing a standard procedure in 3 dimensions we define a new wavefunction $\chi(r)$

$$\Psi(r) = r^{-(N-1)/2} \chi(r) \tag{5}$$

IOP Conf. Series: Journal of Physics: Conf. Series **1143** (2018) 012001 doi:10.1088/1742-6596/1143/1/012001

which now satisfies

$$\left[-\frac{1}{2N^2}\frac{d^2}{dr^2}+\frac{(N-1)(N-3)}{8N^2r^2}+V(r)\right]\chi(r)=\frac{E}{N}\chi(r) \tag{6}$$

This is exactly like the eigenvalue equation of a one dimensional Schrodinger problem. Something interesting happens in the large N limit. The one dimensional problem now has a N independent effective potential

$$V_{eff}(r)=\frac{1}{8r}+V(r) \tag{7}$$

The only place where N appears is the kinetic term of the radial hamiltonian, where it plays the role of $1/\hbar$. Therefore $1/N$ plays the role of a coupling in this effective theory of radial variables. The $N \to \infty$ limit is the *classical limit.* In fact the spectrum of this theory in this limit can be obtained from a classical calculation based on the potential (7).

This general pattern continues to work for any large-N quantum field theory : $N = \infty$ is a classical limit, and this can be seen by re-writing the theory in terms of appropriate invariant variables. The problem is that it is not possible to perform this change of variables for most interesting theories, like QCD. For pure $SU(N)$ Yang-Mills theories, the appropriate gauge invariant variables are Wilson loops

$$W(C)=\mathrm{Tr}P\exp[i\int_C A\cdot dl] \tag{8}$$

the integral is over the closed loop C. At large N it is easy to see that these factorize, i.e.

$$< W(C_1)W(C_2) >=< W(C_1) >< W(C_2) > +O(1/N^2) \tag{9}$$

which is characteristic of a classical limit.

2.2. Matrix Quantum Mechanics

In the early 1980's the problem of obtaining an effective action for Wilson loops was a popular activity, and people started to do "warm up problems". One of them later turned out to be of great importance in the development of the ideas behind the holographic correspondence.

The basic fields of $SU(N)$ Yang-Mills theory are matrix valued fields $(A_\mu)(\vec{x},t)_{ij}$ which are functions of the spatial coordinates $\vec{x}$ and time t. Consider a toy model obtained from this by forgetting about spatial dependence and restricting to only one Lorentz component. The dynamical variables of this toy model would be a single matrix $M_{ij}(t)$ - i.e. quantum mechanics of a single hermitian matrix. The $U(N)$ invariant hamiltonian would be of the form

$$H=-\sum_{ij}\frac{\partial^2}{\partial M_{ij}\partial M_{ij}}+V(M) \tag{10}$$

where the potential $V(M)$ is invariant, e.g. containing sum of terms like $\mathrm{Tr}(M^n)$. We want to study the dynamics of this theory in the singlet sector. Just as in the single particle quantum mechanics problem we therefore need to go over to invariant variables. In this case the invariant variable is the density of eigenvalues

$$\rho(x)\equiv\frac{1}{N}\sum_{i=1}^{N}\delta(\lambda_i-x) \tag{11}$$

This is the analog of the radial variable in the single particle quantum mechanics problem discussed in the previous subsection. When N is large $\rho(x)$ is a smooth function restricted by the condition $\int dx\rho(x) = 1$. In the singlet sector quantum mechanics of a $N \times N$ matrix, $M_{ij}(t)$ can be described by the quantum dynamics of $\rho(x,t)$. It turns out that one can perform this change of variables to ρ and one can write down the hamiltonian for ρ and its canonically conjugate momentum π_ρ. In this hamiltonian $1/N^2$ appears as a Planck constant, just as $1/N$ appeared as the Planck constant for the single particle quantum mechanics problem. Therefore in the large N expansion one can treat the theory in a semiclassical fashion. One finds the classical solution of the equation of motion, and treats the fluctuation around the classical solution quantum mechanically - just as one would perform a WKB expansion in usual quantum mechanics.

Motivated by developments in String Theory, this problem was revisited in the early 1990's. The result was rather surprising : the theory of the fluctuations of ρ turned out to be a quantum field theory in $1+1$ dimensions, with x playing the role of a space dimension. There is a completely equivalent description of the quantum mechanics of a single matrix with N^2 degrees of freedom - a $0+1$ dimensional theory in terms of a single field in $1+1$ dimensions. The matrix quantum mechanics of course did not have any "space". The theory of the density of eigenvalues had a "space" - this came from the mathematical space of eigenvalues, but rather surprisingly this behaves as a real space for all purposes. In fact this $1+1$ dimensional theory is the field theory of strings in $1+1$ dimensions, whose only propagating mode is in fact a massless scalar - this is now identified with $\rho(x,t)$.

String Theory is an ultraviolet completion of General Relativity. The quantized quadrapole moment of an oscillating closed string is a graviton. However, the normal modes of $1+1$ dimensional string theory does not have dynamical gravity in the sense that there are no gravitational waves in these low dimensions. However there are static gravitational forces - which in this theory have a finite range. It turns out that these gravitational forces can be deciphered from the theory of eigenvalues.

This is the holographic correspondence in its simplest form : a theory with N^2 degrees of freedom with no "space" and just time is equivalent to a theory in $1+1$ dimension and the latter theory does contain gravity. In this case, the latter theory can be explicitly constructed starting from the matrix model. One might hope that there is a similar story in higher dimensions : however in higher dimensions we have an infinite number of matrices (one matrix at every point) which do not commute - so they cannot be simultaneously digaonalized and there is no simple way to rewrite the theory in terms of invariant variables. In fact the invariant variables are Wilson loops - these are stringlike objects - and that is where we started the story anyway. Therefore if we could derive an effective theory of Wilson loops, one would get a theory of strings in higher number of dimensions - and if we start with the right large-N theory this string theory will contain gravity. That would be a description of gravity in terms of a conventional quantum field theory. The lesson of the lower dimensional example is that this theory of gravity necessarily lives in higher number of dimensions.

Despite many attempts, writing down a field theory of Wilson loops starting from Yang-Mills theory turned out to be impossible. Progress came in the mid 1990's from a rather different direction - the physics of black holes.

3. Black Holes and Strings

In classical General Relativity a black hole is an object which is surrounded by a *horizon*. Nothing can come out of the horizon. In fact once inside the horizon, any object will inevitably fall into the singularity characterized by large space-time curvatures. An important aspect of the horizon which will be important in the following is that any signal emitted from a point near the horizon and travelling to the asymptotic region far away suffers a huge redshift. As the emission point approaches the horizon, this redshift becomes infinite. This is also the reason why an infalling object appears to an asymptotic observer to slow down infinitely fast as it

IOP Conf. Series: Journal of Physics: Conf. Series **1143** (2018) 012001 doi:10.1088/1742-6596/1143/1/012001

approaches the horizon.

While the horizon is certainly a point of no return, the local curvature and tidal forces near the horizon for a very massive black hole are small. This prompted Hawking to apply standard quantum field theory methods to processes near the horizon. The result was surprising : due to particle creation, a black hole was found to emit radiation, which is thermal to a very good approximation. This tied up with earlier observations due to Bekenstein that black holes should be associated with an entropy and a temperature.

The entropy and temperature of every black hole have universal geometric expressions. The entropy (called the Bekenstein-Hawking entropy) is given by

$$S_{BH} = \frac{A_H}{4G_N} \tag{12}$$

in natural units, where A_H is the area of the horizon and G_N is the Newton's constant. This relationship holds for all black holes in all dimensions. The temperature (as e.g. obtained from the properties of Hawking radiation) is given by

$$T_H = \frac{\kappa}{2\pi} \tag{13}$$

where κ is the surface gravity at the horizon. Finally the decay rate of the black hole into some scalar with energy ω is given by

$$\Gamma(\omega) = \frac{\sigma(\omega)}{e^{\frac{\omega}{T_H}-1}} \tag{14}$$

where $\sigma(\omega)$ is the corresponding classical absorption cross-section by the black hole. For minimally coupled massless scalars, a general theorem ensures that in the limit of small ω one has $\sigma(\omega) \to A_H$.

Bekenstein pushed this even further : he argued that if we consider a theory containing dynamical gravity in a large box, the *maximum* possible entropy of *anything* inside the box is bounded by the surface area of the box - rather than the volume of the box

$$S \leq \frac{A_{boundary}}{4G_N} \tag{15}$$

This is strange, since usually we learn that entropy is extensive and should be proportional to volume. In the early 1990's 't Hooft and Susskind provided a radical interpretation of this fact : they argued that a gravitational theory in a large region of space has a completely equivalent description in terms of a non-gravitational theory which lives on the boundary. Since there is no dynamical gravity in the latter theory, its entropy should be proportional to *its* "volume" - which is of course the boundary area of the gravitational theory. This would provide a natural explanation of Bekenstein's bound. However it took a few more dramatic developments to realize this idea concretely.

3.1. Microscopic origin of Black Hole Thermodynamics and Strings

Usually thermodynamics has a basis in Statistical Mechanics of microscopic constituents. The thermal nature of black holes indicates that there should be a microscopic origin as well. For example, the black hole entropy should be related to the number of states (at a given energy) by Boltzmann formula

$$S = \log(\Omega(E)) \tag{16}$$

However, no hair theorems in General Relativity tell us that for a given mass, charge and angular momentum there is basically one state of a black hole. Therefore in this framework the

IOP Conf. Series: Journal of Physics: Conf. Series **1143** (2018) 012001 doi:10.1088/1742-6596/1143/1/012001

appearance of a large entropy appears puzzling. Some physicists believed that maybe there is no microscopic origin of black hole thermodynamics. This would mean that there is a loss of information at the fundamental level - something which is clearly inconsistent with quantum mechanics. In particular this would mean that time evolution in the presence of gravity and black holes is not unitary. Developments in String Theory led to a rather different direction.

Originally String Theory was described as a theory of infinitely thin strings, just as quantum field theory is thought of as a theory of point particles. However particles are not the only excitations in a typical quantum field theory - there are extended objects like strings, membranes etc. which appear as states. In a similar fashion a String Theory is not just a theory of strings. Rather there are other extended objects of different dimensionalities - like membranes (or 2-branes) which are two dimensional. Consistent string theories live in ten dimensions, six of which are supposed to be curled up into tiny compact spaces - so there are many possibilities (3-branes, 4 branes, $\cdots$ 9-branes). Indeed, there are such objects. One class of these are called D-branes of different dimensionalities. In superstring theories N such D-branes parallel to each other can form a threshold bound state. The remarkable thing about these is that their low energy dynamics is described by a $SU(N)$ gauge theory which lives on the worldvolume of these D-branes.

On the other hand, D-branes have a tension and therefore they should produce gravitational fields. When the number of D-branes are large, this gravitational field is macroscopic, just like the gravitational field of a star. Under suitable circumstances these D-branes will collapse under mutual gravitational attraction and form black holes.

D-branes can therefore provide a microscopic description of black holes, pretty much like a collection of electrons can provide a microscopic description of an object with a large charge. In this latter case, we know in principle how to describe the collection of electrons and use this to describe e.g. the motion of a test charge. However when the number of such electrons become huge, this is not a useful way to approach the problem. Rather it is much better to describe the collection of electrons as a classical charge distribution, use Maxwell's equations to determine the fields produced and then use the fields to study the motion of the test charge. Similarly, if we consider a collection of D-branes there is microscopic description in terms of a non-abelian gauge theory on the worldvolume - and if we knew well how to calculate in this theory we would have known in principle to use this to study other objects near these D-branes. However as the number of these branes become very large, this strategy is senseless. Instead we should describe this by the gravitational field produced - which we should obtain as solutions to the equations of Einstein's General Relativity. The latter is what we would normally call a black hole. The D-branes carry charges for some gauge fields - so these black holes are charged.

Just as we can calculate the entropy of a box of gas by counting the number of states of molecules which make up the gas, we should be able to calculate the entropy of a black hole by counting the number of states of the collection of D-branes. This would provide a microscopic basis for black hole thermodynamics which otherwise looks strange. The problem is that in the regime of parameters in which the collection of D-branes look like a large classical black hole, the Yang-Mills theory describing the dynamics of D-branes is strongly coupled - and in general we do not know how to calculate things in strongly coupled field theories. There are exceptions : these are field theories which have enough supersymmetry - in this case certain quantities can be calculated even in strong coupling. The corresponding black hole is an *extremal* black hole - these are black holes with maximal charge for a given mass. They have zero temperature but nevertheless large horizon area and therefore large Bekenstein-Hawking entropy. In the microscopic description in terms of a gauge theory on the branes, this is a ground state in the given charge sector - but a highly degenerate ground state. It turns out that for a large class of such black holes, the ground state degeneracy can be computed reliably even when the theory on the brane is strongly coupled. The logarithm of the number of states is found to be in precise agreement with the Bekenstein-Hawking entropy. This classic result of Strominger and Vafa which builds on earlier work by Sen was the first indication that these D brane configurations are indeed the microscopic description of black holes.

4. The AdS/CFT correspondence

Extremal black holes do not emit Hawking radiation since they have a zero temperature. To understand the nature of Hawking radiation, we need to go off extremality. While it is difficult to deal with black holes which are far from extremality, it is possible to work with near-extremal black holes which have a small temperature. Once again microscopic calculations show that the temperature and the entropy are precisely reproduced.

In the microscopic picture, Hawking radiation corresponds to annihilation of the quanta of the D-brane gauge theory into a mode, while absorption corresponds to the reverse process. The relation (14) is then a statement of detailed balance. This corss-section was computed in the microscopic theory using standard field theory methods and the result was in precise agreement with the grey body factors obtained by a classical absorption calculation.

Maldacena interpreted the agreement of the absorption calculations in the microscopic theory and in classical General Relativity to be indicative of a *duality* of these two descriptions. This is most transparent when one considers a stack of N D-branes which are three dimensional extended objects, $D3$ branes. The microscopic theory for this is a $3+1$ dimensional supersymmetric gauge theory with a gauge group $SU(N)$. The gravitational description of this stack yields a space-time metric

$$ds^2 = (1 + \frac{R^4}{r^4})^{-1/2}[-dt^2 + d\vec{x}^2] + (1 + \frac{R^4}{r^4})^{1/2}[dr^2 + r^2 d\Omega_5^2] \qquad (17)$$

where the branes are extended in the directions labelled by the three coordinates $\vec{x}$. The coordinate r is the radial coordinate in the space transverse to the branes : together with five angles this forms the six dimensional transverse space. $d\Omega_5^2$ is the line element on a unit 5-sphere. The length scale R is given by

$$R^4 = 4\pi l_s^4 g_s N \qquad (18)$$

Here l_s is the string length, and g_s is the dimensionless string coupling. This is related to the ten dimensional Newton constant by

$$G_{10} = 8\pi l_s^8 g_s^2 \qquad (19)$$

Maldacena's argument went as follows. Consider the classical absorption of a graviton by this stack of D3 branes. Now think of dividing the spacetime into two parts : a part which is close to the horizon, and a part which is far from the horizon. In each of these regions we can consider gravitons - the near horizon gravitons and the asymptotic gravitons. One can now think of the absorption process as a conversion of the asymptotic gravitons into near horizon gravitons. The absorption cross-section turns out to poroportional to the square of the energy -this means that at low energies these two regions decouple.

The same absorption cross-section results from a calculation in the microscopic theory. In this calculation absorption may be thought of a conversion of asymptotic gravitons into the excitations of the brane system - the latter being described by the supersymmetric Yang-Mills theory. The agreement of the two calculations then suggest that these excitations must be an alternative description of the near horizon gravitons. Actually there is a little bit more. Since we are working with a string theory, these gravitons are modes of closed strings. Normally all the higher stringy modes are massive particles and decouple at low energies. However when the string lives in a region close to the horizon this is not quite true - because of the large redshift all these modes appear as low energy modes to the asymptotic observer. So the conjecture is that

IOP Conf. Series: Journal of Physics: Conf. Series **1143** (2018) 012001 doi:10.1088/1742-6596/1143/1/012001

the near-horizon strings provide an alternative and equivalent description of the gauge theory modes on the D-branes.

In the coordinate system used in (17) the horizon is at $r = 0$. However this is not a point. If we expand the metric in a power series in R/r we get the near-horizon metric as

$$ds^2 = \frac{r^2}{R^2}[-dt^2 + d\vec{x}^2] + R^2\frac{dr^2}{r^2} + R^2 d\Omega_5^2 \tag{20}$$

In flat space the radii of spheres surrounding a point approach zero as we approach the point. This is not what happens here. Rather, the radius of the 5-spheres approach a constant, R -the geometry resembles a throat. The other five directions $(t, \vec{x}, r)$ decouple from the sphere - they form what is called an anti-De Sitter space-time - a spacetime which has constant negative curvature, $1/R^2$. The space-time is then written as $AdS_5 \times S^5$. On the other hand, the theory on the 3 branes is a theory which is invariant under conformal transformations - a conformal field theory. This is why this correspondence is called the AdS/CFT correspondence. The correspondence relates a $3+1$ dimensional quantum field theory with no dynamical gravity with a $9+1$ dimensional theory with dynamical gravity.

The two theories which are conjectured to be dual to each other have two dimensionless parameters each. The "bulk" theory in AdS has the string coupling g_s and the scale of the curvature of the space-time, R (as well as the radius of the S^5) in units of the string length l_s. On the Yang-Mills side there are two parameters, N and the Yang Mills coupling g_{YM}. The correspondence holds when

$$g_s = g_{YM} \qquad \left(\frac{R}{l_s}\right) = 4\pi g_{YM}^2 N \tag{21}$$

The bulk theory is weakly coupled when $g_s = g_{YM}$ is small - this is when the theory is semiclassical. However the gravitational field is nontrivial when R/l_s is finite - from the second relation above this means that we need to take $N \to \infty$. Thus semiclassical string theory in $AdS_5 \times S^5$ is a good description of the Yang-Mills in the limit $N \to \infty$ and $g_{YM} \to 0$ with $g_{YM}^2 N =$ fixed. This is precisely 't Hooft's large N limit we discussed in previous sections. In this limit the real coupling of the Yang Mills theory is not g_{YM}^2, but $g_{YM}^2 N = \lambda$, which is called the 't Hooft coupling. And indeed in this limit we end up with a theory in higher number of dimensions. And, as we expected, this latter theory is a theory of strings. In this theory of strings, $1/N$ acts as a coupling constant.

If this was the end of the story, things would have been rather difficult. since we really do not know how to calculate things in the full string theory in such backgrounds. However (21) shows that when the 't Hooft coupling is large $g_{YM}^2 N \gg 1$, $R \ll l_s$. This means that the curvatures are much smaller than the string scale. This is the limit where we can ignore the fact that we are dealing with a theory of strings rather than usual quantum fields. The description is now in terms of *classical* field theory which contains gravity described by General Relativity. This, then, is the "holy grail" - we found a regime where a quantum field theory of the usual kind describes General Relativity in higher dimensions. Significantly this quantum field theory is necessarily strongly coupled.

The precise correspondence between the gauge theory and the theory containing gravity in this limit was found by Witten and by Gubser, Klebanov and Polyakov. In this correspondence, the gauge theory is defined on the boundary of AdS space-time - just as conjectured by 't Hooft and Susskind. This then is a concrete realization of the holographic correspondence.

5. Epilogue

While the AdS/CFT correspondence has provided major insight into many puzzling aspects of black holes, it is fair to say that the basic issue - the black hole information problem - has not

IOP Conf. Series: Journal of Physics: Conf. Series **1143** (2018) 012001 doi:10.1088/1742-6596/1143/1/012001

been completely understood. This is a very active field of research at this moment - and most discussions are in the context of this correspondence.

The holographic correspondence has been shown to work for a large variety of quantum field theories, most of which come from String Theory. However it is believed to be a lot more general. In fact it is most widely used in the other direction : to understand strongly coupled field theories which appear in particle physics, condensed matter physics and cold atom physics in terms of gravity. A key fact which makes this possible is that the gravitational dual of a finite temperature field theory happens to be a black hole inside AdS space-time. Field theories are notoriously difficult to study when they are strongly coupled. However if we assume that they have a gravitational dual, it is precisely in this strong coupling regime that the gravitational dual becomes classical. Therefore difficult questions in the field theory become questions in classical General Relativity. This has led to considerable insight into a large variety of phenomena, opening up a new way of thinking about old problems in hydrodynamics, superconductivity, quantum quench and thermalization.

6. Acknowledgements

I would like to thank the organizers of this conference for inviting me to present these lectures, and the participants for stimulating questions and comments. This work was partially supported by a National Science Foundation grant NSF-PHY-1521045.

Further Reading

[1] For relevant aspects of large-N expansion : S. Coleman, "1/N" , https://lib-extopc.kek.jp/preprints/PDF/1980/8005/8005163.pdf; E. Witten, "The 1 / N Expansion In Atomic And Particle Physics," NATO Sci. Ser. B **59**, 403 (1980), https://lib-extopc.kek.jp/preprints/PDF/1980/8002/8002242.pdf

[2] For Matrix quantum mechanics and relation to 1+1 dimensional string theory : I. R. Klebanov, "String theory in two-dimensions," In *Trieste 1991, Proceedings, String theory and quantum gravity '91* 30-101 and Princeton Univ. - PUPT-1271 (91/07,rec.Oct.) 72 p [hep-th/9108019]; S. R. Das, "The one-dimensional matrix model and string theory," In *Trieste 1992, Proceedings, String theory and quantum gravity '92* 172-211 [hep-th/9211085].

[3] For black holes in string theory : J. M. Maldacena, "Black holes in string theory"; hep-th/9607235.S. R. Das and S. D. Mathur, "The quantum physics of black holes: Results from string theory" Ann. Rev. Nucl. Part. Sci. **50**, 153 (2000) doi:10.1146/annurev.nucl.50.1.153 [gr-qc/0105063];J. R. David, G. Mandal and S. R. Wadia, "Microscopic formulation of black holes in string theory" Phys. Rept. **369**, 549 (2002) doi:10.1016/S0370-1573(02)00271-5 [hep-th/0203048].

[4] For AdS/CFT : I. R. Klebanov, "TASI lectures: Introduction to the AdS / CFT correspondence," hep-th/0009139.

[5] For applications to hydrodynamics, QCD, condensed matter and quantum quench : V. E. Hubeny, S. Minwalla and M. Rangamani, "The fluid/gravity correspondence," arXiv:1107.5780 [hep-th]; S. S. Gubser and A. Karch, "From gauge-string duality to strong interactions: A Pedestrian's Guide," Ann. Rev. Nucl. Part. Sci. **59**, 145 (2009) doi:10.1146/annurev.nucl.010909.083602 [arXiv:0901.0935 [hep-th]]. S. A. Hartnoll, "Lectures on holographic methods for condensed matter physics," Class. Quant. Grav. **26**, 224002 (2009) doi:10.1088/0264-9381/26/22/224002 [arXiv:0903.3246 [hep-th]] S. R. Das, "Old and New Scaling Laws in Quantum Quench," PTEP **2016**, no. 12, 12C107 (2016) doi:10.1093/ptep/ptw146 [arXiv:1608.04407 [hep-th]].

IOP Conf. Series: Journal of Physics: Conf. Series **1143** (2018) 012002 doi:10.1088/1742-6596/1143/1/012002

Compact stars in the pseudo-complex general relativity

C. A. Zen Vasconcellos[1 2], J. E. S. Costa[1], D. Hadjimichef[1], M. V. T. Machado[1], F. Köpp[1], G. L. Volkmer[1], M. Razeira[3]

Abstract. The field equations of the pseudo-complex General Relativity (pc-GR) have an extra term, of repulsive character, which may halt the gravitational attractive collapse of matter distributions in the evolution process of compact stars. This additional extra term simulates the presence of dark energy in the Universe. On the first part of this contribution we consider the effect of this additional term of pc-GR on the structure of a white dwarf. Observations of type Ia supernova (SNe Ia) admit white dwarfs with masses as high as $2.3M_{\odot} - 2.6M_{\odot}$, surpassing this way the mass limit established by Chandrasekhar ($1.44M_{\odot}$). In a more conventional theoretical description of a white dwarf, the electron degeneracy pressure ultimately stabilizes the star against gravitational collapse and establishes this way the Chandrasekhar maximum limit of the stelar mass; this mechanism can not explain however values of masses more expressive like those mentioned above. We investigate here a possible mechanism for carrying out such a hypothesis: the combination of the electron degeneracy and dark energy internal pressures. In this study we use a very simple model for the white dwarf mater which consists of nucleons and degenerate electrons, held together by the presence of the gravitational interaction and superimposed to a repulsive background of dark energy. In the pc-GR formalism, the corresponding modified Tolman-Oppenheimer-Volkoff (TOV) equations are solved and the mass-radius relations as well as the maximum mass of the white dwarf star are determined for different parameter configurations. In the second part of this contribution we explore the presence of this additional term and study the role of dark energy in the structure of neutron stars composed by nucleons, hyperons, mesons, and weakly interacting massive fermion dark matter particles (WIMPs) held together by the presence of the nuclear force and the gravitational interaction, superimposed to the repulsive background of dark energy. To describe the hadron-lepton sector we consider three different effective nuclear models, Zimanyi-Moszkowski, Boguta-Bodmer, and the analytic parametrized coupling model, which we extend to consider, in the baryonic sector, the presence of the whole fundamental baryon octet. By solving the corresponding pc-GR Tolman-Oppenheimer-Volkoff (TOV) equations we estimate the maximum gravitational mass of neutron stars.

1. Introduction

A white dwarf represents the final stage of the evolution process of stars with up to ~ 10 solar masses. After the hydrogen-fusing period which transforms its hydrogen content in heavier elements through different cycles of thermonuclear reactions, the star passes by a series of thermal pulses before definitively ejecting its outer layers. The remaining core is a white dwarf

[1] Instituto de Física, Universidade Federal do Rio Grande do Sul (UFRGS), CEP 91501-970 Porto Alegre, RS, Brazil.
[2] Author to whom any correspondence should be addressed - cesarzen@cesarzen.com.
[3] Universidade Federal do Pampa (UNIPAMPA), CEP 96570-000, Caçapava do Sul, RS, Brazil.

with around 0.6 solar mass contained into a spherical volume with cents of the solar diameter of about two times the Earth diameter, one of the most dense and compact forms of matter. In a more conventional approach, white dwarfs are called degenerate stars, in the sense that by taking into account that the material in a white dwarf no longer undergoes fusion reactions, the star has no longer internal sources of energy, and as a result the electron degeneracy pressure mainly supports the star matter against gravitational collapse.

With respect to the internal structure of a conventional description of a white dwarf, it contains a core of electron-degenerate matter covered by a gaseous envelop, which has less than 1% of the total mass. It is the initial mass of the progenitor star that determines the dominant thermonuclear reaction cycles in each stage of its evolution and, therefore, the mass and composition of the core in the white dwarf stage. In a typical white dwarf, the core is composed by a mixture of carbon and oxygen above which lies a shell of helium with a outer thin layer of hydrogen. White dwarfs with helium core have lower masses ($M < 0.5M_{\odot}$) and most of them are the result of mass-transfer episodes during the evolution of binary systems or of strong mass-loss episodes during the red giant branch (RGB) evolution of single stars [1]. Additionally, more massive progenitors produce white dwarfs with cores with heavier elements, as iron, for example.

Although most of the mass of a white dwarf (99% or more) is concentrated in the core, the envelop corresponds to a significant part of the stellar volume. The envelope thickness depends on the relative concentration of H+He, on the surface temperature and gravity. In hot white dwarfs with low mass, for instance, the envelop thickness can reach around 1/3 of the stellar radius. For massive white dwarfs, the envelop is very thin and its thickness is independent of temperature.

Spectroscopic observations of a white dwarf may reveal not only its surface chemical composition but also its surface temperature and gravity ($g = GM/R^2$). By assuming a model for its internal structure, it is possible to estimate the mass and radius of a white dwarf. However, the direct observational determination of the mass and radius of a white dwarf is only possible for binary systems.

White dwarf mass-radius relations are illustrated in Fig. 1. The points correspond to white dwarfs identified in the Data Release 4 of the Sloan Digital Sky Survey (SDSS) and spectroscopically classified as DAs, i.e., white dwarfs with hydrogen in the outer layer [2]. Their masses and radii were calculated assuming a C+O core at a ratio of 50% for each component. The continuous curves indicate the zero-temperature mass-radius relations for cores with different compositions calculated considering star models with a hydrogen envelope [3]. Sirius B, Procyon B and 40 Eri B are examples of white dwarfs in binary systems with well-determined distances [4] and with mass and radius directly measured, with no assumption on the core compositions. For the particular case of Procyon B, the mass-radius relation is not compatible with a typical C+O core, suggesting a core with a denser composition, such as iron.

Stars with masses four to eight times larger than the mass of the sun suffer a supernova explosion process which ejects their outer layers, leaving behind a small, dense core that continues to collapse. Gravity presses the material upon itself so strongly that protons and electrons combine to make neutrons and neutrinos (neutronization process), this being the main aspect that gave rise to the designation neutron stars to these compact objects: the neutrinos produced in the inverse beta decay process usually escape from the contracting core but the neutrons pack closer together until their density becomes equivalent to that of an atomic nucleus.

The composition of a neutron stars is however much more complex, involving different layers of nuclear matter with the presence of the fundamental baryon octet, mesons, leptons, kaon and pion condensates and, in the center of the star, a deconfined composition of quarks and gluons. If the remaining core of the star is less than about 3 solar masses, the particles of the system exert a quantum pressure which is capable of supporting the star. For masses larger than this

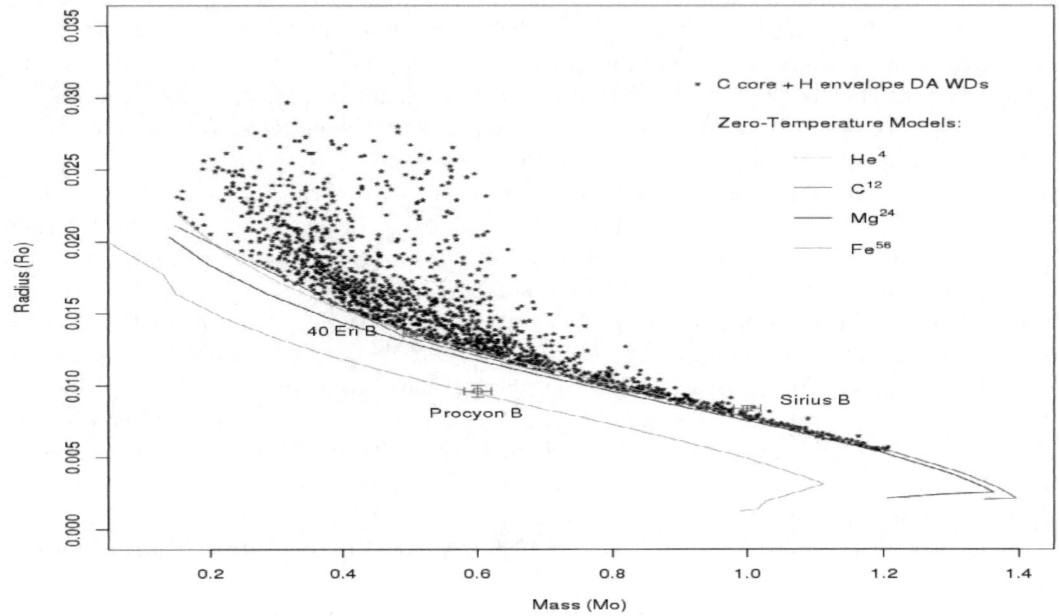

Figure 1. White dwarf mass-radius relations (in terms of solar mass and solar radius, respectively). The continuous curves are the zero-temperature mass-radius relations calculated for different core composition. The black points indicate the mass-radius of DA white dwarfs from the SDSS DR4, assuming a C+O core with an hydrogen envelope. The green points indicate mass-radius for three white dwarfs in binary systems.

amount, the internal pressure of the system of particles cannot support the star against gravity and the system collapses into a stellar black hole.

Neutron stars may be seen as a pulsar if their internal magnetic field are favorably aligned with their spin axis. Neutron stars moreover pack their mass inside a 20-kilometer diameter and on average, gravity on a neutron star is around 2 billion times stronger than gravity on Earth.

2. About the masses of white dwarfs and neutron stars

One of the most important and controversial topics related to white dwarfs and neutron stars is about their maximum masses.

Following the formulation of the Fermi-Dirac statistics, Fowler [5] treated the electron gas in the Sirius-B white dwarf as a degenerate non-relativistic gas and found no limiting mass for the star. Anderson [6] and Stoner [7] considered a relativistic treatment for the electron gas and found a limiting density although their treatment was heuristic. Chandrasekhar [8, 9, 10] obtained the limiting mass as $0.91M_{\odot}$ initially by using a relativistic treatment for the degenerate electron gas. Subsequently, he succeeded in formulating the theory of white dwarfs to full generality employing Newtonian gravity and an equation of state valid for the entire range of electron velocities (including relativistic velocities) for the degenerate Fermi gas. He has obtained this way the equations of hydrostatic equilibrium of the star, in the form of Lane-Emden equations with index 3. Chandrasekhar then solved the corresponding differential equations numerically to find the limiting mass of $1.44M_{\odot}$ for the white dwarf. Chandrasekhar [11] also considered the problem in the general relativistic framework to study the instability of a radially pulsating white dwarf star and obtained the critical mass as $1.4176M_{\odot}$. Anand [12] studied the

IOP Conf. Series: Journal of Physics: Conf. Series **1143** (2018) 012002 doi:10.1088/1742-6596/1143/1/012002

effect of rotation on a white dwarf star and showed that the value of the limiting mass increases to $1.704M_\odot$. Qualitative arguments given by Landau & Lifshitz [13] suggest that the inter-particle Coulomb interaction is negligible in a white dwarf star. Using the method of Bohm & Pines [14], Singh [15] showed that the correction to the electron density due to electron-electron interaction is small and can be treated as negligible. Salpeter [16] reconsidered the problem of taking into account Coulomb effects, Thomas-Fermi corrections, the contribution of exchange and correlation energies in white dwarfs and he showed that the corresponding equation of state departs measurably from the ideal degenerate case.

A wide range of mass distributions of a large number of white dwarfs stars, including low and large massive stars, were plotted by Bergeron et al. [17] et al. and Kepler et al. [18]. The most massive non-magnetic white dwarf observed was LHS4033 (see Bergeron et al. [17], Kepler et al. [18] and Dahn et al. [19]) which was predicted to have an oxygen-neon core with a mass in the range of $1.318M_\odot - 1.335M_\odot$.

Recent observations of type Ia supernova (SNe Ia) admit white dwarfs with masses as high as $2.3M_\odot - 2.6M_\odot$. Howell et al. [20] argued that the over-luminosity and low expansion velocities around the SN 2003fg white dwarf could be explained assuming the star to have a mass larger than $1.44M_\odot$. Hicken et al. [21] presented SN2006gz as a possible SNe Ia candidate that was identified with similar properties. Scalzo et al. [22] estimated the total mass of the SN 2007if progenitor to be in the range $2.2M_\odot - 2.6M_\odot$. Silverman et al. [23] suggested another member of SNe Ia class, SN2009dc, with similar peculiarities, possibly formed from the merger of two white dwarfs.

Similarly to white dwarfs, one of the most controversial and intriguing questions in modern astrophysics concerns the existence of a maximum mass of a neutron star. The establishment of this threshold and the controversies correlated with this conception followed historically a logical line of thought quite similar to the corresponding determination of the maximum mass of the white dwarfs.

In 1932, Lev Landau [24] mentioned the possible existence of dense stars which would look like giant nuclei and which can be regarded today as an early theoretical prediction or anticipation of the existence of neutron stars. He showed that hydrostatic equilibrium of these stars, in which the gravity contraction is supported by the pressure of degenerate electrons, can only be reached if the mass of the star do not exceed $1.5M_\odot$. His predictions in this regard occurred before the discovery of the neutron by James Chadwick [25]. Landau assumed that protons and electrons constitute atomic nuclei in stable configurations. After the discovery of the neutrons, an adaptation of his calculations provided a limit of $1.8M_\odot$ for the maximum mass of a neutron star.

In 2010, Paul Demorest and colleagues [26], using observational results for the Shapiro delay, measured the mass of the millisecond pulsar PSR J16142230. The result obtained was $1.97 \pm 0.04M_\odot$, a value substantially higher than any previously measured neutron star mass, placing this way strong constraints on the interior composition of neutron stars. Moreover, in 2013, John Antoniadis and colleagues [27] measured the mass of the millisecond pulsar PSR J0348+0432 to be $2.01 \pm 0.04M_\odot$, based on the spectroscopy of its white dwarf companion. Furthermore, this result allowed, for the first time, a test of general relativity predictions for compact stars.

In the following, as pointed out in the abstract of this contribution we consider a possible mechanism for the generation of such high masses, that is, we consider a treatment of the matter contained in white dwarfs and neutron stars in the scope of the pseudo complex general relativity.

3. The pseudo-complex general relativity

It was shown by Hess, Schäfer and Greiner [28, 29], within the pseudo-complex general relativity (pc-GR) theory, that the resulting field equations include an extra term associated to

IOP Conf. Series: Journal of Physics: Conf. Series **1143** (2018) 012002 doi:10.1088/1742-6596/1143/1/012002

the nature of spacetime itself, of repulsive character, that may sustain the collapse of matter distributions avoiding for instance the standard general relativity predictions of black hole creation as the final stage of the evolution of compact stars and space-time singularities. This additional term arises from micro-scale phenomena due to vacuum fluctuations which simulate the presence of dark energy in the Universe. In order to proceed within a familiar GR framework, similarly to the original formulation, we represent this term by an energy-momentum tensor $(T^{DE})_{\mu\nu}$:

$$G_{\mu\nu} = 8\pi(T^{DE})_{\mu\nu} + 8\pi(T^{\mathcal{M}})_{\mu\nu}\,; \tag{1}$$

thus, in this expression $(T^{DE})_{\mu\nu}$ represents the dark energy contribution and $(T^{\mathcal{M}})_{\mu\nu}$ is the energy-momentum tensor for the white dwarf or neutron star matter content. The corresponding equations in the pc-GR are

$$G_{\mu\nu} = R_{\mu\nu} - \frac{1}{2}g_{\mu\nu}\mathcal{R} \quad \epsilon\ \mathcal{P}^0\,, \tag{2}$$

with $\mathcal{P}^0$ being a zero divisor quantity (for the details see Refs. [28, 29]); in this expression $G_{\mu\nu}$ is the pseudo-complex Einstein tensor, $R_{\mu\nu}$ is the Ricci tensor, defined in the same way as in the standard GR, with the difference that now it is pseudo-complex, and $\mathcal{R}$ is the Riemann curvature (also known as scalar curvature).

4. Equations of State

4.1. White dwarf sector

We describe the nucleons and electrons sector in our simple model of the white dwarf considering a massive star in the final stage of its evolutionary process, i.e., the stage of exhaustion of its nuclear fuel; in this stage the star begins to cool and contract [30]. A more complete description of a white dwarf may be found in the Ref. [31]. When the stellar temperature reaches a sufficient low level, the stellar electrons will occupy the lowest available energy levels. According to the Pauli Principle, there will be two electrons for each energy level due to the spin multiplicity. The number of energy levels per unit volume for the electrons with momenta between k and $k + dk$ is equal to $4\pi k^2\, dk/(2\pi^- h^3)$.

Each fermion of the system (electrons and nucleons) contribute to the energy density (ε_i) , pressure (p_i) and number density (ρ_i) (with $i = e, n$), according to (with $c = \hbar = 1$):

$$\begin{aligned} \varepsilon_i &= \frac{\gamma_i}{2\pi^2}\int_0^{k_{Bi}} \sqrt{k_i^2 + m_i^2}\, d^3k\,, \\ p_i &= \frac{1}{3}\frac{\gamma_i}{2\pi^2}\int_0^{k_{Bi}} \frac{k_i^2}{\sqrt{k_i^2 + m_i^2}}\, d^3k\,, \\ \rho_i &= \frac{\gamma_i}{2\pi^2}\int_0^{k_{Bi}} k_i^2\, d^3k\,, \end{aligned} \tag{3}$$

where $\gamma_i = 2$ is the degeneracy of each momentum state due to the two spin projections of the fermions.

After integration of these equations we arrive to the following solutions for the equations of energy and pressure of the system in the low density regime $k_i << m_i$ (see Ref. [32]):

$$\varepsilon_i \approx \frac{m_i^4}{\pi^2}\left[\frac{1}{3}(\frac{k_i}{m_i})^3 + \frac{1}{10}(\frac{k_i}{m_i})^5 - \frac{1}{56}(\frac{k_i}{m_i})^7 + \frac{1}{144}(\frac{k_i}{m_i})^9\right] \approx \rho_i m_i + \frac{(3\pi^2\rho_i)^{5/3}}{10\pi^2 m_i}\,; \tag{4}$$

$$p_i \approx \frac{m_i^4}{3\pi^2}\left[\frac{1}{5}(\frac{k_i}{m_i})^5 - \frac{1}{14}(\frac{k_i}{m_i})^7 + \frac{1}{24}(\frac{k_i}{m_i})^9\right] \approx \frac{(3\pi^2\rho_i)^{5/3}}{15\pi^2 m_i}\,. \tag{5}$$

Since $m_N >> m_e$, the nucleons and dark matter do not contribute appreciably to the pressure, so we can use the following approximation for the equation of state of the system:

$$\varepsilon_{\mathcal{M}} = \rho_e \rho_N \nu \,, \tag{6}$$

for the energy density; in this expression, $\nu = \frac{\rho_n+\rho_p}{\rho_e}$ is a parameter that represents the nucleon-electron ratio of the dominant nuclear species and which must be determined by the equilibrium condition for a neutral Fermi gas ($\rho_n = \rho_p(\nu - 1)$);

$$p_i = \frac{1}{3}\frac{\gamma_i}{2\pi^2}\int_0^{k_{Bi}} \frac{k_i^2}{\sqrt{k_i^2 + m_i^2}} d^3k \,, \tag{7}$$

for the pressure of the system.

4.2. Baryon-meson and lepton sector

The equation of state in the baryon-meson sector may be synthesized as $p_{NM} = p_{NM}(\varepsilon_{NM})$ where

$$\begin{aligned} \varepsilon_{NM} &= \frac{1}{2}m_\sigma^2\sigma_0^2 + \frac{1}{2}m_{\sigma*}^2\sigma_0^{*2} + \frac{1}{2}m_\omega^2\omega_0^2 + \frac{1}{2}m_\phi^2\phi_0^2 + \frac{1}{2}m_\rho^2\varrho_{03}^2 + \frac{1}{2}m_\delta^2\delta_3^2 + \frac{bM}{3}\sigma_0^3 + \frac{c}{4}\sigma_0^4 \\ &+ \frac{1}{\pi^2}\sum_B \int_0^{k_{F,B}} k^2 dk\sqrt{k^2 + M_{B\zeta}^{*2}} + \frac{1}{\pi^2}\sum_l \int_0^{k_{F,l}} k^2 dk \sqrt{k^2 + m_l^2} \,, \end{aligned} \tag{8}$$

which describes the internal energy density of nuclear matter (NM), and

$$\begin{aligned} p_{NM} &= -\frac{1}{2}m_\sigma^2\sigma_0^2 - \frac{1}{2}m_{\sigma*}^2\sigma_0^{*2} + \frac{1}{2}m_\omega^2\omega_0^2 + \frac{1}{2}m_\phi^2\phi_0^2 + \frac{1}{2}m_\rho^2\varrho_{03}^2 - \frac{1}{2}m_\delta^2\delta_3^2 + \frac{bM}{3}\sigma_0^3 + \frac{c}{4}\sigma_0^4 \\ &+ \frac{1}{3}\frac{1}{\pi^2}\sum_B \int_0^{k_{F,B}} \frac{k^4 dk}{\sqrt{k^2 + M_{B\zeta}^{*2}}} + \frac{1}{3}\frac{1}{\pi^2}\sum_l \int_0^{k_{F,l}} \frac{k^4 dk}{\sqrt{k^2 + m_l^2}} \,, \end{aligned} \tag{9}$$

which represents the internal pressure of nuclear matter.

5. Dark matter-dark energy sector

Following Ref. [28, 29] we consider an isotropic fluid model for both the dark matter and the dark energy densities and we decouple the physical Higgs from the dark Higgs. This choice also decouples dark matter from ordinary matter and transforms the fermion-singlet system, with mass m_χ to a non-interacting Fermi gas of WIMPs. There remains a residual interaction of the χ singlet with the dark Higgs. However, after symmetry breaking in the dark sector, the χ singlet acquires a mass $m_\chi = m_{\chi_0} + g_\varphi x_0$ (for the details see [33]). We shall treat m_χ as a free parameter, studying the impact of different χ-mass values on the exotic stellar quantities.

The equation of state for a free gas of dark fermions at zero temperature $p_{DM}(\varepsilon_{DM})$ can be calculated via explicit expressions for the energy density

$$\varepsilon_{DM} = \frac{1}{\pi^2}\int_0^{k_F} dk\, k^2 \sqrt{m_\chi^2 + k^2} \,, \tag{10}$$

and pressure

$$p_{DM} = \frac{1}{3\pi^2}\int_0^{k_F} dk \frac{k^4}{\sqrt{m_\chi^2 + k^2}} \,. \tag{11}$$

Inside the star, the numerical implementation of the system of differential equations demands the specification of the dark matter (DM) equation of state and its relation to the dark energy (DE) sector. Following Ref. [28, 29] we consider an isotropic fluid model for both the dark matter and the dark energy densities and we assume a linear coupling between ϵ_{DM} and ϵ_{DE}:

$$\epsilon_{DE} = \alpha\epsilon_{DM}. \tag{12}$$

The central pressure of the DE-term is assumed to be related to the central energy density of the WIMPs, ϵ_0, through $p_{DE_C} = |\alpha|\,\epsilon_0$, with a standard value $\epsilon_0 =$ 141 MeV/fm^3.

6. Modified TOV Equations

For simplicity we consider a spherically symmetric and static metric for the star:

$$ds^2 = -e^{\nu(r)}dt^2 + e^{\mu(r)}dr^2 + r^2\Big(\sin^2(\theta)d\phi^2 + d\theta^2\Big). \tag{13}$$

Moreover, in the following, both $(T^{DE})_{\mu\nu}$ and $(T^{\mathcal{M}})_{\mu\nu}$ will be treated as isotropic perfect fluids,

$$T_{\mu\nu} = \Big(\epsilon + p\Big)u_\mu u_\nu + p g_{\mu\nu}\,,$$

where p is the pressure and ϵ is the energy density of the fluid, and u_μ are the components of the four velocity of the fluid. Solving the field equations and assuming that the fluid components interact only gravitationally, the hydrostatic pc-GR TOV equilibrium equations can be written as (see Hess, Schäfer & Greiner [28, 29]):

$$\frac{dp_{\mathcal{M}}}{dr} = -\frac{1}{r}\left(\frac{\epsilon_{\mathcal{M}}(r) + p_{\mathcal{M}}(r)}{r - 2m_{\mathcal{M}}(r) + 2m_{DE}(r)}\right)\left(m_{\mathcal{M}}(r) - m_{DE}(r) + 4\pi r^3\Big[p_{DE}(r) + p_{\mathcal{M}}(r)\Big]\right); \tag{14}$$

$$\frac{dp_{DE}}{dr} = -\frac{1}{r}\left(\frac{\epsilon_{DE}(r) + p_{DE}(r)}{r - 2m_{\mathcal{M}}(r) + 2m_{DE}(r)}\right)\left(m_{\mathcal{M}}(r) - m_{DE}(r) + 4\pi r^3\Big[p_{DE}(r) + p_{\mathcal{M}}(r)\Big]\right), \tag{15}$$

where:

$$m_{\mathcal{M}}(r) = 4\pi\int_0^r r'^2\epsilon_{\mathcal{M}}(r')dr'\,, \tag{16}$$

and

$$m_{DE}(r) = -4\pi\int_0^r r'^2\epsilon_{DE}(r')dr'\,, \tag{17}$$

represent the corresponding mass functions.

7. Results and Conclusions

In this work we have analyzed the role in the stellar stability of white dwarfs and neutron stars of the additional term of the pseudo-complex general relativity, associated to the nature of spacetime itself and which simulates, according to Hess & Greiner [28, 29], the presence of dark energy in the Universe.

Our main results for the mass-radius relations of white dwarfs are summarized in Fig. (2). For $\alpha = 0$ our predictions for the maximum mass of white dwarfs are in good agreement with the classical results of Subrahmanyan Chandrasekhar (see for instance [31]) using the conventional formulation of General Relativity. Our calculations for other values of the alpha parameter predict values of maximum masses of white dwarfs much larger than the results observed so far. Obviously, these results lack an observational confirmation.

IOP Conf. Series: Journal of Physics: Conf. Series **1143** (2018) 012002 doi:10.1088/1742-6596/1143/1/012002

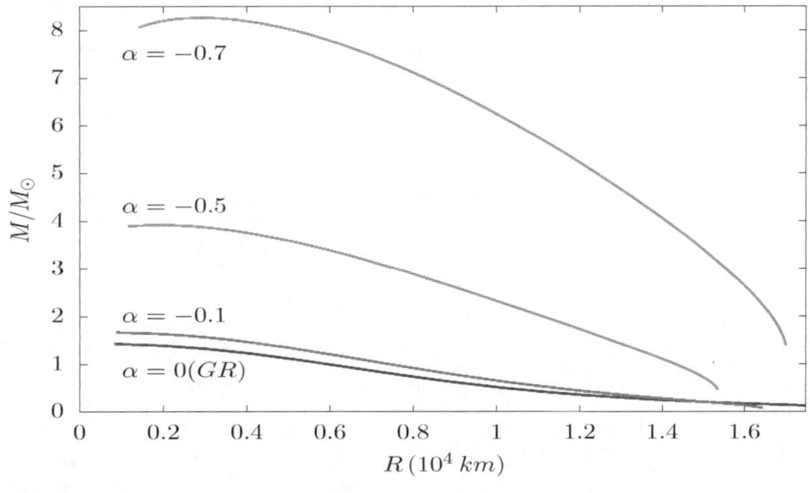

Figure 2. The figure shows the α-dependent predictions of our calculations for the mass-radius relations of the white dwarf. Our results go beyond the conventional treatments using the Einstein's equations of general relativity as we consider here the formalism of pc-GR and the effects of the presence of dark energy in the realization of the equilibrium conditions of the white dwarf star.

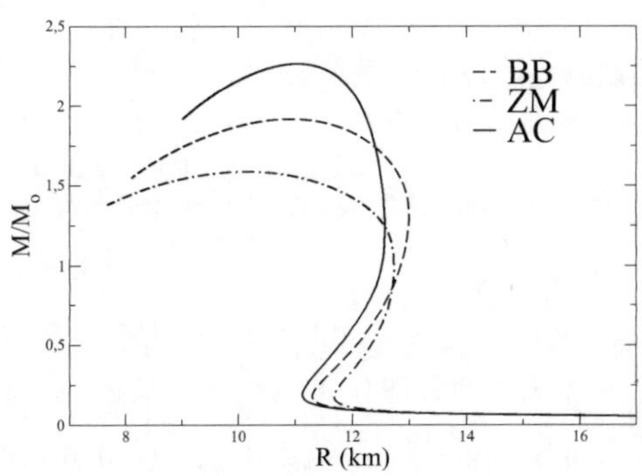

Figure 3. M $\times$ R relation of neutron stars for the models BB, ZM and AC without the presence of dark matter in the framework of the GR formalism.

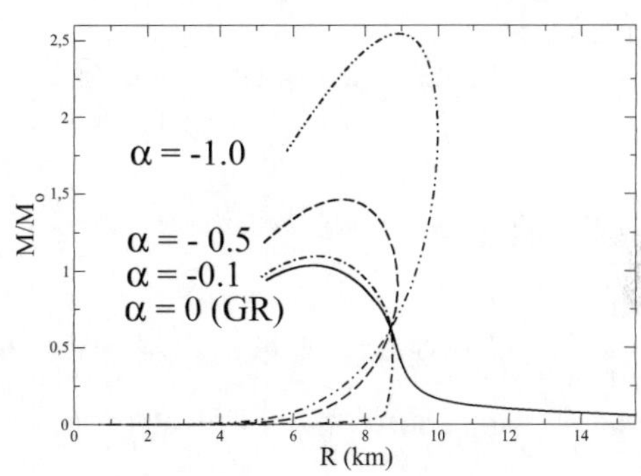

Figure 4. M $\times$ R relation of neutron stars for the BB model considering the presence of dark matter in the system and the pc-GR formalism.

For neutron stars, our main results for the mass-radius relations are summarized in Figs. (3) - (6). For $\alpha = 0$, which characterizes a more conventional formalism of QHD + GR, i.e., without the presence of dark energy, and using Einstein's equations of General Relativity, — although considering the presence of a less conventional ingredient in the interior of a neutron stars, i.e., dark matter—, our results are in very good agreement with the observational results of Refs. [26, 27]. However, for $\alpha \neq 0$, the presence of dark matter in the interior of a neutron star superimposed to a repulsive background of dark energy, due to the additional repulsion term of the field equations as predicted by pc-GR, causes the stellar mass to grow expressively, and even reach a value close to the mass limit predicted by Rhoades and Ruffini [34] of $2.9M_\odot$. It is

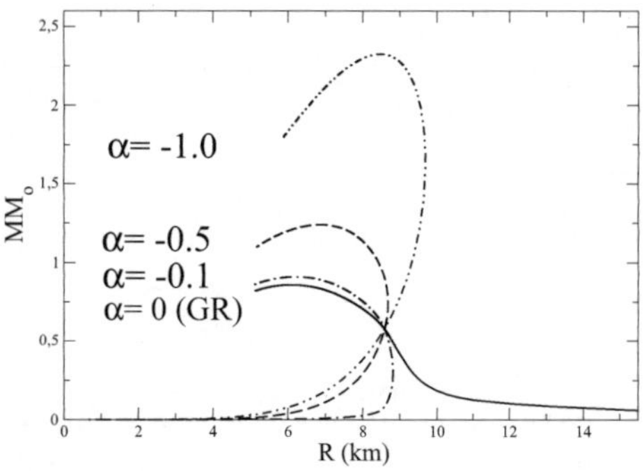

Figure 5. M × R relation of neutron stars for the ZM model considering the presence of dark matter in the system and the pc-GR formalism.

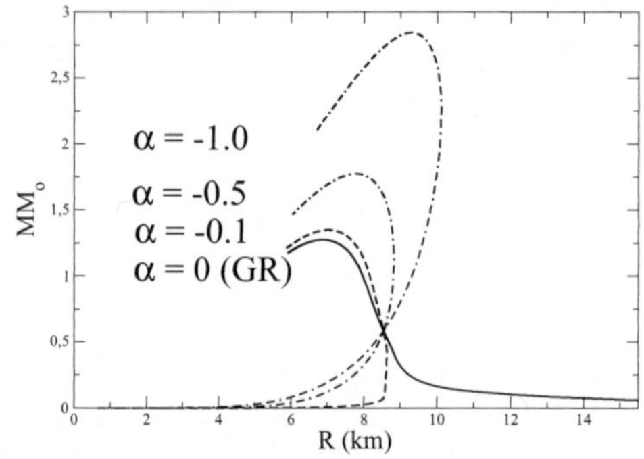

Figure 6. M × R relation of neutron stars for the AC model considering the presence of dark matter in the system and the pc-GR formalism.

important to emphasize that our model for neutron stars considers the presence of hyperons that as is well known soften the equation of state of nuclear matter thus contributing to the decrease of the maximum mass of a neutron star. However, the effect of the additional repulsion due to the presence of vacuum fluctuations in the Universe, as predicted by pc-GR, are dominant when compared to the softening of the EoS by the presence of hyperons in the configuration of the maximum mass of a neutron star.

The results of mss-radius relations for white dwarfs and neutron stars indicate that pc-GR may represent a very interesting theoretical alternative in the determination of stable configurations of compact stars. Further work along this line is projected by our theoretical research group.

References

[1] Althaus, L.G.; Panei, J.A.; Romero, A. D. et al: 2009, *Astronomy and Astrophysics* **502**, 207
[2] Eisenstein, D.J., Liebert, J., Harris, H. et al: 2006, *Astrophysics Journal Suppl.* **167**, 40
[3] Hamada, T. and Salpeter, E.E.: 1961, *Astrophysical Journal* **134**, 683
[4] Provencal, J.L., Shipman, H.L., Hg, E., and Thejll, P.: 1998, *The Astrophysical Journal* **494** (2), 759; 176
[5] Fowler, R.H.: 1926, *Mon. Not. R. Astron. Soc.* **87**, 114
[6] Anderson, W.: 1929, *Z. Phys.* **54**, 433
[7] Stoner, E.C.: 1930, *Phil. Mag.* **9**, 944
[8] Chandrasekhar, S.: 1931, *Astrophys. J.* **74**, 81
[9] Chandrasekhar, S.: 1935, *Mon. Not. Astron. Soc.* **95**, 207
[10] Chandrasekhar, S.: 1939 (Cambridge: University of Cambridge Press, USA)
[11] Chandrasekhar, S.: 1964, *Astrophys. J.* **146**, 417
[12] Anand, S.P.S.: 1965, *Proc. Nat. Acad. Sci.* **54**, 23
[13] Landau, L.D. & Lifshitz, E.M.: 1980, *Statistical Physics, 3rd edn.* (Amsterdam: Butterworth-Heinemann/Elsevier, Holland)
[14] Bohm, D. & Pines, D.: 1951, *Phys. Rev.* **82**, 625
[15] Singh, R.P.: 1957, *Astrophys. J.* **126**, 213
[16] Salpeter, E.E.: 1961, *Astrophys. J.* **134**, 669
[17] Bergeron, P., Gianninas, A., & Boudreault, S.: 2007, *Proc. 15th European Workshop on White Dwarfs*, ed. Napiwotzki, R. and Burleigh, M. (San Francisco, CA: ASP) **372**, 29
[18] Kepler, S.O., Kleinman, S.J., Nitta, A., et al.: 2007, *Mon. Not. R. Astron. Soc.* **375**, 1315
[19] Dahn, C.C., Bergeron, P., Liebert, J., et al.: 2004, *Astrophys. J.* **605**, 400
[20] Howell, A., Sullivan, M., Nugent, P.E., et al.: 2006, *Nature* **443, 308**
[21] Hicken, M., Garnavich, P.M., Prieto, J.L., et al.: 2007, *Astrophys. J. Lett.* **669**, L17

[22] Scalzo, R.A., Aldering, G., Antilogus, P., et al.: 2010, *Astrophys. J.* **713**, 1073
[23] Silverman, J.M., Ganeshalingam, M., Li, W., et al.: 2011, *Mon. Not. R. Astron. Soc.* **410**, 585
[24] Landau, L.D.: 1932, *Phys. Z. Sowjetunion* **1**, 285
[25] Chadwick, J.: 1932, *PRSL A* **136**, 692
[26] Demorest, P., Pennucci, T., Ransom, S., Roberts, M., and Hessels, J.: 2010, *Nature* **467**, 1081
[27] Antoniadis, J., Freire, P.C.C., Wex, N. et al.: 2013, *Science* **340**, 6131, 448
[28] Hess, P.O, Schäfer, M., and Greiner, W.: 2016, in *Pseudo-Complex General Relativity* (Cham: Springer International Publishing AG, Switzerland)
[29] Hess, P.O and Greiner, W.: 2016, in *Centennial of General Relativity: A Celebration*, Edited by C. A. Zen Vasconcellos (Singapore: World Scientific Publishing), p. 97
[30] Weinberg, S.: 1972, in *Gravitation and Cosmology* (New York: Wiley, USA)
[31] Carvalho, S.M., Rotondo, M., Rueda, J.A., and Ruffini, R.: 2014, *Phys. Rev. C* **89** (1), id.015801
[32] Glendening, N.K.: 1997, in *Compact Stars: Nuclear Physics, Particle Physics, General Relativity* (New York: Spinger, USA)
[33] Hadjimichef, D., Volkmer, G.L., Gomes, R.O., and Vasconcellos, C.A.Z., in *Homage Book to Walter Greiner* (World Scientific, Singapore, 2017) (to be published)
[34] Rhoades, Jr., C.E. and Ruffini, R.: 1974, *Phys. Rev. Lett.* **32**, 324

IOP Conf. Series: Journal of Physics: Conf. Series **1143** (2018) 012003 doi:10.1088/1742-6596/1143/1/012003

ARAPUCA light trap for large liquid argon time projection chambers

H da Motta[1,1], A A Machado[2,2], L Paulucci[2], E Segreto[3,2], A Fauth[3], B Gelli[3], M R Guzzo[3], G Marques[3], D Galante[4], V Teixeira[4], W Araujo[4], C Ambrósio[4], M Bissiano[4], F Marinho[5]

[1]Centro Brasileiro de Pesquisas Físicas (CBPF), R. Dr. Xavier Sigaud, 150 - Urca, Rio de Janeiro - RJ, 22290-180, Brazil
[2]Universidade Federal do ABC (UFABC), Av. dos Estados, 5001, Santo André, SP, 09210-170, Brazil
[3]Instituto de Física Gleb Wataghin, Universidade Estadual de Campinas - Unicamp, Rua Sergio Buarque de Holanda, No 777, CEP 13083-859 Campinas, SP, Brazil
[4]Laboratório NAcional de Luz Síncroton (LNLS), Rua Giuseppe Máximo Scolfaro, 10000 - Guará, Campinas - SP, 13083-100, Brazil
[5]Universidade Federal de São Carlos,
Rodovia Anhanguera, km 174, 13604-900, Araras, SP, Brazil

E-mail: helio@cbpf.br, ana.amelia@ufabc.edu.br, segreto@ifi.unicamp.br, fauth@ifi.unicamp.br, gelli@ifi.unicamp.br, marina.r.guzzo@gmail.com, gabrielmarques3197@gmail.com, douglas.galante@lnls.br, veronica.teixeira@lnls.br, w.teleco@gmail.com, camili.ambrosio@lnls.br, marcos.errada@lnls.br, fmarinho@ufscar.br

Abstract. ARAPUCA is a totally innovative device for liquid argon scintillation light detection. It is composed of a passive light collector and of active devices. The active devices are standard SiPMs that operate at liquid argon temperature, while the passive collector is a photon trap that allows the collection of light with extremely high efficiency. The total detection efficiency of the device can be tuned by modifying the ratio between the area of the active components (SiPM) and that of the optical window. Few arrays of ARAPUCAs will be installed inside the prototype of the Deep Underground Neutrino Experiment - protoDUNE - and their performances will be compared with those of more standard solutions based on guiding bars. The results of the most recent tests of ARAPUCAs in a liquid argon environment, which led to the actual design for the protoDUNE, will be reported together with the proposal of a photon detection system for the Deep Underground Neutrino Experiment based on ARAPUCAs combined with dielectric mirror foils coated by wavelength-shifter.

1. Introduction

Neutrino physics is actually recognized to be one of the most important tool to investigate new physics beyond the Standard Model. One of the most relevant experiments in this field is the Deep Underground Neutrino Experiment (DUNE), which is aimed to address some of the open

[1] Speaker
[2] Corresponding author

IOP Conf. Series: Journal of Physics: Conf. Series **1143** (2018) 012003 doi:10.1088/1742-6596/1143/1/012003

questions in this field, namely the measurement of the CP violating phase in the leptonic sector, the hierarchy of neutrino masses and the θ_{23} octant.

DUNE foresees the realization of a neutrino beam and of a near detector, both located at the Fermi National Accelerator Laboratory (Fermilab), Illinois, and a far detector based on the technology of Liquid Argon (LAr) TPC that will be installed at the Sanford Underground Research Facility in South Dakota, 1,300 km far away. This is the ideal distance since it will give the highest sensitivity in measuring the CP violating phase and in determining the mass hierarchy [1].

Moreover, the huge active mass of the far detector will allow the development a rich program of non accelerator physics, that includes the search for proton decay, the detection of supernova neutrinos and of atmospheric neutrinos.

The experimental technique foresees the detection of LAr scintillation light for various purposes. LAr is known to be an abundant scintillator, emitting 40 photons per KeV of deposited energy by minimum ionizing particles. Scintillation photons are emitted through the transition of the Ar_2^* excimer to the ground dissociative state made of two separate argon atoms. Unfortunately, photons are emitted in the Vacuum Ultra Violet (VUV) with a wavelength of 128 nm, where the vast majority of commercial photo-sensors is not sensitive. The most common and viable way to detect LAr VUV photons is to downconvert their wavelengths to the visible through the use of wavelength shifting compounds which are coupled to standard (cryogenic) photo-sensitive devices with glass or fused silica window.

The scintillation light detection system for a massive LArTPC, like the one foreseen for the DUNE far detector, is very challenging since it needs to meet the discording requirements of having an extremely large coverage - given the internal dimensions of the detector - and a good detection efficiency, at the level of percent - to meet the physics goals of the experiment - while keeping the costs at an acceptable level.

One recent development for VUV photon detection in LAr is represented by the ARAPUCA device [2], whose basic design is shown on figure 1. It is constituted by a light collector coupled to silicon photo-sensors (SiPM) used to read-out the collected photons. This configuration allows to increase the effective area of the SiPM array by a factor which can be higher than ten in optimal conditions.

The core of the ARAPUCA concept is represented by the light collector. It consists of a box cavity with internal highly reflective walls from which photons can reflect back and forth until being detected by the SiPM array (or absorbed). The acceptance window of the box is made by a short-pass dichroic filter[3] coated with two different wavelength shifters, one on each side.

The external shifter converts the VUV photon coming from the LAr active volume to a wavelength below the cut-off, allowing it to cross the filter, while the internal one absorbs again the converted photons and shifts its wavelength above the cut-off, where the filter is totally reflective. Since the box has reflective internal surfaces, the photon gets trapped inside it.

For use in a liquid argon (LAr) environment the two wavelength shifting layers can be optimized as follows. Scintillation light is emitted by LAr with a wavelength of 127 nm. The first shifter is para-terphenyl (pTP) [3], deposited on the external side of the acceptance window, which converts the liquid argon scintillation light to a photon of wavelength that peaks around 350 nm, below the filter cutoff, chosen to be $\sim$ 400 nm. After passing through the dichroic filter, the photon reaches the second shifter, TetraPhenyl-Butadiene (TPB) which shifts it to the visible region (peak $\sim$ 430 nm) [4]. When entered the box, the photon will likely reflect a few times (including on the dichroic filter surface) before being detected by the SiPM array.

[3] This particular type of filter has the properties of being highly transparent to photons with wavelength below a given cut-off, while being highly reflective to photons with wavelength above the same cut-off.

IOP Conf. Series: Journal of Physics: Conf. Series **1143** (2018) 012003 doi:10.1088/1742-6596/1143/1/012003

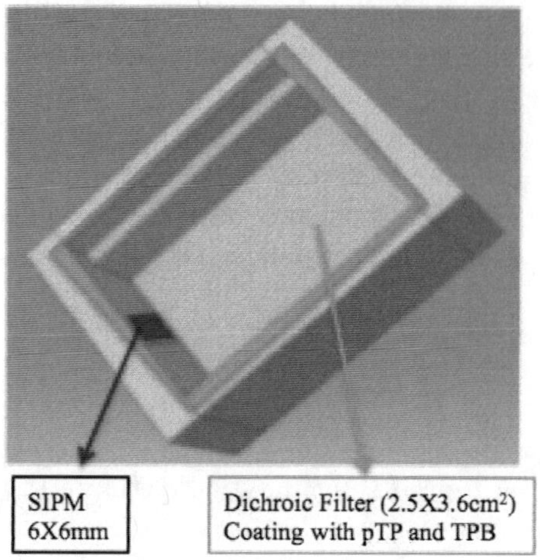

Figure 1. Schematic diagram of the ARAPUCA.

ARAPUCA provides a reasonably high light detection efficiency (few percent) on large areas (hundreds of cm^2) with a limited coverage of active components (SiPM). These properties make it particularly suited to be used in large liquid argon Time Projection Chambers, like the ones of the DUNE experiment.

2. Experimental Setup

The tests of the ARAPUCA using liquid argon (LAr) were performed at the facilities of the Toroidal Grating Monochromator (TGM) beamline of the Brazilian Synchrotron Light Laboratory (LNLS). In order to observe the ARAPUCA's efficiency, the sensor was installed in a vacuum tight stainless steel cylinder filled with ultra pure LAr. Before filling, the chamber is evacuated down to a residual pressure of 10^{-6} mbar to avoid argon contamination. An alpha source was installed in front of the acceptance window of the ARAPUCA at a distance of 4 cm, in correspondence of its geometrical center. The radioactive source is constituted by an ^{238}U-Al alloy in the form of a metallic foil, which emits α particles with a continuous spectrum with an end-point of 4.267 MeV. The energy loss of the energetic particle in LAr was responsible for the production of the scintillation ligth, which was collected and detected by the ARAPUCA. A schematic of the experimental setup is shown on figure 2.

The alpha particle source has been characterized with the robust method of alpha particle spectrometry in a solid state detector. It consists in exposing the passive plastic detector to the radioactive source. The emitted alpha particles interact with the polymeric chains, causing damages and creating the so called etch pits. These pits have been demonstrated to be compatible with those produced by the α particles emitted by an ^{238}U source of infinite thickness. The signal coming from the SiPM is digitized (Acqiris, DP235 Dual-Channel PCI Digitizer Card - 1GS/s, 8 bit dynamic range) and stored on a computer.

3. Data Analysis

Recorded waveforms have been individually analyzed. In order to determine the Single Electron Response (SER) of the SiPM in LAr, single photon signals have been searched in the tail of the scintillation signals. A hit-find algorithm has been implemented, which searches for single, isolated peaks and integrates its charge over 300 ns. The histogram of the integrals for a large number of isolated peaks is shown in figure 3, right.

IOP Conf. Series: Journal of Physics: Conf. Series **1143** (2018) 012003 doi:10.1088/1742-6596/1143/1/012003

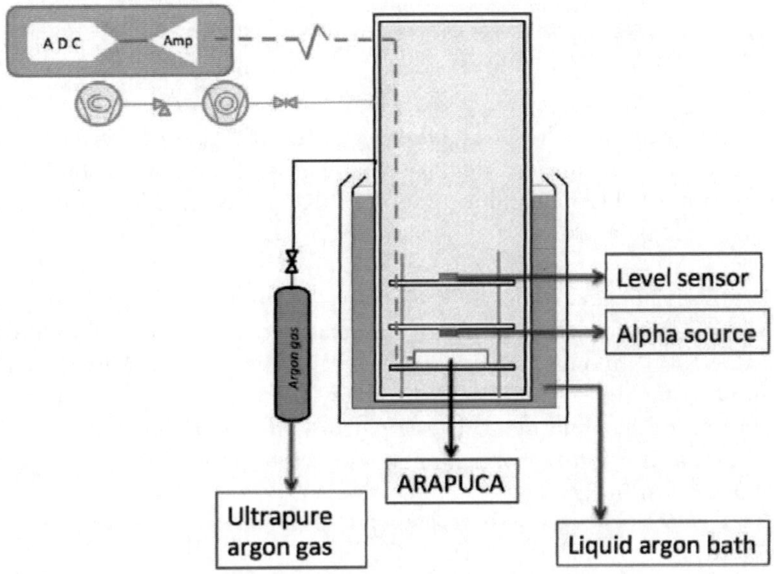

Figure 2. Schematic diagram of the vacuum chamber, complete with gas system and the ARAPUCA assembly, inside a dewar filled with liquid argon.

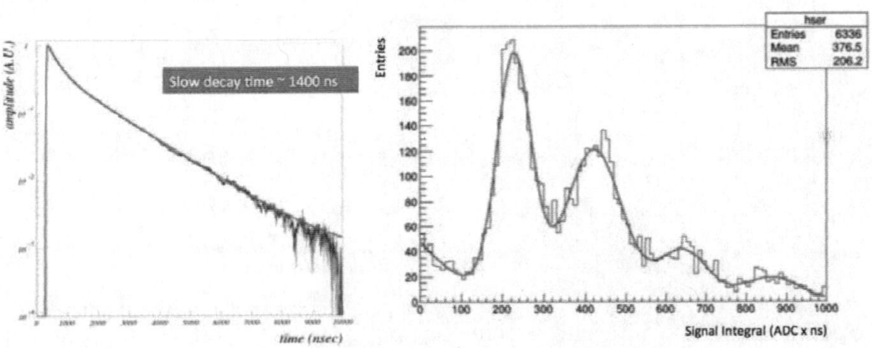

Figure 3. Left: Average waveform detected by the ARAPUCA, fast and slow components of the scintillation light are clearly visible. The fast component is degraded by the slow decay time of the SiPM. The slow component has a decay time of about 1400 ns, as expected for ultra clean argon. Right: Single Electron Response of the SiPM used to read-out the photons trapped inside the ARAPUCA box. The first peak is due to genuine one photon singal, while the others are due to multiple photons. Red line is the result of the fit of the histogram with a function which is constituted by the sum of four non-independent gaussians (see text).

The first peak is produced by one photon signals, while the others by multiple photons (3 more peaks are visible, due to two, three and four photons). The distribution has been fitted with a function made by four Gaussians. The mean value and standard deviation of the Gaussian distribution related to the single photon have been left as free parameters, while the mean value and standard deviation of the Gaussians distributions of multi-photons peaks are not free and are given by the relations:

$$\bar{x}_n = n \times x_0 \quad (1)$$

$$\sigma_n = \sqrt{n} \times \sigma_0 \quad (2)$$

where $\bar{x}_n$ is the mean value of the peak relative to the nth photon, x_0 is the mean value of the one photon peak, σ_n is the standard deviation of the peak relative to the nth photon and σ_0 is the standard deviation of the one photon gaussian distribution. The normalization factor of each one of the peaks has been left as free parameter.

The result of the fit is shown in figure 3, right with a red line. x_0 represents the value of the integral of a signal produced by one single photon and is the overall-calibration factor of the light sensor. The total amount of photons detected by the SiPM in coincidence with an ionization event induced by the α particles emitted by the source is determined by integrating the signal from t_0 (time corresponding to the trigger of the digitizing board) up to $t_0 + 10\ \mu s$ and dividing it by the calibration factor x_0. The spectrum of the number of photons detected by the ARAPUCA is shown in figure 4.

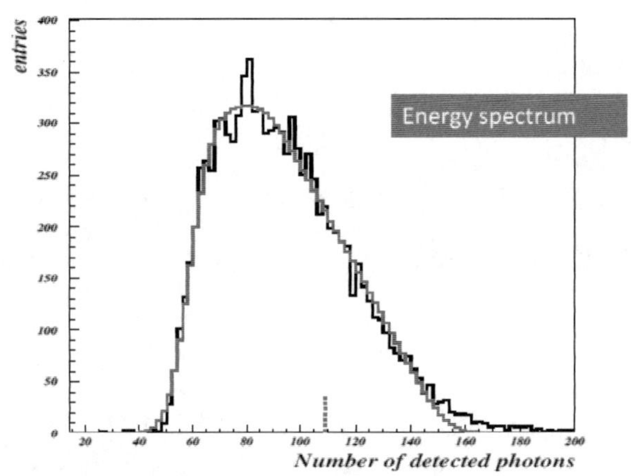

Figure 4. Spectrum of the integrated scintillation signals (black line). The experimental spectrum has been fitted with the source energy spectrum predicted by a dedicated Monte Carlo simulation (red line).

The spectrum has been fitted with a function which describes the expected energy spectrum of the residual energy of the α particles emerging from the source, determined through a dedicated Monte Carlo simulation. In fact, the number of photons detected by the ARAPUCA is proportional to the total number of photons produced by the α particle, through the fraction of the solid angle subtended by the ARAPUCA acceptance window, which is in turn proportional to the α's residual energy. The end point of the spectrum in terms of photo-electrons comes from the fitting procedure and lies approximately in the middle of the high energy descending part of the spectrum, indicated by a red line in figure 4. Its value results to be of 109 ± 1 photo-electrons. The number of photons impinging on the acceptance window of the ARAPUCA can be calculated through simple geometrical considerations. The total amount of photons emitted by α paricles with energies corresponding to the end point of the spectrum is given by:

IOP Conf. Series: Journal of Physics: Conf. Series **1143** (2018) 012003 doi:10.1088/1742-6596/1143/1/012003

$$N_{ph} = N^0_{Lar} \times 4.3\ MeV \times Q_\alpha \tag{3}$$

where N_0 is the ideal photon yield in LAr that is equal to $\sim 5.1 \times 10^4$ [6] photons/MeV and Q_α is the quenching factor for α particles, which is known to be 0.7 [7]. Photons are emitted isotropically and the fraction which hit the ARAPUCA window is given by the solid angle it subtends. It can be analytically calculated because of the very simple geometry of the system with the equation:

$$\Omega = 4 \arctan\left(\frac{ab}{2d\sqrt{4d^2 + a^2 + b^2}}\right) \tag{4}$$

where a and b are the linear dimensions of the acceptance window (2.5 x 3.6 cm^2) and d is the distance of the source (4 cm). The solid angle subtended by the ARAPUCA window is approximately 0.49 rad and the fraction f of photons impinging on the ARAPUCA window is 3.9 %. The detection efficiency of the ARAPUCA is estimated by dividing the number of photo-electrons detected in correspondence of the end point of the α spectrum by the number of photon impinging on its window, N$_{ph}$ $\times$ f, and results to be:

$$\epsilon = 1.8\% \pm 0.3\% \tag{5}$$

The average waveform of the signals is shown in figure 3, left. Fast and slow components of scintillation light are clearly visible. The fast component decay time is dominated by the SiPM response, while the slow one by the scintillation light whose value of 1400ns is compatible with pure liquid argon [8] [9].

4. Monte Carlo Simulations

The Geant4 toolkit [5] was used to simulate the whole device. The reflection properties of the teflon used in the box were implemented along with the ones for the three layers of the acceptance window. A view of the whole simulated device is seen in figure 5. The dimensions and geometry were chosen to reproduce the prototype. It has internal dimensions of 36 mm x 25 mm x 6 mm (length $\times$ width $\times$ height) and a teflon thickness of 1 cm with 95% reflectivity for visible light. The SiPM has an active surface of 6 x 6 mm^2 and a quantum efficiency of 25%. Roughly half of the incident photons are emitted back when converted by the PTP while the remaining photons are effectively trapped inside the ARAPUCA. Some of them reach the SiPM after a few reflections (2-3) while the remaining are either detected after many reflections or absorbed by the surface. The simulation predicted a detection efficiency of 1.7 ± 0.1 %, well compatible with the experimental result.

5. Conclusions

The measured efficiency of the ARAPUCA with 9 cm^2 active area was $1.8\% \pm 0.3\%$. This value is in good agreement with the Monte Carlo simulation, which gives a value of $1.7\% \pm 0.1\%$. The ARAPUCA device promises to have a high detection efficiency for low energy events (supernova neutrino and proton decay) in huge LArTPC detectors (two of the three main goals of DUNE experiment). R&D for ARAPUCA is ongoing, not only in Brazil but also in USA, Europe and other Latin America Countries.

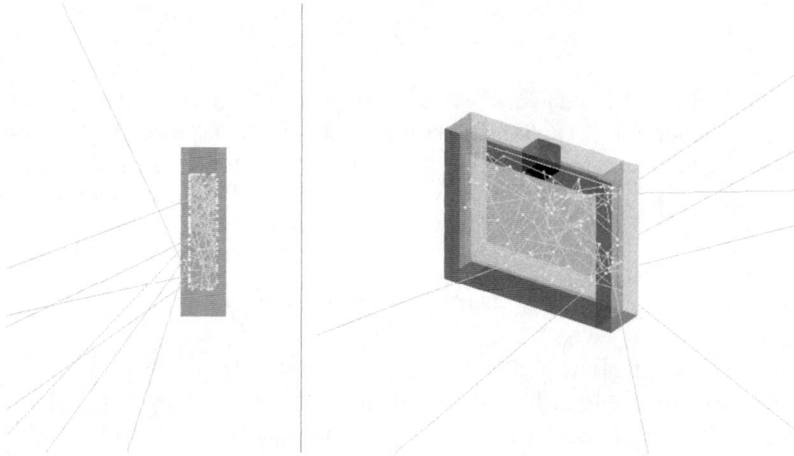

Figure 5. Two views of the simulated ARAPUCA device: lateral (left) and rotated (right). The SiPM (blue) is located on the top. The box is made of highly reflective teflon (gray). Photons are shown as green lines.

References

[1] DUNE/LBNE/LBNF Collaboration, *Baseline optimization for the measurement of CP violation and mass hierarchy in a long-baseline neutrino oscillation experiment*, arXiv:1311.0212 [hep-ex], 2013.
[2] A.A. Machado and E. Segreto, Journal of Instrumentation, 11, C02004 (2016).
[3] T.A. DeVol et al., Evaluation of p-terphenyl and 2,2" dimethyl-p-terphenyl as wavelength shifters for barium fluoride, Nucl. Instrum. Meth. A 327 (1993) 354.
[4] R. Francini et al., VUV-Vis optical characterization of Tetraphenyl-butadiene films on glass and specular reflector substrates from room to liquid Argon temperature, JINST 8 P09006 (2013).
[5] S. Agostinelli et al., Nuclear Instruments and Methods in Physics Research Section A, 506, 250-303 (2003).
[6] T. Doke et al., Nuclear Instrument and Methods in Physics Research Section A, 291 617-620 (1990).
[7] T. Doke et al., Nuclear Instrument and Methods in Physics Research Section A, 269 291 (1988).
[8] R. Acciarri et al. 2010 JINST 5 P06003.
[9] R. Acciarri et al. 2010 JINST 5 P05003.

IOP Conf. Series: Journal of Physics: Conf. Series **1143** (2018) 012004 doi:10.1088/1742-6596/1143/1/012004

Modified non-linear Schrödinger models, infinite tower of conserved charges and dark solitons

H Blas[1], **M Zambrano**[2]

[1] Instituto de Física
Universidade Federal de Mato Grosso
Bairro Boa Esperança, Cep 78060-900, Cuiabá - MT - Brazil

[2] Instituto de Ciências Matemáticas e de Computação; ICMC/USP;
Universidade de São Paulo,
Caixa Postal 668, CEP 13560-970, São Carlos-SP, Brazil

E-mail: blas@ufmt.br

Abstract. We show that the quasi-integrability concept holds for the modified defocusing NLS model with dark soliton solutions and it exhibits the new feature of an infinite sequence of alternating conserved and asymptotically conserved charges. For the special case of two dark soliton solutions, where the field components are eigenstates of a space-reflection symmetry, the first four and the sequence of even order charges are exactly conserved in the scattering process of the solitons. We perform extensive numerical simulations and consider the scattering of dark solitons for the cubic-quintic NLS model with potential $V = \eta I^2 - \frac{\epsilon}{6} I^3$ and the saturable type potential satisfying $V'[I] = 2\eta I - \frac{\epsilon I^q}{1+I^q}$, $q \in \mathbb{Z}_+$, with a deformation parameter $\epsilon \in \mathbb{R}$ and $I = |\psi|^2$. The saturable NLS supports elastic scattering of two soliton solutions for a wide range of values of $\{\eta, \epsilon, q\}$. Our results may find potential applications in several areas of non-linear science, such as the Bose-Einstein condensation.

1. Introduction

Certain non-linear field theory models with relevant physical applications possess soliton-like solutions and it is difficult to know *a priori* if they are in fact true solitons [1]. The so-called quasi-integrability concept has recently been put forward in the context of certain deformations of the integrable sine-Gordon (SG), Bullough-Dodd (BD) and non-linear Schrödinger (NLS) models [2, 3, 4]. The idea is that many non-integrable theories possess solitary wave solutions that behave similar to solitons, i.e. the scattering of such solitons basically preserve their shapes and velocities.

In this context, our aim is to predict the results of solitary wave collisions and test the quasi-

IOP Conf. Series: Journal of Physics: Conf. Series **1143** (2018) 012004 doi:10.1088/1742-6596/1143/1/012004

integrability concept in deformed defocusing NLS models. The deformed focusing NLS with bright soliton solutions and the structures responsible for the phenomenon of quasi-integrability have been discussed in [3]. It has been shown that this model possesses an infinite number of asymptotically conserved charges. On the other hand, the deformed defocusing NLS with dark soliton solutions also presents the above remarkable properties, however in this case we can say even more [4]. There are some soliton like solutions which present a special space-reflection symmetry for any time, so this property implies the exact conservation of the sequence of the even order charges, which in the absence of this special symmetry would be conserved only asymptotically, as in the models studied in [2, 3]. However, it seems to be that such parity property is a sufficient but not a necessary condition in order to have the sequence of exactly conserved charges. In fact, it has been shown, by numerical simulations, that there are certain soliton like configurations without this symmetry which also exhibit such conserved charges [4]. So, these properties constitute the distinguishing new features associated to the deformed defocusing NLS with dark soliton solutions, as compared to the previous quasi-integrable models [2, 3].

In order to simulate the dark soliton collisions the so-called time-splitting cosine pseudo-spectral finite difference (TSCP) and the time-splitting finite difference with transformation (TSFD-T) methods [5] have been used. In fact, these methods allowed us to improve in several orders of magnitude the accuracy in the computation of the charges and anomalies presented in [3].

The paper is organized as follows. In the next section we discuss the defocusing NLS model, its 2-dark soliton solutions and the space-reflection parity symmetry. In section 3, we discuss the first conservation laws. In section 4 we construct analytically the quasi-conserved charges and anomalies based on analytical results. It is introduced an anomalous zero-curvature equation based on the $sl(2)$ Kac-Moody (KM) loop algebra. It is defined the first non-trivial quasi-conservation law. In section 5 we present the results of our numerical simulations and the computations of the anomaly $\beta^{(4)}$ for dark soliton scattering of the modified NLS model. We present the results of the deformations with potentials $V = \eta|\psi|^4 - \frac{\epsilon}{6}|\psi|^6$ and $V'[I] = 2\eta I - \epsilon\frac{I^q}{1+I^q}$. In section 6 we present some conclusions.

2. The non-linear Schrödinger model (NLS)

The NLS model has many applications in non-linear physics: optics, condensed matter (e.g. Bose-Einsten condensation-BEC), plasma physics, hydrodynamics, 2D gravity, analogs of gravitation in BECs, etc. The NLS model is a well known integrable system defined by

$$i\frac{\partial}{\partial t}\psi(x,t) + \frac{\partial^2}{\partial x^2}\psi(x,t) - 2\eta\,|\psi(x,t)|^2\,\psi(x,t) = 0, \quad \psi(x,t) \in C. \tag{2.1}$$

Let us summarize some properties of the eq. (2.1):
- For $\eta < 0$, one has the **focusing** NLS, which supports bright soliton solutions.
- For $\eta > 0$, it is the **defocusing** NLS, which possesses dark soliton solutions [6].
- The model (2.1) is an integrable system with reach mathematical and physical structures: N-soliton solutions, infinite number of conserved charges, solvable by the inverse scattering method (IST), Backlund transformations, (bi-tri-) Hamiltonian formulation , etc.

2.1. (Defocusing) Non-linear Schrödinger model and dark solitons

The 1-dark soliton of the defocusing NLS eq. (2.1) becomes [4, 7]

$$\psi(x,t) = \frac{|\psi_0|}{2} \exp\left[i(kx + wt + x_0)\right] \left\{1 + y + (y-1) \tanh\left[\frac{\theta(x,t) + \bar{\theta}(x,t)}{2}\right]\right\} \tag{2.2}$$

This solution possesses *three* arbitrary real parameters, say $|\psi_0|, k$ and the phase associated to $y = e^{2i\delta}$. The intensity function $|\psi|$ moves at the velocity $v = -2p_I, (p \equiv p_R + ip_I)$, which is the velocity of the dark soliton. The dark soliton approaches constant amplitude $|\psi_0|$ as $|x| \to \pm\infty$. As x varies from $-\infty$ to $+\infty$ the soliton acquires a 2δ phase. We restrict $-\pi/2 < \delta < \pi/2$. Moreover, at the center of the soliton, one has that the intensity becomes $|\psi|_{center} = |\psi_0| \cos\delta$. This center intensity is lower than the asymptotic amplitude $|\psi_0|$ and this property characterizes a dark soliton. Notice that this center intensity is controlled by the parameter δ; i.e. this parameter defines the "darkness" of the soliton.

2.2. Space-reflection parity transformation

Let us write the function ψ as $\psi = \sqrt{R}e^{i\frac{\varphi}{2}}$ such that $\varphi/2 = (kx + wt + x_0) + \phi + \delta$. It is interesting to see the behaviour of the fields R and ϕ for the two-soliton solutions possessing a space-reflection symmetric amplitude $R(-\widetilde{x}, \widetilde{t}) = R(\widetilde{x}, \widetilde{t})$ and phase $\phi(-\widetilde{x}, \widetilde{t}) = \phi(\widetilde{x}, \widetilde{t})$ for any given shifted space and time, $\widetilde{x} = x - x_\rho$, and $\widetilde{t} = t - t_\rho$. Consider the next 2-dark soliton solution [4, 7]

$$\begin{aligned} R(\widetilde{x}, \widetilde{t}) &= \frac{1}{4}|\psi_0|^2 \frac{\mathcal{N}(x_1, t_1)[\mathcal{N}(x_1, t_1)]^\star}{\mathcal{D}^2}; \quad x_1 \equiv 2p_{1R}\widetilde{x}, \; t_1 \equiv 4p_{1R}p_{1I}\widetilde{t} \\ &= \frac{1}{2}|\psi_0|^2 \times \\ & \frac{e^{-\Delta} + \cos 4\delta_1 + e^{-\Delta}\cosh 2x_1 + 4e^{-\Delta/2}\cos 2\delta_1 \cosh t_1 \cosh x_1 + \cosh 2t_1}{[e^{-\Delta/2}\cosh x_1 + \cosh t_1]^2} \end{aligned} \tag{2.3}$$

$$\phi(\widetilde{x}, \widetilde{t}) = \arctan\left[\frac{\sin 2\delta_1 \sinh t_1}{e^{-\Delta/2}\cosh x_1 + \cos 2\delta_1 \cosh t_1}\right]. \tag{2.4}$$

$$x_0 \equiv \frac{-\Delta - 2Re(\theta_o^1) - 2Re(\theta_o^2)}{4p_{1R}}, \; t_o \equiv \frac{Re(\theta_o^2) - Re(\theta_o^1)}{8p_{1R}p_{1I}}$$

In Fig. 1 we plot the functions R and ϕ for three successive times. Note that they are symmetric under the reflection $\widetilde{x} \to -\widetilde{x}$ for any value of the variable $\widetilde{t} = t$.

Therefore, it is clear that the above solution satisfies

$$\mathcal{P}_x: \; R(-\widetilde{x}, \widetilde{t}) \to R(\widetilde{x}, \widetilde{t}), \quad \phi(-\widetilde{x}, \widetilde{t}) \to \phi(\widetilde{x}, \widetilde{t}). \tag{2.5}$$

This type of solution will be associated to the even order ($n \geq 4$) exactly conserved charges as discussed below. In our numerical simulations we will verify the properties for this kind of solutions of deformed NLS models, as well as for configurations without this type of symmetry.

IOP Conf. Series: Journal of Physics: Conf. Series **1143** (2018) 012004 doi:10.1088/1742-6596/1143/1/012004

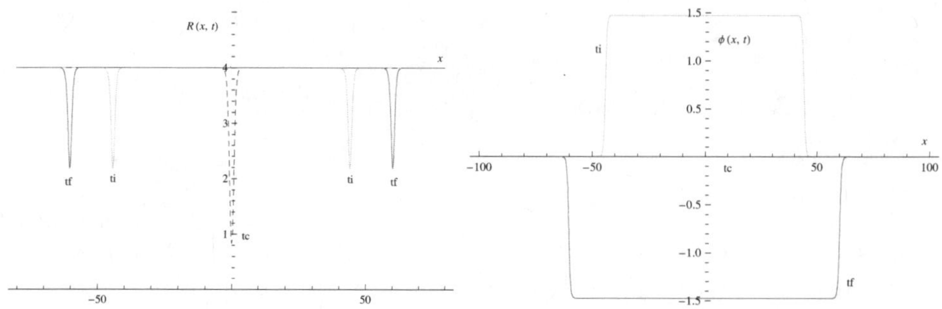

Figure 1. (color online) The amplitude $R = |\psi|^2$ (a) and phase ϕ (b) ($\psi = \sqrt{R}e^{i\phi}$) with space-reflection symmetries for 2-solitons sent at $v_1 = -v_2 = 2, v_s = 4$, $|\psi_0| = 2$, $\eta = 1$, for initial t_i (green), collision t_c (dashed) and final t_f (red) times, respectively. At t_c the solitons have merged and their phases cancel each other. After collision, t_f, each soliton again reveals its phase rise or fall.

3. Conservation laws

The infinite number of conserved charges of the integrable NLS model satisfy

$$\frac{dQ^{(n)}}{dt} = 0, \;\; n = 0, 1, 2, 3, ... \tag{3.6}$$

The first conserved charges of the NLS model become

$$Q^{(0)} = -i \int dx \Big[\frac{\bar{\psi}\partial_x\psi - \psi\partial_x\bar{\psi}}{2|\psi|^2}\Big], \tag{3.7}$$

$$Q^{(1)} = 2\eta \int_{-\infty}^{+\infty} dx \, |\psi|^2 \tag{3.8}$$

$$Q^{(2)} = 2\eta \int_{-\infty}^{+\infty} dx \, \Big[\bar{\psi}\partial_x\psi - \psi\partial_x\bar{\psi}\Big], \tag{3.9}$$

$$Q^{(3)} = \int_{-\infty}^{+\infty} dx \, \Big[|\partial_x\psi|^2 + \frac{1}{2}\eta|\psi|^4\Big]. \tag{3.10}$$

These charges correspond to the conservation of the physical quantities : $Q^{(0)}$ = phase; $Q^{(1)}$ = normalization; $Q^{(2)}$ =momentum; $Q^{(3)}$ =energy. In the case of dark solitons some charges need to be renormalized; e.g. $Q^{(1)} = 2\int_{-\infty}^{+\infty} dx\, |\psi|^2(1 - \frac{|\psi_0|^2}{|\psi|^2})$. Since NLS is an integrable system, it possesses infinite number of conserved charges $Q^{(n)}, n = 0, 1, 2, ...$ and N-dark soliton solutions.

4. Quasi-conserved charges and anomalies: analytical results

4.1. (Anomalous) zero-curvature equation in the sl(2) Kac-Moody (KM) loop algebra

The analytical study of the properties of the deformed NLS model (2.1) can be done using the well known techniques from the integrable field theories. We follow the developments and notations put forward in [2, 3] on quasi-integrability. Then, one considers an anomalous zero curvature representation of the deformed NLS model (2.1) with the deformed gauge connections (Lax pair) $\{A_x, A_t\}$ given for the NLS complex field $\psi(\bar{\psi})$ and deformed potential $V(|\psi|^2)$

$$\begin{aligned}
A_x &= -i\,T_3^1 + \bar{\gamma}\bar{\psi}\,T_+^0 + \gamma\psi\,T_-^0; \\
A_t &= i\,T_3^2 + i\frac{\delta V}{\delta|\psi|^2}\,T_3^0 - (\bar{\gamma}\bar{\psi}\,T_+^1 + \gamma\psi\,T_-^1) - i(\bar{\gamma}\partial_x\bar{\psi}\,T_+^0 - \gamma\partial_x\psi\,T_-^0), \\
F_{xt} &\equiv \partial_t A_x - \partial_x A_t + [A_x\,, A_t] \\
&= XT_3^0 + i\bar{\gamma}\Big[-i\partial_t\bar{\psi} + \partial_x^2\bar{\psi} - \bar{\psi}\frac{\delta V}{\delta|\psi|^2}\Big]T_+^0 - i\gamma\Big[i\partial_t\psi + \partial_x^2\psi - \psi\frac{\delta V}{\delta|\psi|^2}\Big]T_-^0 \\
X &\equiv -i\partial_x\Big[\frac{\delta V}{\delta|\psi|^2} - 2\eta|\psi|^2\Big], \quad \eta \equiv \bar{\gamma}\gamma.
\end{aligned}$$

We have used the $\hat{sl}(2)$ affine KM algebra generators: $\{T_\pm^n, T_3^n\}$. The anomaly is defined by X, and the non-vanishing curvature by F_{xt}. Therefore, for a deformation of the NLS model defined by

$$i\partial_t\psi + \partial_x^2\psi - \psi\frac{\delta V}{\delta|\psi|^2} = 0, \tag{4.11}$$

one has that $F_{xt} \neq 0$. For the usual NLS one has: $V(|\psi|^2) = \bar{\gamma}\gamma\,(|\psi|^2)^2$, which implies the vanishing of the anomaly $X \equiv 0$. Therefore, one has the usual result $F_{xt} \equiv 0$, for NLS field configurations.

In order to study the quasi-conservation laws it has been applied the abelianization procedure, which amounts to gauge transform $A_\mu \to a_\mu = gA_\mu g^{-1} + \partial_\mu g\,g^{-1}$ such that the next anomalous conservation laws arise [3]

$$\frac{\partial}{\partial t}a_x^{(3,-n)} - \frac{\partial}{\partial x}a_t^{(3,-n)} = X\alpha^{(3,-n)}; \quad n = 0, 1, 2, ... \tag{4.12}$$

So, one has asymptotically conserved charges

$$\frac{dQ^{(n)}}{dt} = \beta^{(n)}; \quad Q^{(n)} = -i\int_{-\infty}^{\infty} dx\, a_x^{(3,-n)}; \quad \beta^{(n)} = -i\int_{-\infty}^{\infty} dx\, X\,\alpha^{(3,-n)}. \tag{4.13}$$

For the usual (focusing or defocusing) NLS one has $\beta^{(n)} = 0 \Rightarrow Q^{(n)} = \text{constant}$, $n = 0, 1, 2, 3, 4, 5, ...$

4.2. First non-trivial quasi-conservation law

Except for relevant renormalization procedures, the form of the first three charges $Q^{(i)}, (i = 0, 1, 2)$ of (3.7)-(3.9) remain the same for the deformed NLS model (2.1). The next order charge $Q^{(3)}$ also requires a renormalization procedure. In fact, the renormalized Hamiltonian associated to the dark soliton itself for $k = 0$ becomes [6]

$$H = \int_{-\infty}^{+\infty} dx\,\Big\{|\partial_x\psi|^2 + \int_{|\psi_0|^2}^{|\psi|^2} dI\Big[V'[I] - V'[|\psi_0|^2]\Big]\Big\}. \tag{4.14}$$

So, the renormalized Hamiltonian of the usual NLS model ($V = \eta|\psi|^4$) can be written as

$$H_{NLS} = \int_{-\infty}^{+\infty} dx \left\{|\partial_x \psi|^2 + \eta(|\psi|^2 - |\psi_0|^2)^2\right\}. \tag{4.15}$$

The next fourth order charge $Q^{(4)}$ presents the first non-trivial anomaly, i.e.

$$\frac{dQ^{(4)}}{dt} = \beta^{(4)}. \tag{4.16}$$

The charge $Q^{(4)}$ and anomaly $\beta^{(4)}$, respectively, in the parametrization $\psi = \sqrt{R}e^{i\varphi/2}$ are

$$Q^{(4)} = \frac{\eta}{4}\int_{-\infty}^{+\infty} dx \left[12\eta R^2 \partial_x \varphi + 3\partial_x \varphi \frac{(\partial_x R)^2}{R} + R[(\partial_x \varphi)^3 - 4\partial_x^3 \varphi]\right] \tag{4.17}$$

$$\beta^{(4)} \equiv -\eta \int_{-\infty}^{+\infty} dx [V''[R] - 2\eta]\left\{\frac{3}{4}\partial_x R^2 (\partial_x \varphi)^2 - \partial_x (\partial_x R)^2 + \frac{3}{2}\frac{(\partial_x R)^3}{R}\right\} \tag{4.18}$$

$$\equiv \int_{-\infty}^{+\infty} dx\, \gamma(x,t), \qquad \gamma(x,t) \equiv \text{anomaly density} \tag{4.19}$$

$$Q^{(4)}(\widetilde{t}_0) - Q^{(4)}(0) = \int_0^{\widetilde{t}_0} \beta^{(4)}(t)\, dt = \int_0^{+\widetilde{t}_0} dt \int_{-\infty}^{+\infty} dx\; \gamma(x,t). \tag{4.20}$$

Notice that for soliton solutions with the symmetry (2.5) the anomaly density of $\gamma(x,t)$ becomes an odd function of x. Therefore, the anomaly $\beta^{(4)}$ would vanish for soliton configurations presenting this type of symmetry. This would imply the exact conservation of the charge $Q^{(4)}$ for this type of soliton configurations.

5. Examples and numerical results

Next we will consider some modified NLS models (4.11) and numerically simulate the behaviour of the quasi-conservation law (4.16) under the scattering of two dark solitons.

5.1. Modified (defocusing $\eta > 0$) non-linear Schrödinger (MNLS)

Next we discuss some particular modifications (defocusing $\eta > 0$) of the NLS equations (MNLS) defined for the deformation parameter ϵ.

Cubic-quintic potential NLS: $V(|\psi|^2) = \eta(|\psi|^2)^2 - \frac{\epsilon}{6}(|\psi|^2)^3$

$$i\frac{\partial}{\partial t}\psi(x,t) + \frac{\partial^2}{\partial x^2}\psi(x,t) - \left[2\eta|\psi(x,t)|^2 - \frac{\epsilon}{2}(|\psi(x,t)|^2)^2\right]\psi(x,t) = 0, \quad \psi(x,t) \in C. \tag{5.21}$$

Saturable-type potential: $V'(|\psi|^2) = 2\eta|\psi|^2 - \frac{\epsilon(|\psi|^2)^q}{1+(|\psi|^2)^q}$, $q \in \mathbb{Z}$

$$i\frac{\partial}{\partial t}\psi(x,t) + \frac{\partial^2}{\partial x^2}\psi(x,t) - \left[2\eta|\psi(x,t)|^2 - \frac{\epsilon(|\psi(x,t)|^2)^q}{1+(|\psi(x,t)|^2)^q}\right]\psi(x,t) = 0. \tag{5.22}$$

Notice that in the both cases one has that in the limit $\epsilon = 0$ one recovers the usual NLS model.

IOP Conf. Series: Journal of Physics: Conf. Series **1143** (2018) 012004 doi:10.1088/1742-6596/1143/1/012004

The cubic-quintic modified NLS system (5.21) possesses the solitary waves

$$\Phi^2_{\pm}(z) = \frac{\xi_1 + r\xi_2 \tanh^2[k^{\pm}(z-z_0)]}{1 + r\tanh^2[k^{\pm}(z-z_0)]} \tag{5.23}$$

$$\Theta_{\pm}(z) = \mp \arctan\left[\sqrt{r\frac{\xi_2}{\xi_1}}\, \tanh[k^{\pm}(z-z_0)]\right] \tag{5.24}$$

where

$$r \equiv \frac{|\psi_0|^2-\xi_1}{\xi_2-|\psi_0|^2}, \quad k^{\pm} \equiv \sqrt{\frac{|\epsilon|}{6}}\sqrt{(\pm)(\xi_2-|\psi_0|^2)(|\psi_0|^2-\xi_1)}, \tag{5.25}$$

$$\xi_1 = \frac{B-\sqrt{B^2-6v^2\epsilon}}{2\epsilon}, \tag{5.26}$$

$$\xi_2 = \frac{B+\sqrt{B^2-6v^2\epsilon}}{2\epsilon}, \quad B \equiv 6\eta - 2\epsilon|\psi_0|^2. \tag{5.27}$$

The signs $\pm$ in $\Phi_{\pm}$ and $\Theta_{\pm}$ correspond to the case $\epsilon > 0$ and $\epsilon < 0$, respectively. The parameter ξ_1 is always positive and ξ_2 becomes positive (negative) for $\epsilon > 0$ ($\epsilon < 0$) . For $\epsilon > 0$ they satisfy $\xi_1 < |\psi_0|^2 < \xi_2$, whereas for $\epsilon < 0$ it holds $\xi_2 < \xi_1 < |\psi_0|^2$. Notice that $k^{\pm}$ characterizes the inverse soliton width and $\sqrt{\xi_1}$ is the minimum intensity (dip) of the dark soliton. The maximum intensity of the dark soliton approaches $\Phi_2(\pm\infty) = |\psi_0|$. Moreover, for $v = 0$ the parameter ξ_1 vanishes and the dark soliton becomes a black soliton.

For the second deformed model (5.22), with $\epsilon \neq 0$, $q \in \mathbb{Z}_+$, we do not know any analytical expression for its solitary wave; however, one can generate numerically a solitary wave.

One can study the dark soliton collisions numerically for the both deformed NLS models and compare the outcomes with the properties for the integrable NLS model. Then, we can take two one dark solitons located some distance apart as the initial condition for our numerical simulations of soliton scattering.

In the Figs. 2-5 we present the numerical simulation of the collision of two dark solitons in the cubic-quintic NLS model (5.21). The two-soliton collision with equal and opposite velocities (equal amplitudes) and parameter $\epsilon = -0.01$ is shown in Fig. 2. The anomaly density $\gamma(x,t)$ in (4.19) has been plotted (Fig. 3) as function of the space variable x for three successive times. The anomaly (4.18) (Fig. 4) and the time integrated anomaly (4.20) as functions of time (Fig. 5) are plotted.

One notices the vanishing of the anomaly, and therefore the exact conservation of the charge $Q^{(4)}$, within numerical accuracy, for this type of soliton configuration. In addition, one has the qualitative realization of the symmetry (2.5) and the confirmation of our expectations for the vanishing of the anomaly, since the anomaly density $\gamma(x,t)$ in (4.19) becomes an odd function for this type of soliton configuration (see Fig. 3).

In the Figs. 6-9 we present the numerical simulation of soliton collisions in the saturable-type potential NLS model (5.22). The two-soliton collision with equal and opposite velocities (equal amplitudes) and parameter $\epsilon = 0.1$ is shown in the Fig. 6. The anomaly density $\gamma(x,t)$ in (4.19) has been plotted as function of the space variable x for successive times (see Fig. 7). The two-soliton

IOP Conf. Series: Journal of Physics: Conf. Series **1143** (2018) 012004 doi:10.1088/1742-6596/1143/1/012004

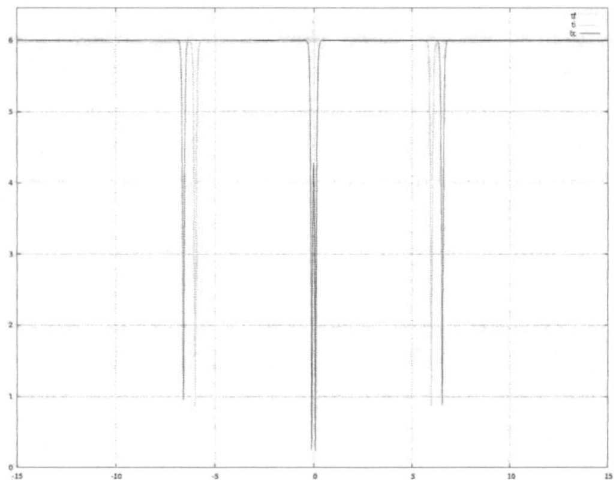

Figure 2. The reflection of two dark solitons of the **cubic-quintic NLS model** is plotted for $\epsilon = -0.01$, $|\psi_0| = 6$, $\eta = 2.5$. The initial gray solitons (t_i =green line) travel in opposite direction with velocity $v \approx 1.97\sqrt{2}$. They partially overlap (t_c= blue line) in their closest approximation and then reflect to each other. The gray solitons after collision are plotted as a red line (t_f). Note that $v < v_s/2$ ($v_s \approx 13.89\sqrt{2}$).

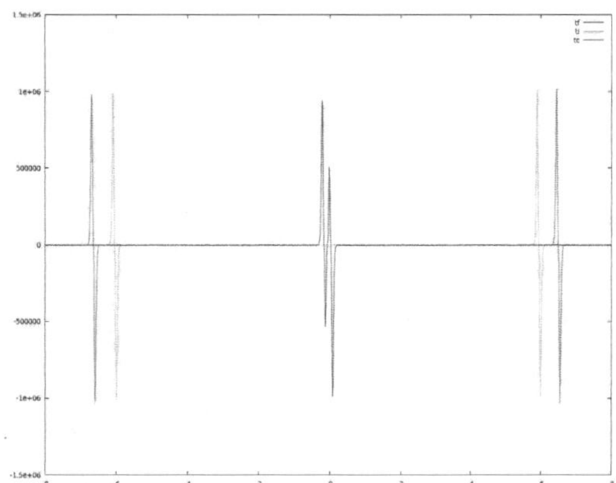

Figure 3. The anomaly density $\gamma(x,t)$ plotted for three successive times (t_i, t_c and t_f).

collision with equal and opposite velocities (equal amplitudes) and parameter $\epsilon = 0.1$. The relevant anomaly (4.18) and the integrated anomaly (4.20) are plotted in the Figs. 8 and 9, respectively.

Similarly, in this case the anomaly vanishes, and the charge $Q^{(4)}$ is conserved, within numerical accuracy, for this type of soliton configuration. In addition, one has the qualitative realization of the symmetry (2.5) and the vanishing of the anomaly, as we have anticipated above based solely on symmetry arguments.

Some comments are in order here: First, the deformed models (5.21)-(5.22) have dark solitary waves and they present the properties of quasi-integrable models. Second, they present infinite tower of exactly conserved charges [4]: $Q^{(0)}, Q^{(1)}, Q^{(2)}, Q^{(3)}, \hat{Q}^{(4)}, \hat{Q}^{(6)}, \hat{Q}^{(8)}, ...$ Third, they present infinite tower of asymptotically-conserved charges: $\hat{Q}^{(5)}, \hat{Q}^{(7)}, \hat{Q}^{(9)}, ...$ such that

$$\frac{d\hat{Q}^{(n')}}{dt} = \beta^{(n')}, \quad \int_{-\infty}^{+\infty} dx\, \beta^{(n')} = 0 \Rightarrow \hat{Q}^{(n')}(x=+\infty) = \hat{Q}^{(n')}(x=-\infty), \;\; n' = 5, 7, 9, ... \qquad (5.28)$$

Fourth, one can argue that they exhibit solitary waves resembling to solitons, i.e. preserve their properties after collision.

IOP Conf. Series: Journal of Physics: Conf. Series **1143** (2018) 012004 doi:10.1088/1742-6596/1143/1/012004

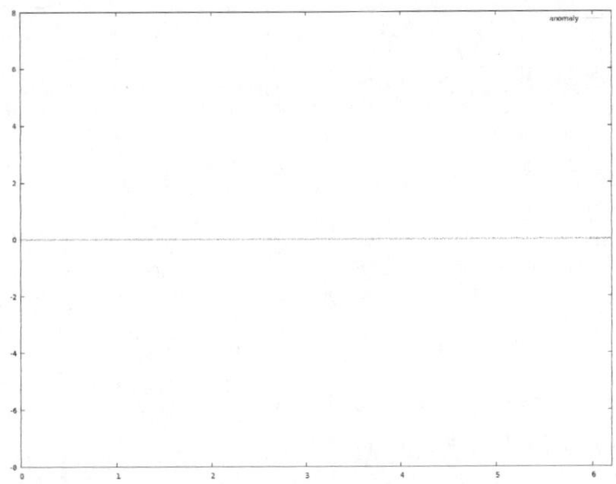

Figure 4. The anomaly $\beta_r^{(4)}(t)$.

Figure 5. The time integrated anomaly $\int^t dt' \beta_r^{(4)}(t')$.

6. Conclusion

The modified defocusing NLS models have been studied analytically and numerically. The charges of the MNLS models split into two subsets:

- A first subset of *asymptotically conserved* charges for 2-dark soliton solutions possessing *space-time* reflection parity symmetry [3].

- A second subset of infinite tower of *exactly conserved* charges for two-dark soliton solutions possessing *space reflection* parity symmetry [4].

We have computed numerically the first non-trivial anomaly $\beta^{(4)}$ of the $Q^{(4)}$ charge quasi-conservation law. We have verified that this anomaly vanishes, and consequently the exact conservation of the charge $Q^{(4)}$ holds for various 2-soliton configurations, within numerical accuracy. The only explanation we have found, so far, for the exact conservation of the even order charges $(n \geq 4)$, is the space-reflection parity symmetry of 2-dark soliton solutions.

In [4] we have found that even for two-soliton solutions with different velocities (different amplitudes) which do not satisfy the space-reflection symmetry, the anomaly $\beta^{(4)}$ vanishes, within the numerical accuracy. Then, we may argue that the parity property (2.5) is not the cause of

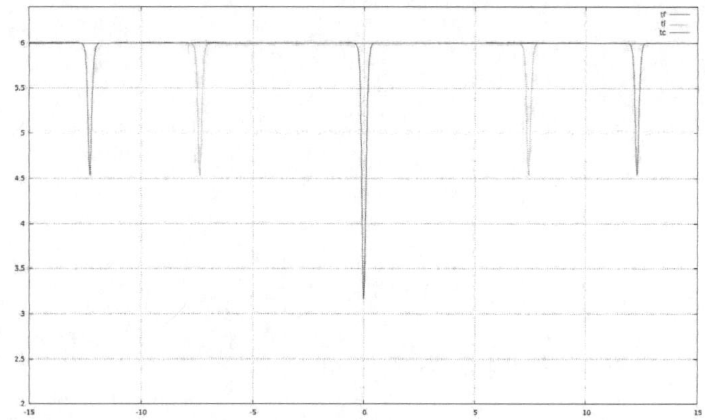

Figure 6. Transmission of two dark solitons of the **saturable NLS model** with $q = 3$ and equal velocities are plotted for $\epsilon = 0.1$, $|\psi_0| = 6$, $\eta = 2.5$. The initial gray solitons (t_i=green line) travel in opposite direction with velocities $v_1 = -6.42\sqrt{2}$ (right soliton), $v_2 = 6.42\sqrt{2}$ (left soliton). They completely overlap (t_c=blue line) and then transmit to each other. The gray solitons after collision are plotted as a red line (t_f). Note that $|v_1| + v_2 > v_s$ ($v_s = 7.1\sqrt{2}$)..

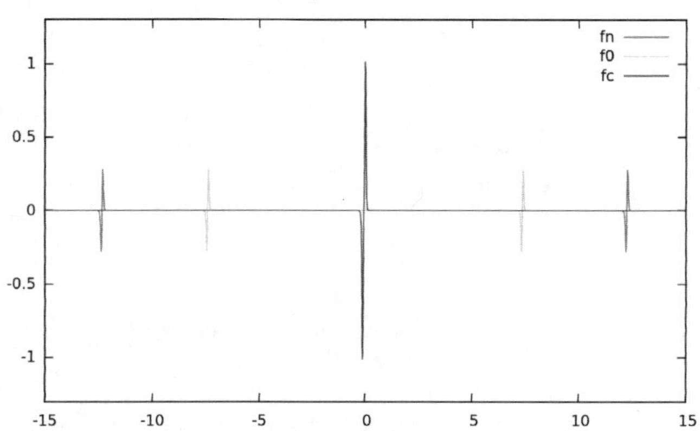

Figure 7. The anomaly density $\gamma(x,t)$ plotted for three successive times (t_i, t_c and t_f).

the exact conservation of the charges, but according to our analytical calculations it is a sufficient condition for these phenomena to happen. Further research work is necessary to settle such questions which involve the non-linear dynamics of the scattering. So, the symmetries involved in the quasi-integrability phenomenon deserve further investigation and they may have relevant applications in many areas of non-linear sciences.

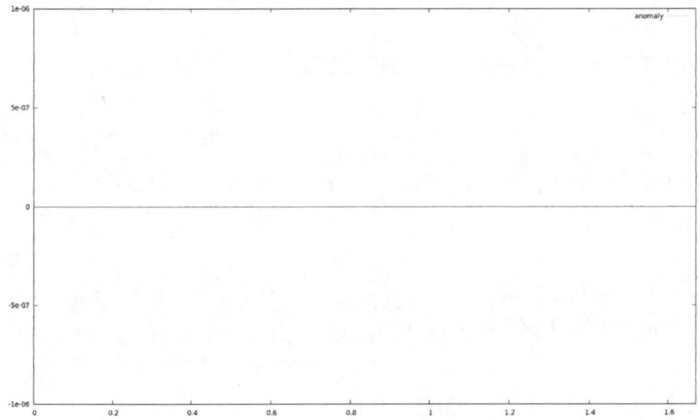

Figure 8. The anomaly $\beta_r^{(4)}(t)$.

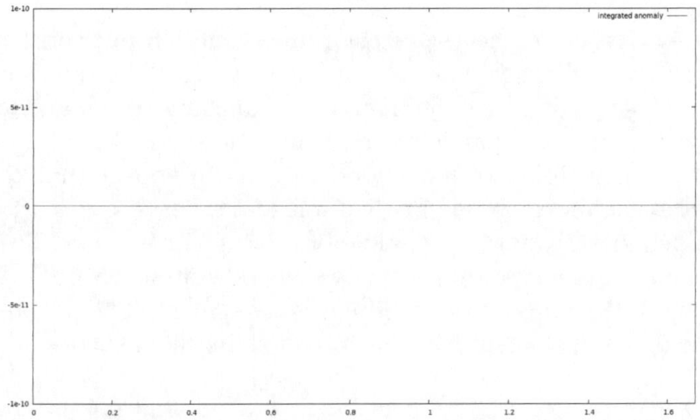

Figure 9. The time integrated anomaly $\int^t dt' \beta_r^{(4)}(t')$.

References

[1] J. Hietarinta, *Hirota's bilinear method and partial integrability*, in *Partially Integrable Equations in Physics*, R. Conte and N. Boccara eds., NATO ASI Series C310, Les Houches France March 21-30 1989 [inSPIRE].

[2] L.A. Ferreira and Wojtek J. Zakrzewski, *JHEP* **05** (2011) 130.
L.A. Ferreira and Wojtek J. Zakrzewski, *JHEP* **01** (2014) 058
V.H. Aurichio and L.A. Ferreira, *JHEP* **03** (2015) 152

[3] L.A. Ferreira, G. Luchini and Wojtek J. Zakrzewski, *JHEP* **09** (2012) 103.

[4] H. Blas and M. Zambrano, *JHEP* **03** (2016) 005.

[5] Weizhu Bao, *Methods and Applications of Analysis* **11** (2004) 001.
Weizhu Bao, Qinglin Tang, Zhiguo Xu, *Journal of Computational Physics* **235** (2013) 423.

[6] Y. S. Kivshar, B. Luther-Davies, *Phys. Reports* **298** (1998) 81.

[7] A. de O. Assunção, H. Blas and M. J. B. F. da Silva, *J. Phys. A: Math. Theor.* **A45** (2012) 085205.
Ohta Y, Wang D-S and Yang J 2011 *Stud. Appl. Math.* **127** (2011) 345.

IOP Conf. Series: Journal of Physics: Conf. Series **1143** (2018) 012005 doi:10.1088/1742-6596/1143/1/012005

Introduction to open string field theory

Carlos Tello
Faculty of Sciences, National University of Engineering, Av. Tupac Amaru, Lima 25, Peru
E-mail: ctelloe@uni.edu.pe

Abstract. We review basic concepts of open string theory. Then we sketch briefly some aspects of string field theory and comment some applications.

1. Introduction

The quantum mechanics and special relativity are the two most fundamental frameworks in high energy physics today. When these two achievements are combined we get the quantum field theory. Using this theory and the concept of gauge symmetry we can build the standard model describing successfully all observed elementary particles and fields. However such elegant synthesis does not include the gravity. In principle, we have good reasons to expect that the higher symmetry we include in our model the more unified description we get.

If we could discover this more unified description we could obtain a complete and deep understanding of details of our universe including quantum gravity. One key step toward such unifying theory is to replace our idea of point particles by one dimensional strings. The purpose of this article is to give an introduction to some elementary concepts of string field theory.

2. Open String

A string moving in a flat Minkowski $D = 26$ dimensional space-time is described by its space-time coordinates $X^\mu(\tau,\ \sigma)$, where τ, σ are the world sheet time-like and spatial-like parameters of the string, and $\eta_{\mu\nu}$ is the space-time Minkowski metric. A good textbook to learn more about string theory can be found in [1].

2.1. Polyakov Action

The dynamic of a relativistic string is determined by the **Nambu-Goto** Action

$$S_{NG}[X] = -\int d\tau d\sigma \sqrt{-\begin{vmatrix} \partial_\tau X^\mu \partial_\tau X_\mu & \partial_\tau X^\mu \partial_\sigma X_\mu \\ \partial_\sigma X^\mu \partial_\tau X_\mu & \partial_\sigma X^\mu \partial_\sigma X_\mu \end{vmatrix}}. \tag{1}$$

This action represent the geometrical area of the worldsheet swept by the evolution of the open string, and it is invariant by reparametrizations of the worldsheet. Since the Hamiltonian of this action is nonlinear the quantization is hard, physicists prefer to use an equivalent action proposed by **Polyakov**

$$S_P[X,g] = \int d\tau d\sigma \sqrt{g} g^{\alpha\beta} \partial_\alpha X^\mu \partial_\beta X_\mu, \tag{2}$$

were $g^{\alpha\beta}$ is the intrinsic metric.

2.2. Symmetries
The Polyakov action has space-time Poincare symmetry, two dimensional worldsheet reparametrization symmetry, and weyl invariance. Using the last two symmetries we can gauge fix so that $g_{\alpha\beta} = h_{\alpha\beta} = diag(-1, 1)$ and obtain the action

$$S[X] = \int d^2 z \partial X^\mu \bar{\partial} X_\mu \tag{3}$$

This action has infinite dimensional two dimensional conformal invariance. This means that from the point of view of the world-sheet the space-time coordinates X^μ are free fields of a conformal field theory. This theory has infinite number of constraints associated to the vanishing of the world-sheet energy-momentum tensor $T = \partial X^\mu \partial X_\mu$. The equation of motion is $\partial\bar{\partial} X = 0$, and it solution is $X(z, \bar{z}) = f(z) + g(\bar{z})$. The boundary conditions for each freely moving endpoints are Neumann-like.

2.3. Quantization
Since the action is quadratic the quantization is simple, and we can use the radial quantization method. The spectrum is discrete with energy eigenvalues $-1,\ 0,\ 1,\ 2, \ldots$. All string states are of the form $|\Psi\rangle = \alpha^{\mu_1}_{-n_1}\alpha^{\mu_2}_{-n_2}\cdots\alpha^{\mu_i}_{-n_i}|0\rangle$, where α^μ_{-n} are creation operators acting on the $SL(2, R)$ vacuum $|0\rangle$. Each elementary particles is identified with an irreducible representation of the Poincare group. In covariant quantization the string physical states must satisfy the **Virasoro** constraints $(L_n - \delta_{n,0})|\Psi\rangle = 0,\ n \geq 0$.

We consider the BRST quantization because very convenient for strings. This method requires to include an anti-commuting $(b,\ c)$ Fadeev-Popov ghosts.

$$S[X, b, c] = \int d^2 z(\partial X^\mu \bar{\partial} X_\mu + b\bar{\partial}c + \bar{b}\partial\bar{c}). \tag{4}$$

Physical states are of the form

$$|\Psi\rangle = \alpha^{\mu_1}_{-n_1}\alpha^{\mu_2}_{-n_2}\cdots\alpha^{\mu_i}_{-n_i} \otimes b_{-m_1}b_{-m_2}\cdots b_{-m_j} \otimes c_{-p_1}c_{-p_2}\cdots c_{-p_k}|0\rangle$$

$\mu_i = 0, 1, 2, \ldots D-1,\ n_i \geq 0,\ \ m_j \geq 2,\ \ p_k \geq -1.$

The BRST operator $Q = \int dzc(T + b\partial c)$ is anticommuting and nilpotent $Q^2 = 0$ only for $D = 26$. This operator is used to select the physical states. A physical state $|\Psi\rangle$ must be annihilated by Q and must not be BRTS trivial. This means $Q|\Psi\rangle = 0$ and $|\Psi\rangle \neq |\lambda\rangle$. In other words, all physical states V belongs to the Q cohomology.

Corresponding to each physical state of the string we have a vertex operator. These vertex operators are used to compute scattering amplitudes. In general there is a standard perturbative prescription to compute this scattering amplitude involves a path integral over all configurations of the worldsheet sum over all world sheet topologies, integration over a moduli space and products of vertex operators. The scattering on-sheel tree level amplitude for four open strings in tachyon states was found by Veneziano. The generalization of this amplitude for n-point open strings in tachyon states was found by Koba-Nielsen.

3. Open String Field

To understand the behavior of strings in non flat space-time backgrounds, to learn about compactification mechanisms, and symmetry breaking etc. we need to go beyond the first quantized perturbative approach. To learn about all these non perturbative phenomena in string theory, we need a second quantized approach

3.1. *Witten String Field*

The vacuum of a string field is the BPZ vacuum $|\Omega\rangle$ defined by $|\Omega\rangle = c_1|0\rangle$, where $|0\rangle$ is the $SL(2,R)$ vacuum, and satisfy $\alpha_n|\Omega\rangle = 0,\ b_n|\Omega\rangle = 0,\ c_n|\Omega\rangle = 0,\ \forall n \geq 1$, and $b_0|\Omega\rangle = 0,\ \langle\Omega|c_0|\Omega\rangle = 1$.

A string field is a functional $A[X,\ b,\ c]$ with ghost number $n_{gh} = 1$. Expanding the first lower levels we have

$$\begin{aligned} A[X(\sigma),\ b(\sigma),\ c(\sigma)] &= [T(x) + A_\mu(x)\alpha^\mu_{-1} + B_\mu(x)\alpha^\mu_{-2} + C_{\mu\nu}(x)\alpha^\mu_{-1}\alpha^\nu_{-1} \\ &\quad + D(x)b_{-1}c_{-1} + \cdots]\,|\Omega\rangle + [U(x)b_{-1} + \cdots]\,c_0|\Omega\rangle \end{aligned}$$

As we can see the string field contains infinite number of space-time fields of arbitrary high integer spin as coefficients. $T(x)$ is the tachyon scalar field, A_μ is the massless Yang-Mills vector potential field, $B_\mu(x)$ is a massive vector field, $C_{\mu\nu}(x)$ is a massive tensor field, $D(x)$ and $U(x)$ are auxiliar scalar fields.

The free string field action is built using the BRST Q operator

$$S[A] = \frac{1}{2}\int d^{26}x\langle A|QA\rangle$$

For simplicity. we omit in this action action a trace over the product of the Chan-Paton matrices associated to each open string corresponding to the gauge group $SO(32)$. This action is invariant by the gauge transformation $\delta A = Q\Lambda$, and its equation of motion is $QA = 0$. In the Siegel's gauge $b_0 A = 0$ this equation of motion is equivalent to $(L_0 - 1)A = 0$. Evaluating the correlation functions in the free string action we obtain

$$S[T, A, U, \ldots] = \int d^{26}x\left[\frac{1}{2}T(+2)T + \frac{1}{2}A_\mu A^\mu + U\partial_\mu A^\mu + U^2 + \ldots\right]$$

The gauge parameter Λ has $n_g = 0$

$$\Lambda = [\lambda(x)b_{-1} + \nu_\mu(x)\alpha^\mu b_{-1} + \ldots]\,|\Omega\rangle + [\omega(x)b_{-1}b_{-2} + \ldots]\,c_0|\Omega\rangle.$$

in components the gauge transformations are $\delta T = 0,\ \delta A_\mu = \partial\lambda,\ \delta U = \frac{1}{2}\lambda$.

In 1986 Witten generalized this action to describe interacting strings [2]

$$S[A] = \int d^{26}x\,\left[\frac{1}{2}\langle A|QA\rangle + \frac{1}{3}\langle A|AA\rangle\right] \qquad (5)$$

gauge invariant under $\delta A = Q\Lambda + A\Lambda - \Lambda A$. This action is a generalization of the Chern-Simons action $S_{CS} = Tr\int d^3x\epsilon^{\mu\nu\rho}\left(\frac{1}{2}A_\mu\partial_\nu A_\rho + \frac{1}{3}A_\mu A_\nu A_\rho\right)$ which is gauge invariant by $\delta A_\mu = \partial_\mu + [A_\mu, \lambda]$.

4. Applications

In a remarkable work Giddings reproduced the Veneziano amplitude using the Witten string field action [3]. He used conformal Riemann-Christoffeld transformations to map the folded Witten vertex into the complex upper half plane and compute the required correlation functions. This work motivated to develop a systematic reformulation of string field theory using conformal field theory by Leclair [4]. An excellent review of conformal field theory can be found in a lecture of Ginsparg [5].

In practical computations in string field theory, all the terms in the action are expressed as n-point correlation functions with a sequence of conformal maps from the upper half plate to wedges

$$\langle A_1, \ldots A_n\rangle = \langle f^{(n)}_1 \circ A_1(0) \ldots f^{(n)}_n \circ A_n(0)\rangle,$$

IOP Conf. Series: Journal of Physics: Conf. Series **1143** (2018) 012005 doi:10.1088/1742-6596/1143/1/012005

$$f_k^{(n)} = e^{\frac{2\pi i(k-1)}{n}} \left(\frac{1+iz}{1-iz}\right)^{2/n}, \quad n \geq 1.$$

More recently, this conformal technique helped to compute string field corrections to Yang-Mills and tachyon condensation. Finally,the quantization of the Witten string field action was completed using Batalin-Vilkovitsky methods. For all these these successful results we believe that string field is a very useful and interesting research area.

5. Conclusion

We explained very briefly some tools, concepts and principles used in open string and open string field theory. A lot of progress has been done using the first quantized approach of strings moving in flat space-time Minkowski background using perturbative methods. New and different kinds of symmetries have been discovered in string theory. On the other hand, open string field theory had a little slow progress in applications, but recently it has stablished as a solid research area.

However open string field theory has the following three problems: It contains a tachyon, it includes only bosonic string states, so it does not consider fermions. Furthermore, since the critical space-time dimension 26 is too high from a phenomenological perspective. In a related paper of this proceeding of the school we will discuss how the incorporation of supersymmetry strongly restrict the string theory in such a way that solves many of these problems.

Acknowledgments

This work had financial support from the Research Institute of the Faculty of Sciences of the National University of Engineering at Lima-Peru.

References

[1] Zwiebach B 2009 *A First Course in String Theory* (Cambridge University Press)
[2] Witten E 1986 *Nucl. Phys.* **B268** 253
[3] Giddings S 1986 *Nucl. Phys.* **B278** 242
[4] Leclair A, Peskin M, and Preitschopf C 1989 *Nucl. Phys.* **B317** 464
[5] Ginsparg P 1989 Applied Conformal Field Theory *Les Houches, Session XLIX* (arXiv:hep-th/9108028)

IOP Conf. Series: Journal of Physics: Conf. Series **1143** (2018) 012006 doi:10.1088/1742-6596/1143/1/012006

Introduction to Open Superstring Field Theory

Carlos Tello

Faculty of Sciences, National University of Engineering, Av. Tupac Amaru, Lima 25, Peru

E-mail: ctelloe@uni.edu.pe

Abstract. We review briefly open superstring theory, and then provide a very short introduction to some basic aspects of open superstring field theory and comment some applications.

1. Introduction

The first quantized approach of open superstring theory has gauge group $SO(32)$ with critical Space-time dimension $D = 10$. This theory is space-time supersymmetric, tachyon free, and describes all known massless bosonic and fermionic particles. Superstring theory has great properties to build better phenomenological models in high energy physics than string theory.

Since our observable physical world is four dimensional we have to consider some compactification of the six extra internal coordinates. The principles that control this compactification is not completely understood yet. In order to make progress in the unification we need to develop non perturbative methods and go beyond the first quantized description aof strings. By implementing supersymmetry in string theory we obtain better convergence of scattering amplitudes and vacuum. The purpose of this article is to provide a short introduction to superstring field theory.

2. Open Superstring

Supersymmetry is a beautiful and useful extension of Poincare algebra that relates fermions an bosons. Supersymmetry makes strong restrictions when we applied it to fundamental strings. In this paper we only discuss descriptions of superstring that have supersymmetry manifest in the world sheet because they are simple to quantize. A great textbook to learn more details about superstrings is [1].

2.1. Ramond Neveu Schwarz Superstring

To the bosonic worldsheet coordinates $X^\mu(\tau,\sigma)$ we incorporate the fermionic coordinates $\psi^\mu(\tau,\sigma)$, where now $\mu = 0, 1, \ldots, 9$. The theory is consistent at quantum level only for $D = 10$ space-time dimensions. Using complex coordinates in the worls-sheet by $z = (\sigma + i\tau)/2$, $\bar{z} = (\sigma - i\tau)/2$, the **RNS** action is

$$S_{RNS}[X,\psi] = \int d^2z(\partial X^\mu \bar{\partial} X_\mu + \psi^\mu \bar{\partial}\psi_\mu + \bar{\psi}^\mu \partial \bar{\psi}_\mu) \tag{1}$$

In superstring theory we have two sectors. The first one is the **Neveu-Schwarz** sector that corresponds to antiperiodic boundary conditions in ψ and contains the space-time bosonic states.

IOP Conf. Series: Journal of Physics: Conf. Series **1143** (2018) 012006 doi:10.1088/1742-6596/1143/1/012006

The second one is the **Ramond** sector that corresponds to periodic boundary conditions in ψ and contains the space-time bosonic states.

2.2. Symmetries
Classically, the RNS action has manifest space-time Poincare symmetry and $N = 1$ world-sheet superconformal symmetry. In RNS superstring the space-time supersymmetry is not manifest, but it this symmetry can be restored at after an appropriate truncation of the quantum spectrum. In this paper we are not consider the Green-Schwarz description that has manifest space-time supersymmetry because its action is nonlinear and hard to quantize.

2.3. Quantization
From the point of view of the world-sheet the variables X^μ and ψ^μ are two dimensional fields with an infinite dimensional superconformal symmetry. In order to quantize such highly constrained system we need both the conformal anti-communting (b, c) ghost system and the superconformal commuting (β, γ) superghost system. The generators of this superconformal symmetry correspond to the **Supervirasoro** constraints and satisfy a superalgebra. The matter contribution of these generators are the world sheet energy momentum tensor T and the world-sheet super-current G. The lowest level of the spectrum is masless and contain the bosonic Yang-Mills gluon and its supersymmetric partner the fermionic gluino.

In 1986. a more convenient superghost variables to build fermionic vertex operators was discovered by Friedan, Martinec, and Shenker [2]. This new system is made of an anti-commuting (η, ξ) superghost system and a chiral boson ϕ. The relation of these variables with the previous one is $\beta = e^{-\phi}\partial\xi$, $\gamma = \eta e^{\phi}$.

The BRST nilpotent operator in terms of the FMS variables is

$$Q = \int dz[c(T + b\partial c - \frac{1}{2}\partial\phi\partial\phi - \partial^2\phi - \eta\partial\xi) + \eta e^{\phi}G^m - b\eta\partial\eta e^{2\phi}]$$

All physical vertex operator V belongs to the Q cohomology. Furthermore, in order to remove the tachyon from the physical spectrum we have to apply the **GSO projection**. The **Space-time supersymmetry** current q^α involve bosonized expressions of ϕ and ψ^μ.

In the first quantized approach we use vertex operators associated to any particular set of superstring states to compute on-shell scattering amplitudes. The prescription to do this computation is essentially perturbative and has been done only for flat Minkowski space-time background.

One subtle aspect in superstring theory is the fact that fermionic vertex operator for Ramond states needs bosonized expresions of ϕ and ψ^μ. Furthermore for each state of the superstring we have an infinite number vertex operator that are physically equivalent. Each vertex operators is characterized by one picture number. Ramond vertex operators have half integer picture values and Neveu-Schwarz vertex operators have integer picture values. Picture number is defined by $n_p = \int dz(\xi\eta - \partial\phi)$ and Picture raising operator is defined by $Z = \{Q, \xi\}$. The Hilbert space of the FMS variables is called the large Hilbert space because it is bigger than the Hilbert space of the (β, γ) system which do not depend of the zero mode ξ_0.

3. Open Superstring Field

In 1986 Witten proposed a action for interacting open superstring field [3]. This cibic action is Chern-Simons-like $S_W[V, \Psi] = \langle V|QV\rangle + \langle V|ZVV\rangle + \langle\Psi|YQ\Psi\rangle + \langle V|\Psi\Psi\rangle$, where V and Ψ are the NS and R superstring fields. But this action turns out to be inconsistent because it requires the use of picture changing operators that are singular when they coincide. Many years physicist were looking for a consistent action in order to fix this technical problem.

IOP Conf. Series: Journal of Physics: Conf. Series **1143** (2018) 012006 doi:10.1088/1742-6596/1143/1/012006

3.1. Berkovits Superstring Field Action

In 1995 Berkovits proposed a WZW-like open superstring field theory action restricted to the Neveu-Schwarz sector [4]. This theory is based on the FMS variables an uses a single superstring field $\Phi[X, \psi, b, c, \xi, \eta, \phi]$ The action is

$$\begin{aligned} S_B[\Phi] &= \frac{1}{2g^2}\Big[\langle\langle e^{-\Phi}G_0^+ e^{\Phi}|e^{-\Phi}\tilde{G}_0^+ e^{\Phi}\rangle\rangle \\ &\quad - \int_0^1 dt\langle\langle e^{-t\Phi}\partial_t e^{t\Phi}|\left\{e^{-t\Phi}G_0^+ e^{t\Phi},\ e^{-t\Phi}\tilde{G}_0^+ e^{t\Phi}\right\}\rangle\rangle\Big] \end{aligned} \tag{2}$$

$\langle\langle\ldots\rangle\rangle$ is correlation in the large Hilbert space, and $G_0^+ = Q$ and $\tilde{G}_0^+ = \eta_0$

3.2. Symmetries

This action is gauge invariant under the gauge transformation

$$\delta e^{\Phi} = (G_0^+\Omega)e^{\Phi} + (\tilde{G}_0^+\tilde{\Omega})e^{\Phi} \tag{3}$$

This action is a generalization of the WZW Wess Zumino Witten Action

$$S[\gamma] = \int_{S^2} dz^2 \left(\gamma^{-1}\partial^{\mu}\gamma\right)\left(\gamma^{-1}\partial^{\mu}\gamma\right) + \int_{B^3} d^3y\epsilon^{ijk}\left(\gamma^{-1}\partial_i\gamma\right)\left[\left(\gamma^{-1}\partial_j\gamma\right),\ \left(\gamma^{-1}\partial_k\gamma\right)\right] \tag{4}$$

The equation of motion is $\eta_0\left(e^{-\Phi}Qe^{\Phi}\right) = 0$.

The string field Φ satisfy the following relations

$$\begin{aligned} \langle\langle\Phi_1\ \Phi_2\ldots\Phi_{n-1}\Phi_n\rangle\rangle &= \langle\langle\Phi_n\ \Phi_1\ldots\Phi_{n-2}\Phi_{n-1}\rangle\rangle \\ \langle\langle\Phi_1\ \Phi_2\ldots\Phi_{n-1}Q\Phi_n\rangle\rangle &= -\langle\langle Q\Phi_n\ \Phi_1\ldots\Phi_{n-2}\Phi_{n-1}\rangle\rangle \\ \langle\langle Q(\cdots)\rangle\rangle &= 0 \\ Q(\Phi_1\Phi_2) &= (Q\Phi_1)\Phi_2 + \Phi_1(Q\Phi_2) \end{aligned}$$

The quantization of the WZW-like action is currently in development using the BV methods. The gauge fixing procedure of the free action has been developed, but the full interacting action needs to be completed.

4. Applications

We sketch here the computation of tree level on-shell amplitudes of four NS superstrings [5].

4.1. Superstring Amplitudes

The linearized gauge transformation is $\delta\Phi = G_0^+\Omega + \tilde{G}^+\tilde{\Omega}$. Two gauge fixing conditions $G_0^-\Phi = b_0\Phi = 0$, $\tilde{G}_0^-\Phi = \xi_0\Phi = 0$. Using this condition we build the appropriate propagator. Then we need to expand the action up to quartic order because they all contributes for the four string amplitudes.

$$S[\Phi] = \langle\langle\frac{1}{2}G_0^+\Phi\tilde{G}_0^+\Phi - \frac{1}{6}\Phi\left\{G_0^+\Phi,\ \tilde{G}_0^+\Phi\right\} - \frac{1}{24}[\Phi,\ G_0^+\Phi][\Phi,\ \tilde{G}_0^+\Phi] + \cdots\rangle\rangle \tag{5}$$

The Giddings conformal transformation is also required to map from the conic-folded Witten vertex worldsheet toward the complex upper-half plane UHP.

The total amplitude has three contributions: (a) Channel-s A_s (one Propagator + two cubic Vertices), (b) Channel-t A_t (one Propagator + two cubic Vertices), and (c) Quartic Vertex A_q (without Propagator) such that $A_4 = A_s + A_t + A_q$ We verified that the first quantized prescription is correctly reproduced

The WZW-Like action has generalized to describe non-BPS D-branes. In order to do that it is considered the $GSO(-)$ projected out states of the superstring which includes the tachyon. This allowed to calculate aproximated computation of tachyon condensation [6].

IOP Conf. Series: Journal of Physics: Conf. Series **1143** (2018) 012006 doi:10.1088/1742-6596/1143/1/012006

5. Conclusions

We introduced some basic ideas about superstring and superstring field theory. The main mathematical tools are superconformal field theory BRST quantization, gauge symmetry, and supersymmetry. After a long time dealing with technical problems related with picture changing operators, the classical superstring field theory action for Neveu Schwarzs has been clearly identified and it is currently being tested.

We expect that a recent proposed open superstring field theory actions that include the Ramond sector will give us better understanding of dualities and other non perturbative effects. And soon we will able to compute better low energy effective actions to super Yang Mills Actions.

Acknowledgments

This work had financial support from the Research Institute of the Faculty of Sciences of the National University of Engineering, Lima-Peru.

References

[1] Polchinski J 1998 *String Theory* vol 2 (Cambridge University Press)
[2] Friedan D, Martinec E, and Shenker S 1986 *Nucl. Phys.* **B271** 93
[3] Witten E 1986 *Nucl. Phys.* **B276** 291
[4] Berkovits N 1985 *Nucl. Phys.* **B450** 90
[5] Berkovits N and Tello C 2000 *Phys. Lett.* **B478** 343
[6] Berkovits N, Sen A, and Zwiebach B 2000 *Nucl. Phys.* **B587** 147

IOP Conf. Series: Journal of Physics: Conf. Series **1143** (2018) 012007 doi:10.1088/1742-6596/1143/1/012007

Sunspot characteristics of active regions NOAA 2268 and NOAA 2305

K. F. Lopez[1], V. S. Cañedo[1,4], D. P. Cabezas[2], and Y. J. Buleje[1,3]

[1] Escuela Académica Profesional de Física, Facultad de Ciencias, Universidad Nacional San Luis Gonzaga de Ica, Av. Los Maestros S/N, Ica, Peru
[2] Kwasan and Hida Observatories, Kyoto University, Yamashina, Kyoto, 607-8471, Japan
[3] Centro de Investigación de la Actividad Solar y sus Efectos sobre la Tierra, Facultad de Ciencias, Universidad Nacional San Luis Gonzaga de Ica, Av. Los Maestros S/N, Ica, Peru

E-mail: karlalopezaraujo131608@gmail.com

Abstract. In this paper we present a study of two solar active regions NOAA 2268 and NOAA 2305 observed on January 28 and 29, and also on March 24 and 25 of 2015, respectively. Such active regions produced multiple solar flares including M-class flares as classified in the soft X-ray scale using GOES observation. We mainly use sunspot sketches obtained with a 15 cm diameter Takahashi telescope located at Universidad Nacional San Luis Gonzaga de Ica, Peru. The active regions were characterized on the basis of the sunspots number and the magnetic configuration. We also derived the physical speed of the spots under study, resulting 1.69 and 1.76 km s^{-1} for the period of January and March, respectively. In addition, considering the calculated speeds we estimated the rotation period of the Sun for the dates mentioned above, which result in 28.9 days for January and 27.4 days for March. The present study is complemented with data taken by the Helioseismic and Magnetic Imager (HMI) instrument onboard the Solar Dynamics Observatory (SDO), which provides high resolution solar images in the continuum at 6173 Å as well as magnetograms.

1. Introduction

The Sun is made of plasma, in its core the energy is produced throughout the nuclear fusion building up helium nuclei from hydrogen nuclei by the mean of proton-proton chain reaction [1]. The energy produced in the core can take as long as approximately 10^7 years of journey through the radiative and convective zones until arrive at the Sun's surface known as the photosphere, where the plasma level is less dense than its interior [1]. Then the radiation passes through the chromosphere, transition region and the corona where the photons stop interacting with the outer atmosphere. This radiation is observed in almost the entire electromagnetic spectrum.

The Sun does not rotate as a rigid sphere, its equator rotates faster than their poles, this is called differential rotation, moreover, it is responsible of continuous twisting of the magnetic field lines. One of the most characteristic manifestations on the solar surface is the presence of sunspots. They are formed in perturbed areas known as active region (ARs), exhibit complex magnetic configuration and variable in time. A sunspot is composed of a dark central core named umbra and it is surrounded by filamentary radial structures called penumbra. The effective

[4] Present address: Escuela de Física, Facultad de Ciencias e Ingeniería, Pontificia Universidad Católica del Perú, Av. Universitaria 1801, San Miguel Lima 32

IOP Conf. Series: Journal of Physics: Conf. Series **1143** (2018) 012007 doi:10.1088/1742-6596/1143/1/012007

temperature of the umbra is approximately 3700 K, $\sim 2100\ K$ cooler than the photosphere [1]. The magnetic field of an umbra is around 2600–2900 Gauss [2], which is equivalent to 4000 times more intense than the value of the Earth's magnetic field, while in the penumbra the magnetic field is in average about 1000 Gauss [1].

In 1844 Heinrich Schwabe noticed a regular variation in the sunspots number, approximately every 10 years. This finding was confirmed by Rudolf Wolf, who in 1849 conducted a more precise estimation and found a variation of 11 years [3]. Every 11 years the Sun moves through a period of fewer, smaller sunspots, prominences, and flares - called "solar minimum" and a period of more, large sunspots, prominences, and flares - called "solar maximum". This activity of minimum and maximum is connected to a period of 22 years known as the Hale's cycle. Wolf suggested that it would be more convenient to identify groups of sunspots rather than individual sunspots. Wolf defined the sunspots number as $R = k(10\ g + n)$, where g is the number of sunspot groups on the solar disk, n the total number of individual spots contained in a sunspot group, and k the correction factor for the observer [3], being k approximately 1.

In the vicinity of the ARs, solar flares may occur releasing large amounts of magnetic energy $10^{26} - 10^{32}$ erg in very short time interval (100–1000 s) [4]. The dissipation of the released magnetic energy leads to the acceleration of non-thermal particles as well as plasma heating in the chromosphere and the corona, resulting in a rapid brightening and simultaneous manifestation in almost the entire electromagnetic spectrum [5]. Solar flares are classified according to the soft X-ray flux variation recorded by the Geostationary Operational Environmental Satellites (GOES). The X-ray fluxes are measured in two bands 0.5–4.0 and 1.0–8.0 Å [6]. The excess of the flux is given in logarithmic scale, meaning levels A and B the lowest which represent mainly "quiet" Sun emissions, levels C, M, and X indicate increases of the flux due to solar activity and very intense events like solar flares. Solar flares can trigger coronal mass ejection (CME) in which large amounts of plasma is ejected into the interplanetary space. When CMEs are directed towards the Earth they are considered potentially geo-effective causing interplanetary shocks and geomagnetic storms, as well as secondary effects on the Earth.

2. Observations and Data

In this work we use mainly sunspot sketches obtained with a 15 cm refractor Takahashi EM-500 telescope, installed at the Facultad de Ciencias, Universidad Nacional San Luis Gonzaga de Ica. This telescope allow us to make observations of the solar photosphere. The data record is based on the projection method, which consists of projecting the solar disk image on a screen where a template of the solar disk of 15 cm diameter is placed, and on which we sketch and record the sunspots (Figure 1). This kind of data acquisition has been carried out regularly for several years and to date we have a large amount of sunspot records. In this paper we focus on the records taken on January 28 and 29 as well as on March 24 and 25 of 2015, respectively.

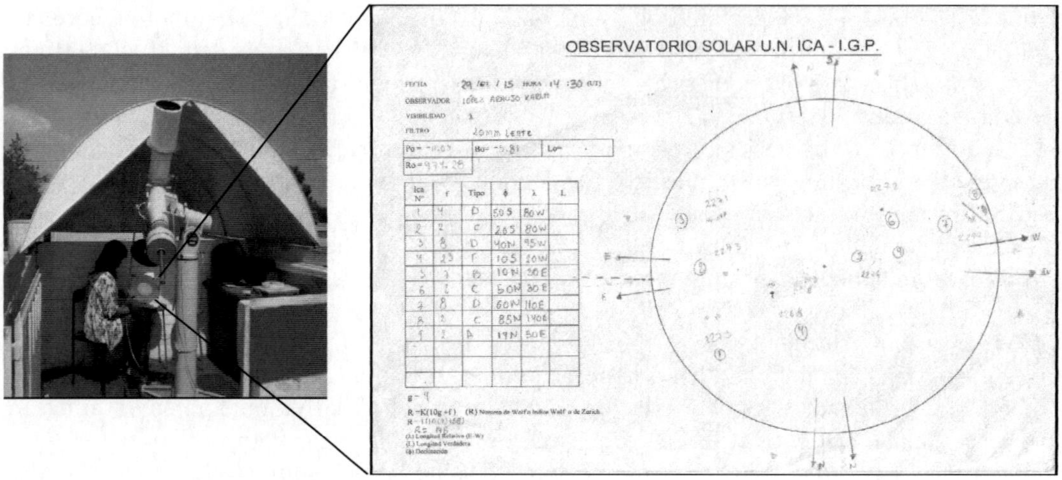

Figure 1. Left: Takahashi telescope located at the Facultad de Ciencias, Universidad Nacional San Luis Gonzaga de Ica, Peru. Right: example of sunspot data collection by using the projection method.

In order to compare our observations, we have used data from the *Helioseismic and Magnetic Imager* instrument on board the *Solar Dynamics Observatory* (SDO) satellite [7]. The HMI instrument provides high resolution observations of the photosphere at wavelength 6173 Å, with a spatial resolution of 0.6" pixel^{-1}. It also provides magnetograms that were used to analyze the magnetic configuration of the ARs. In addition, we have taken extreme-ultraviolet (EUV) data at 211 and 193 Å from the *Atmospheric Imaging Assembly* (AIA) on board the SDO, which allowed us to follow the morphological evolution of the solar flares associated with the ARs under study.

To verify the intensity and number of flares produced by the ARs of our interest, we used GOES satellite data which provides soft X-rays information from the Sun. In our study we use data in the channels 0.5 - 4.0 Å (3.0 - 25 keV) and 1.0 - 8.0 Å (1.5 - 1.0 keV). In panels (a) of Figures 2 and 3 we present soft X-ray emissions for the period of January 23 to 30 and March 23 to 31 of 2015, respectively. In January emission enhancement were observed in soft X-ray due to several M-class flares that took place between January 28 to 30. In contrast during March the Sun was less active exhibiting mainly C-class flares and only one M-class on March 25. In panels (b) of Figures 2 and 3 we also show EUV band images of the flares during their maximum phase. The EUV images show emissions of the hot corona at temperature of about 2 MK.

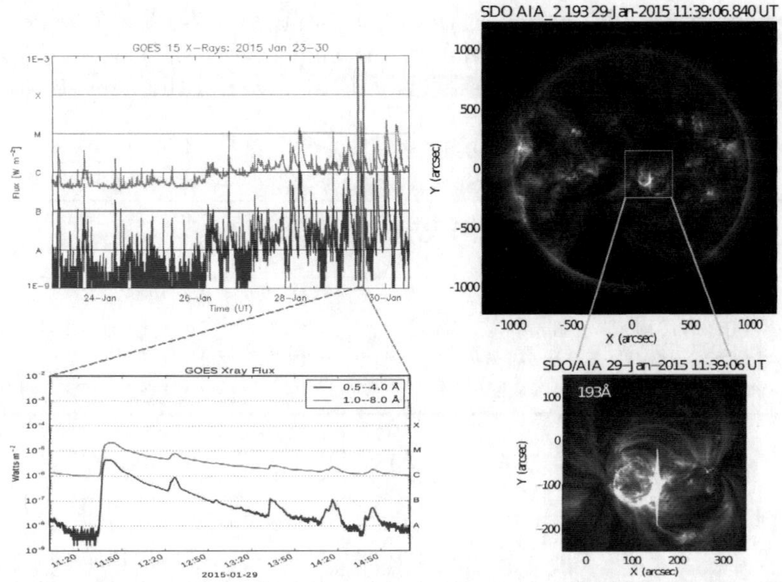

(a) (b)

Figure 2. Upper left panel: GOES X-ray emission at 0.5-4.0 and 1.0-8.0 Å, corresponding to January 23-30 of 2015. In the bottom panel it is also shown an enlarged view of the emission for January 29. It can be see an increase of the solar activity during the last days of this month. Right panels: extreme-ultraviolet image at 193 Å showing the active region NOAA-2268 during the M-class flare occurred on January 29 of 2015 at 11:39:00 UT.

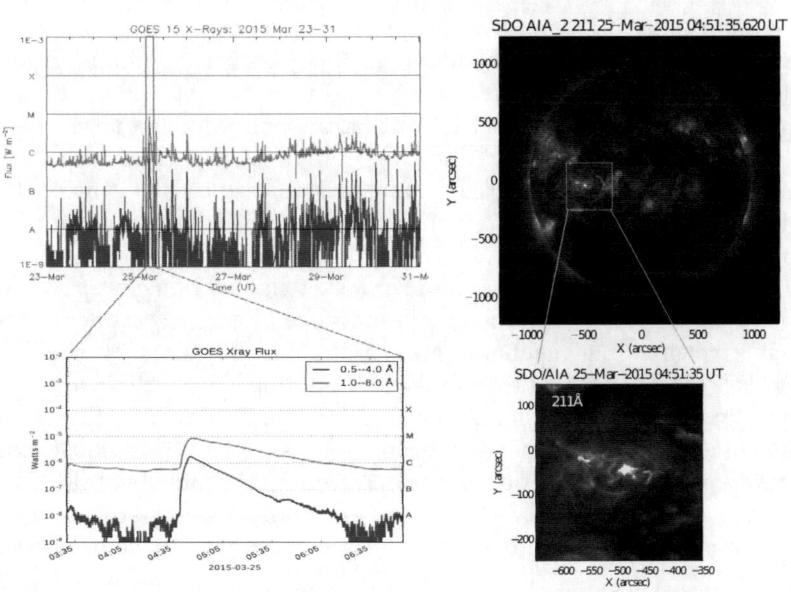

(a) (b)

Figure 3. Same as in Figure 2 but for the period of March 23-31, 2015. During this period the solar activity shows low or moderate levels. Right panels: extreme-ultraviolet image showing the active region NOAA-2305 at 211 $\AA$ during the M-class flare occurred on March 25 of 2015 at 04:51:01 UT.

IOP Conf. Series: Journal of Physics: Conf. Series **1143** (2018) 012007 doi:10.1088/1742-6596/1143/1/012007

In table 1 we summarize the characteristics of the studied ARs. We present the locations in heliographic coordinates, the sunspots number, the magnetic classification, and the number of flares, respectively. Details of the analyzed sunspots are given in the next section.

Table 1. Summary and characteristics of the studied active regions.

Date	Time (UT)	Lat. CMD[7]	NOAA[5]	CIASEST[6]	Zurich Class	Magnetic Class	Flares class C	M	X
28/01/2015	15:41:00	S10W05	29	13	E	BG[8]	3	1	-
29/01/2015	14:30:00	S10W20	35	23	F	BG	2	1	-
24/03/2015	15:44:00	S08E31	15	20	E	BGD[9]	2	-	-
25/03/2015	15:39:00	S08E19	26	26	E	BGD		1	-

3. Analysis and Results

In this section we present the results of our analysis, considering sunspot sketch data obtained with a refractor telescope. The morphology of the sketched sunspots are used to classify them following the Zurich catalog, as well as to determine the magnetic configuration. We also compare the sunspot records with observations by the HMI instrument aboard the SDO satellite. On the other hand, some physical parameters derived from the sketched data are also presented, such as the physical speed of sunspots and the rotation period of the Sun.

3.1. Sunspots characteristics and magnetic configuration

We characterize briefly our sunspot sketches obtained with the projection method, and compare with the observational report provided by the Space Weather Prediction Center (SWPC) of the NOAA. We determine the Wolf's number considering the sunspots number in our observation. According to the Zurich catalog, the sunspots can be classified as E and F because they are bipolar sunspots surrounded by several penumbrae. In relation to the magnetic configuration, the sunspots are classified as beta-delta (BD) during the period of January, whereas beta-delta-gamma (BGD) for the period of March. The BG configuration is characterized by the mixture of polarities in a dominantly bipolar distribution, while in case of BDG the configuration of the sunspots exhibit two cores sharing a common penumbra. This latter known as delta spots are more prolific to trigger solar flares [6][8].

In left panels of Figure (4) we present our sunspot sketches on January 29 and March 25, 2015, respectively. For comparison it is also shown observations of the photosphere at 6173 Å (middle panels) and their corresponding magnetograms (right panels) taken by the HMI instrument. From the figure we see that our sunspot sketches obtained with the projection method are in good agreement in terms of morphology and sunspots number, with those observed in HMI images. It is also noted that the magnetograms for the dates mentioned above exhibit a complex magnetic configuration, particularly on January 29, where the positive polarity (white region) is surrounded by the negative polarity (black region). Indeed, because the complexity and the existence of delta sunspots in the studied ARs, several solar flares have occurred (see Table 1).

[5] National Oceanic and Atmospheric Administration
[6] Centro de Investigación de la Actividad Solar y sus Efectos sobre la Tierra
[7] Central Meridian Distance
[8] Beta-Gamma
[9] Beta-Gamma-Delta

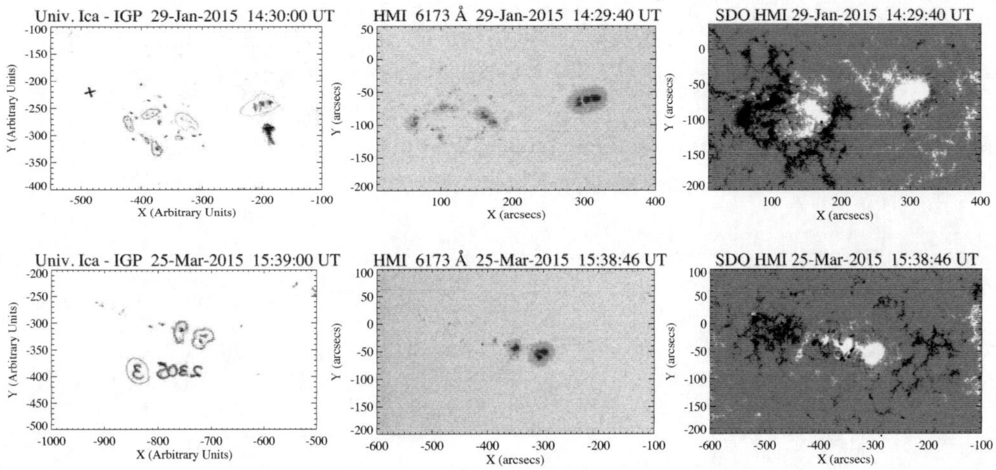

Figure 4. Left panels: sunspot records obtained with the projection method on January 29 and March 25, 2015. Middle and right panels: images observed in the continuum (photosphere, 6173 Å) by HMI onboard the SDO satellite and their corresponding magnetograms (positive polarity: white color; negative polarity: black color).

3.2. Physical speed of the sunspots

Taking into account the sunspot sketches obtained on January 28 and 29, 2015 at t_1=14:30 and at t_2=15:41 UT, respectively, we proceed to calculate the physical speed v considering the following relation of proportionality as in [9],

$$\frac{dcl}{D} = \frac{mdcl}{mde} \tag{1}$$

here D denotes the "real" diameter or the angular size of the Sun, being equal to two times the solar radius ($2R_s$), where $R_s \approx 974.3$ arc-seconds (6.82×10^5 km) as estimated from the ephemerides provided by the *Solarsoft*[10]. On the other hand, dcl is the "real" diameter of the latitude in which the sunspots would be located. Similarly, mde and $mdcl$ represent the "apparent" diameter or solar equator and the "apparent" latitude in which the sunspot is observed, both variables estimated from our records resulting in 15.0 and 14.9 cm, respectively. So, equation (1) is reduced to $dcl = 0.99D$; taking for $D = 1.36 \times 10^6$ km we obtain the value of $dcl = 1.35 \times 10^6$ km.

In Figure (5) a schematic picture of the sunspot trajectory is shown, in which the sunspots depicted by the dark structures on the disk travel a distance d from day-1 to day-2. This represents the "real" distance traveled by the sunspots from January 28 to 29. We measure d by using the following relation

$$\frac{d}{dcl} = \frac{md}{mdcl} \tag{2}$$

where md is the apparent distance traveled by the sunspot in our sketch, being equal to 1.7 cm. Having calculated dcl previously, we get $d = 1.54 \times 10^5$ km. Next, the physical speed is derived as follows

[10] The Solarsoft system is a set of integrated software libraries, data bases, and utilities which provide data analysis enviroment for solar physics

$$v = \frac{d}{\Delta t} \tag{3}$$

In our analysis, $\Delta t = 24$ hours$+(t_2 - t_1)$ is the time spent by the sunspot for traveling from the initial position at t_1 to the final position arriving at t_2, that is, $\Delta t = 90660$ seconds. Therefore the physical speed v is 1.69 km s^{-1}.

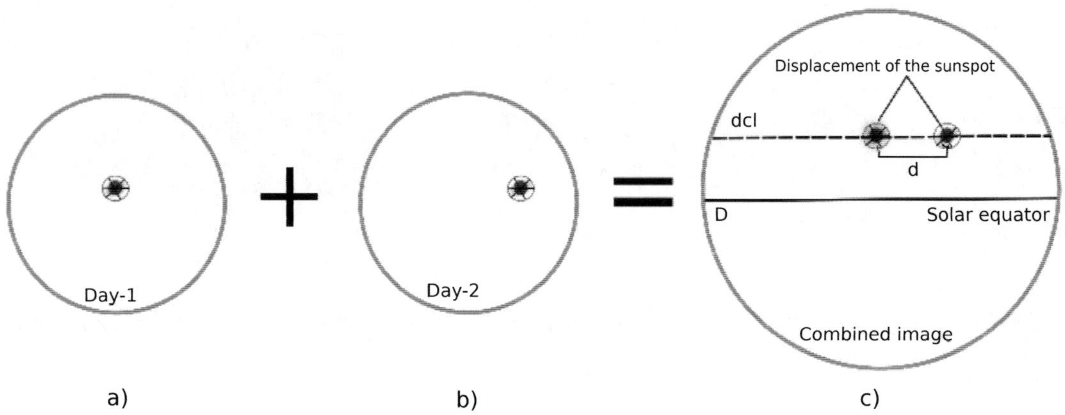

Figure 5. Schematic picture of the procedure we used for the calculation of sunspot physical speeds and rotation period of the Sun. Panel (a) shows the position of the sunspot at time t_1, panel (b) shows the position of the same sunspot at time t_2, in panel (c) we show a combined image showing the sunspot displacement d from t_1 to t_2.

3.3. Rotation period of the Sun

The rotation period was calculated tracing the sunspot trajectory. Following a standard procedure we estimate (i) the distance c traveled by the sunspot during a solar rotation that in analogy it would describe a circumference, (ii) such a distance could be calculated using the circumference of the circle $c = 2\pi r_s$, here r_s is the radius of the circumference where the sunspot is located. In our case the previous equation reduces to $c = \pi dcl$, since dcl represents the diameter of the latitude where the sunspot is located at, then we obtain $c = 4.24 \times 10^6$ km. Therefore, the rotation period is

$$T = \frac{c}{v} = \frac{4.24 \times 10^6 \text{ km}}{1.69 \text{ km s}^{-1}} = 2.50 \times 10^6 \text{ seconds} = 28.9 \text{ days} \tag{4}$$

Similar analysis has been done for the sunspot records taken on Mach 24 and 25, considering the following parameters: $\Delta t = 86100$ seconds, $md = 1.7$ cm, $mdcl = 14.9$ cm and $R_s \approx 962.91$ arc-seconds (6.74×10^5 km). Using equation (3) we obtained the physical speed v=1.76 km s^{-1}. Regarding the rotation period, we got $T = 27.4$ days.

4. Summary and Conclusions

We studied the ARs NOAA 2268 and NOAA 2305 during the periods January 28 to 29, and March 24 to 25 of 2015, respectively. From the GOES soft X-ray observations we identified several solar flares associated with the studied ARs, including M and C-class flares. In the magnetograms we confirmed the complex magnetic configuration of the ARs, in consequence the number of observed flares for the studied period.

By using our sunspot records obtained with the projection method, we derived the physical speed of the sunspots. From these results we also estimated the rotation period of the Sun. Our results are quite similar with the standard values. On the other hand, the morphological characteristics we found in our sunspot observations are very close to that observed in the HMI continuum images. Finally, despite the projection method to record sunspots is not considered a sophisticated technique nowadays, still it can be used to calculated basic physical parameters of the Sun such as we have presented in this work.

5. Acknowledgments

We are grateful to Dr. Nestor Vargas Maya and the staff members of the Facultad de Ciencias, Universidad Nacional San Luis Gonzaga de Ica, for all their support during the data adquisition with the Takahashi telescope. Special thanks to Dr. Mutsumi Ishitsuka for the motivation and his great efforts in promoting and establishing collaboration with our university in the field of Astronomy and Solar Physics, indeed this work has been done thanks to his encouragement. The authors also wish to thank the Facultad de Ciencias, Universidad Nacional de Ingenieria, for giving us the opportunity to present this work in the XVI Meeting of Physics. Finally, our acknowledgement to the SDO team for providing high-quality solar data.

6. Reference lists

[1] Priest, E. R. 1982, Solar Magnetohydrodynamics (Dordrecht: Reidel)
[2] Borrero, J.M., Ichimoto, K., 2011, Magnetic Structure of the Sunspots, Living Reviews in Solar Physics
[3] Hathaway, D.H. 2010, The Solar Cycle, Living Reviews in Solar Physics
[4] Tajima, T. & Shibata, K. 1997, Plasma Astrophysics (Massachusetts: Addison-Wesley)
[5] Aschwanden, M. J., 2004, Physics of the Solar Corona (Springer: Praxis Publishing)
[6] National Oceanic and Atmospheric Administration (NOAA) 2012, The Preliminary Report and Forecast of Solar Geophysical Data
[7] Pesnell, W. D., Thompson, B. J., & Chamberlin, P. C. 2012, Solar Physics, 275, 3
[8] Antiochos, S. K. 1998, The Magnetic Topology of Solar Eruptions, The Astrophysical Journal, 502, L181-184
[9] Solar-B Education and Public Outreach 2000, Touch the Sun, Chabot Space & Science Center

IOP Conf. Series: Journal of Physics: Conf. Series **1143** (2018) 012008 doi:10.1088/1742-6596/1143/1/012008

Do we live in an under-dense region?

Esly Calcina[1] **and Julio C. Tello**[1]

[1]Faculty of Sciences, Escuela Profesional de Física, Universidad Nacional de Ingeniería; Av. Tupac Amaru 210, Lima 25, Apartado 31-139, Lima, Peru

E-mail: ecalcinas@uni.pe; jtellog@uni.edu.pe

Abstract. According to the Cosmological Principle, matter structures and radiation are expected to be homogeneously and isotropically distributed in the universe at sufficiently large scales. An open question in observational cosmology is to estimate the size of such scale from the data. In the case of the cosmic background radiation, it seems clear that such scale corresponds to the horizon scale at matter-radiation decoupling, that is, at $\sim 1°$, but the problem is still open for the size of the homogeneity scale. But, due to the growth of structures in the evolving universe the homogeneity scale depends on the redshift of the data in analyses. Here we study this problem for data in the local universe by analyzing the 21cm HI-line sources using the public catalog ALFALFA, with median redshift $\langle z \rangle = 0.025$. We use the *scaled counts-in-spheres* method to establish the approximate size of the homogeneity scale in the local universe. In this analysis one compares the data sample with respect to a randomly generated homogenous sample. Defining the scale of transition to homogeneity as the scale at which the *scaled counts-in-spheres* estimator reaches the limiting value 1 within 1%, we find that this transition scale is $r_H \simeq 69$ Mpc, in excellent accordance with what is expected considering that the HI-line sources come from low-mass blue galaxies which have an anti-bias b with respect to the matter fluctuation field (with $b \simeq 0.48 - 0.68$ depending on the composition of the sample), from which one expects a transition scale in the local universe of $56 < r_H < 79$ Mpc.

1. Introduction

Since the Hubble results [1], the astronomical community believed that the universe expands decelerated. However, the discovery in 1998 [2] that the universe is currently in an accelerating expansion phase produced a challenge to modern cosmology. According to Einstein equations, if the background space-time is Friedmann-Robertson-Walker (FRW), the source for the acceleration should be a fluid with the unfamiliar property that it exerts a negative pressure over itself, continuously increasing the universe expansion rate. But the problem is that a real fluid with such properties is unknown; this fluid is termed the *dark energy*.

Since then, alternative scenarios have been proposed in the literature to explain the accelerating expansion of the universe without appealing to the (existence of the) dark energy. One possible solution claims that, if the local universe is an under-dense region then the dimm light (photons traveling along the intergalactic media) coming from type Ia supernovae (SNIa) do not need to be explained as emitted from an extremely far object, as it would be if the universe expands accelerated, instead they could have been emitted from a nearer SNIa but travelled at different speeds when inside and outside the void region. This alternative approach is known as Lemaître-Tolman-Bondi (LTB) due to the general relativity metric developed by these authors with which one constructs locally under-dense (or void) region (i.e., an inhomogeneous space) surrounded by a FRW spacetime [3].

IOP Conf. Series: Journal of Physics: Conf. Series **1143** (2018) 012008 doi:10.1088/1742-6596/1143/1/012008

The Cosmological Principle assumes that matter structures [4, 5] and radiation [6, 7, 8, 9, 10] are expected to be homogeneously and isotropically distributed in the universe at sufficiently large scales (see, e.g., [11, 12] for anisotropic features at large-angles in the cosmic background radiation or [13] for anisotropic radiation in compact spaces). Efforts are been done to estimate the size of such scale from the data [14, 15, 16, 17, 18]. Such analyses are particularly interesting due to the possibility that the LTB hypothesis can be rigorously tested using several cosmological probes, including publicly available astronomical data from the local universe like the ALFALFA catalog. This is what we planned to analyze here. In section 2 we present the data utilized in our analyses, while in section 3 we detail our methodology, and in section 4 we present our preliminary results. Finally, in section 5 we draw our conclusions and final remarks.

2. The ALFALFA catalog

The Arecibo Legacy Fast ALFA survey [19, 20] consists on astronomical data of the HI emission sources. At the time these analyses were done the publicly available survey was **alpha.70** (or α.70). In order to avoid systematic errors, noise, and non-extragalactic emission sources we have chosen CODE 1 detection sources, which is the subset of data suggested by the collaboration [20]. This dataset comprises 18, 984 extragalactic objects, with median redshift $\langle z \rangle = 0.025$. Every source is regarded as an object which spherical coordinates are given by the adopted distance, right ascension and declination(see Figures 1, 2, and 3).

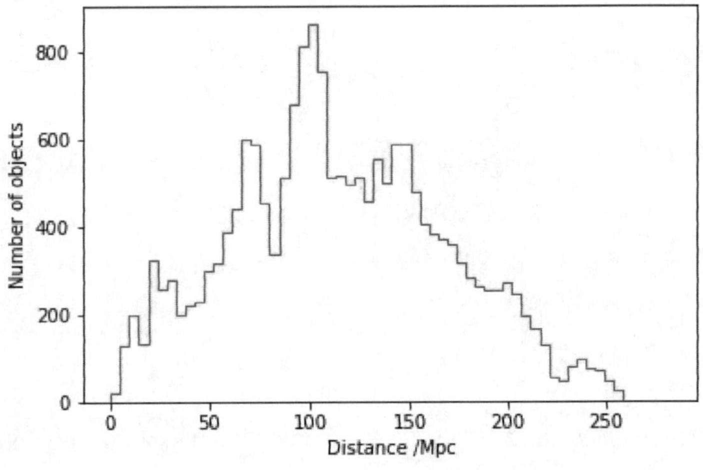

Figure 1. Distance histogram distribution for the extragalactic objects with CODE 1 in the catalog α.70.

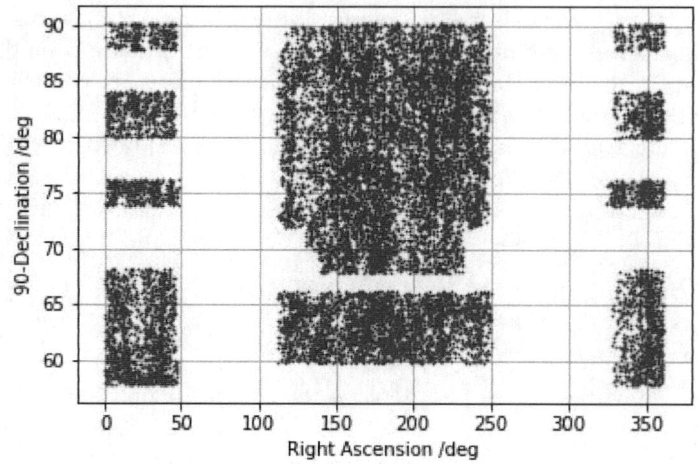

Figure 2. Angular distribution for the extragalactic dataset, CODE 1, in the catalog $\alpha.70$.

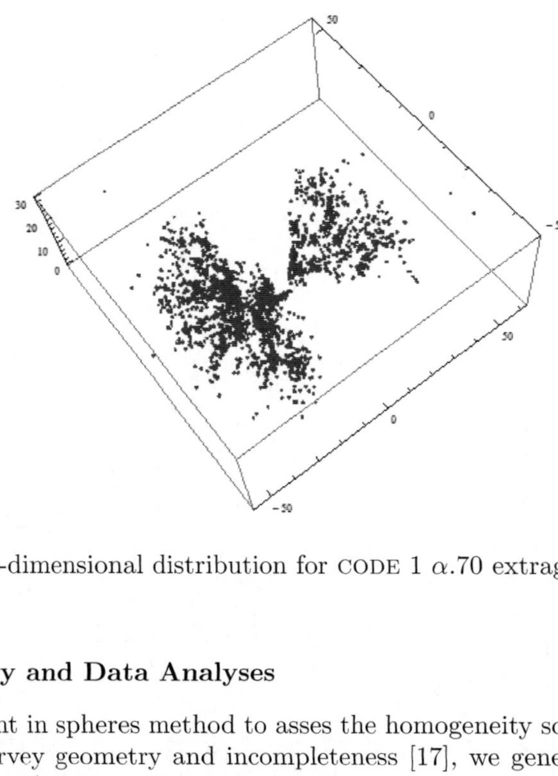

Figure 3. 3-dimensional distribution for CODE 1 $\alpha.70$ extragalactic objects

3. The Methodology and Data Analyses

We use the scaled count in spheres method to asses the homogeneity scale in the local universe. Considering to the survey geometry and incompleteness [17], we generated a randomly set of data (see Figures 4 and 5), in the same region of the survey.

The *scaled count-in-spheres* $\mathcal{N}_{center}(r)$ is calculated through extragalactic data and random generated data according to

$$\mathcal{N}_{center}(r) = \frac{1}{N} \sum_{i}^{N} \frac{n_{gal}^{i}(< r)}{n_{rand}^{i}(< r)}, \tag{1}$$

IOP Conf. Series: Journal of Physics: Conf. Series **1143** (2018) 012008 doi:10.1088/1742-6596/1143/1/012008

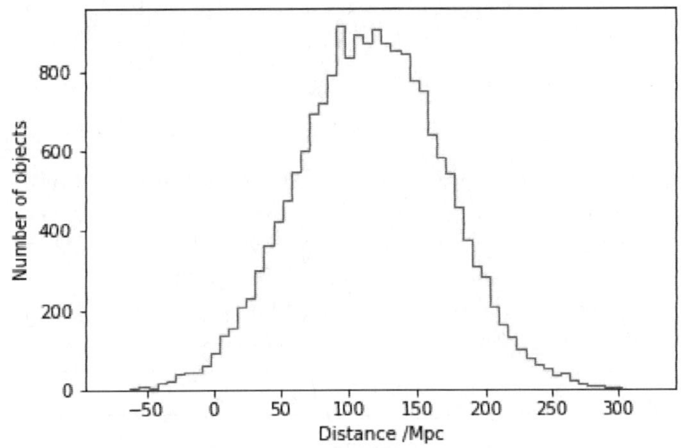

Figure 4. Distance histogram distribution for the random catalog. For the small fraction of negative distances we consider their absolute value.

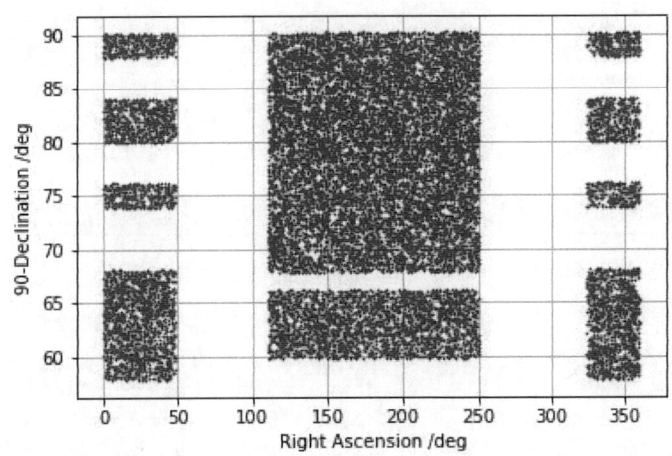

Figure 5. Angular distribution of the objects in the simulated random catalog.

where $n^i_{gal}(< r)$ is the number of extragalactic objects inside a sphere of radius r centered in the position of the i-extragalactic object. In the same way, $n^i_{rand}(< r)$ is the number of random generated objects. N is the total number of objects, that is 18, 984.

The fractal dimension is defined as follows [17]

$$\mathcal{D}_2(r) \equiv \frac{d \ln \mathcal{N}(< r)}{d \ln r} + 3 \,. \tag{2}$$

For a homogeneous distribution $\mathcal{N}(r)$ is proportional to r^3, thus $\mathcal{D}_2(r) = 3$, while for a fractal distribution the fractal dimension is a number between 2 and 3. It is worth to mention here that there are several approaches to define $\mathcal{N}$, as can be seen, e.g., in reference [18], one of these methods being the calculation of $\mathcal{N}_{center}$ used here [15, 17, 18]. In particular, one can also use the two-point correlation function to define an estimator for the scaled count-in-spheres $\mathcal{N}(r)$ (for other applications of the two-point correlation function, see e.g., [21, 22]).

4. Results

The *scaled count-in-spheres* methodology was applied to the sample of the ALFALFA.70 survey, a public catalog recently released [19] of 21cm HI-line extra-galactic sources. Our preliminary results are summarized in figure 6, where the curve goes down very fast at 100 Mpc in accordance with simulations done for matter fluctuations where the scale expected is $\sim$ 100 Mpc [23]. One also observes that the scaled count-in-spheres curve oscillates, probably due to under- and over-densities, but then obtain a level of stability at $\sim$ 260 Mpc.

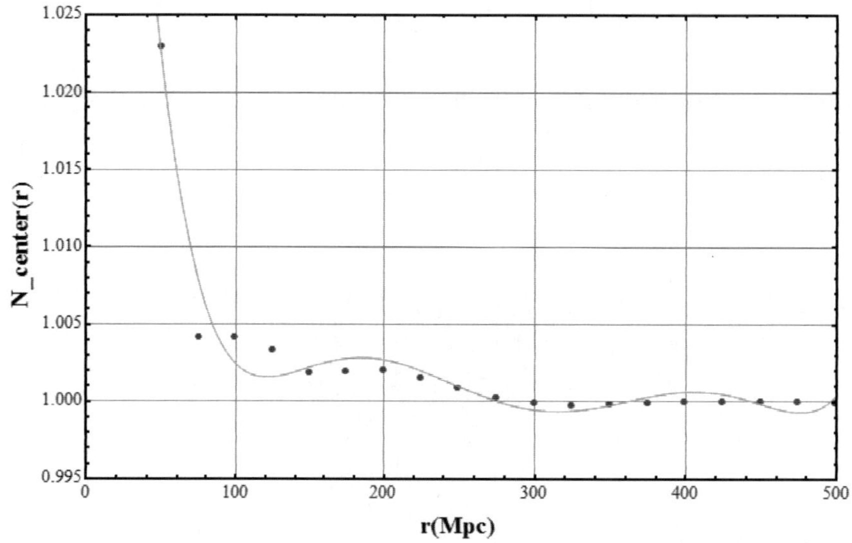

Figure 6. Scaled count-in-spheres *versus* radial distance. In this plot we consider distances from 50 to 500 Mpc, there are 19 bins (dot points): the inicial point is $r_0 = 50$ Mpc, and from this we consider 18 equal bins of size 25 Mpc. The best-fit to the data corresponds to a polynomial of degree six, and the intersection with the horizontal red line (that represents the constant value $N_{center} = 1.01$, that is, the threshold below which one considers that the homogeneity is attained) occurs at the scale $r_H = 69.0072$ Mpc.

5. Conclusions anf Final remarks

Our preliminary analyses are concentrated in Figure 6, where we show that the intersection of the green line, which is the best-fit of the data (dot points), with the red line (that represents the constant value $N_{center} = 1.01$, which is the threshold below which the homogeneity is achieved) defines the homogeneity scale [17]. This intersection occurs at the scale 69.0072 Mpc, a value that is in good accordance with the value obtained for the galaxy distribution in the SDSS DR6 [14]. In figure 6 we also observe some oscillations at distances $300 - 400$ Mpc are possibly due to the presence of over-density regions (galaxy clusters?) while oscillations at scales $200-300$ Mpc possibly due to under-density regions.

Moreover, considering that the HI-line sources come from low-mass blue galaxies which have an anti-bias b with respect to the matter fluctuation field (with $b \simeq 0.48 - 0.68$ depending on the composition of the sample), from which one deduces that the transition scale expected in the local universe for our sample is $56 < r_H < 79$ Mpc [17].

Finally, although these analyses are preliminary we can say that, despite the presence of regions with over and above the mean number density of extra-galactic objects, the 3-dimensional scale of homogeneity is achieved at the predicted scale [17], i.e. $r_H = 69.0072$ Mpc, confirming the consistency of the CP through the ALFALFA local universe dataset.

Acknowledgments

The authors thank Camila Novaes and Felipe Avila for insightful discussions and suggestions. EC acknowledges a fellowship of the CONCYTEC, the Peruvian agency for Science and Technology.

References

[1] Hubble E and Humason M L 1931 *The Astrophysical Journal* **74** 43
[2] Perlmutter S, et al. 1999 *The Astrophysics Journal* **517** 565; arXiv:astro-ph/9812133
[3] Ellis G F R, Maartens R and MacCallum M A H 2012 *Relativistic Cosmology* (Cambridge: Cambridge University Press)
[4] Bernui A, Ferreira I S and Wuensche C A 2008 *The Astrophysical Journal* **673** 968; arXiv:0710.1695
[5] Marques G A, Novaes C P, Bernui A, Ferreira I S 2018 *Monthly Notices of the Royal Astronomical Society* **473** 165; arXiv:1708.09793
[6] Scharf C A, et al. 2000 *The Astrophysical Journal* **544** 49
[7] Planck Collaboration, Planck 2015 results XVI, 2016 *Astronomy and Astrophysics* **594** A16; arXiv:1506.07135
[8] Bernui A 2009 *Physical Review* D **80** 123010; arXiv:0912.1147
[9] Abramo L R, Bernui A and Pereira T S 2009 *Journal of Cosmology and Astroparticle Physics* **12** 013; arXiv:0909.5395
[10] Novaes C P, Bernui A, Marques G A and Ferreira I S 2016 *Monthly Notices of the Royal Astronomical Society* **461** 1363; arXiv:1606.04075
[11] Bernui A, Villela T, Wuensche C A, Leonardi R and Ferreira I 2006 *Astronomy and Astrophysics* **454** 409; astro-ph/0601593
[12] Bernui A, Oliveira A F and Pereira T S 2014 *Journal of Cosmology and Astroparticle Physics* **10** 041; arXiv:1404.2936
[13] Bernui A, Gomero G I, Rebouças M J and Teixeira A F F 1998 *Phys Rev* D **57** 4699; gr-qc/9802031
[14] Sarkar P, Yadav J, Pandey B and Bharadwaj S 2009 *Monthly Notices of the Royal Astronomical Society* **399** L128
[15] Scrimgeour M I 2012 *Monthly Notices of the Royal Astronomical Society* **425** 116; arXiv:1205.6812
[16] Laurent P, et al. 2016 *Journal of Cosmology and Astroparticle Physics* **11** 060; arXiv:1602.09010
[17] Ntelis P, et al. 2017 *Journal of Cosmology and Astroparticle Physics* **06** 019; arXiv:1702.02159
[18] Gonçalves R S, et al. 2017; arXiv:1710.02496
[19] ALFALFA Project Team 2012, The Arecibo Legacy Fast ALFA Survey. Cornell University, Nueva York, available in: http://egg.astro.cornell.edu/alfalfa/data/index.php
[20] Haynes M P, et al. 2011, *The Astronomical Journal* **142** 170; astro-ph/1109.0027
[21] Carvalho G C, Bernui A, Benetti M, Carvalho J C, Alcaniz J S 2016 *Physical Review* D **93** no.2, 023530 arXiv:1507.08972
[22] de Carvalho E, Bernui A, Carvalho G C, Novaes C P 2017; arXiv:1709.00113
[23] Koo D C and Kron R G 1986 *Observational Cosmology: Proceedings of the 124th Symposium of the International Astronomical Union*, Beijing, China

IOP Conf. Series: Journal of Physics: Conf. Series **1143** (2018) 012009 doi:10.1088/1742-6596/1143/1/012009

SEARCH FOR ECLIPSING BINARY STARS IN THE DIRECTION OF GALAXY BULGE WITH DATA OF OGLE-II AND MACHO CATALOGUES

E. Urday E.[1], J. Tello G.[2]

Faculty of Science, National University of Enginnering, Av.T´upac Amaru 210, Lima, Perú

E-mail: llue_130@hotmail.com[1], jtellog@uni.edu.pe[2]

Abstract. It is known that the parameters of a simple star (radii, temperatures, masses) are obtained from the study of eclipsing binary systems. With the eclipsing binaries of the OGLE catalogue, we look for their counterparts in the MACHO catalogue, that is, the same binaries that are contained in the second catalog. As a result, we will generate our own catalogue of eclipsing binaries with photometric information in three filters I, R and V.

1. Introduction

The Optical Gravitational Lensing Experiment(OGLE) is an astronomical project created to discover dark matter using the gravitational microlensing technique[1]. This technique consists of observing continuously and systematically different regions of the sky where the microlenses could be produced. The microlens have a high probability of occurring in the galaxies Large and Small Magellanic Clouds and in the bulge of our Galaxy, due to the large number of stars in those regions. As a result of using this technique, in addition to observing the microlenses, data were obtained on the variation of the luminous flux of variable stars (pulsating stars and eclipsing binary systems). The OGLE project, in its second stage of operation (called OGLE-II), made measurements in the direction Galactic bulge, and as a result, data of 200000 variable stars were obtained[2]. The sky region measured by OGLE was divided into 49 fields and their results are contained in a catalogue with information about each field of the sky, position of the stars in that field, observation time and variation of stellar magnitude in filter I.

On the other hand, the collaboration of Massive Compact Objects of the Halo (collaboration MACHO) is a project made to prove that a significant fraction of dark matter in the halo of the Milky Way is composed of brown dwarfs or planets [3]. For this reason, they observed for several years in the direction of the galaxy Greater Cloud of Magellan and the bulge of our Galaxy. Like the OGLE catalogue, the results of their observations consist of information on the position of the measured stars and variation of the luminous flux, but in two filters R and V. The observed region was divided into 94 fields and the measured objects contain the stars variables.

2. Methodology

An eclipsing binary system is a system composed of 2 stars gravitationally linked and whose orbital plane is close to the observer's visual line, so, in each orbital cycle, 2 eclipses occur: first

IOP Conf. Series: Journal of Physics: Conf. Series **1143** (2018) 012009 doi:10.1088/1742-6596/1143/1/012009

when one of the components is interposed between the observer and the second star and then, at another time, when the second component eclipses the first.

The variation of the light flux in a given filter is called the light curve. In the case of an eclipsing binary system, the more light curve are measured in different filters, the more system parameters can be determined.

The OGLE catalogue presents light curves of variable stars measured with a single filter. Our work consists of using the OGLE-II data of variable stars in the direction of the bulge of our Galaxy with the purpose of separating from them a type of variable stars: the eclipsing binary stars. We finely seek to identify these systems in the MACHO catalogue to obtain their counterparts. As a result, we generated our own catalogue of eclipsing binaries with light curves in three I, R and V filters.

As a first step we must select the eclipsing binaries from the OGLE catalogue. Next, we will look for these same objects in the MACHO catalogue (that is, we will find the counterparts of the eclipsing binaries of the OGLE catalogue in the MACHO catalogue).

In Figure 1 we present a scheme of the procedure followed to obtain the MACHO counterparts of an OGLE binary star.

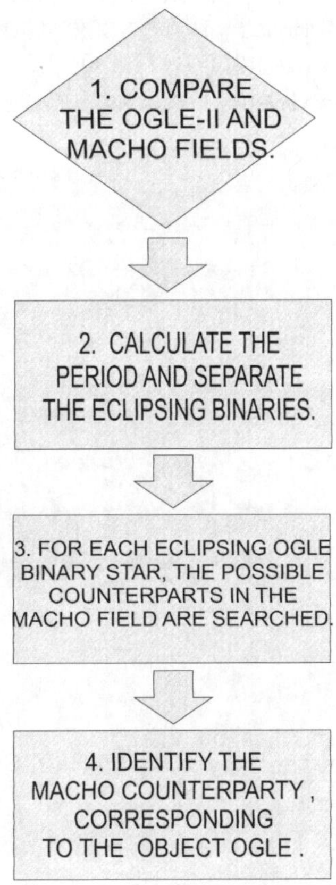

Figure 1. Scheme that describes the steps to follow to obtain the MACHO counterpart of an OGLE binary star.

The procedure applied consists of the following steps:

IOP Conf. Series: Journal of Physics: Conf. Series **1143** (2018) 012009 doi:10.1088/1742-6596/1143/1/012009

2.1. Compare the OGLE-II and MACHO fields.

The OGLE II and MACHO catalogues are examined to determine which fields of a catalogue are closest to each other. For example, we find that the center of the field 10 of OGLE called BUL-SC10 has equatorial coordinates close to field 309 of MACHO. Table 1 shows the equatorial celestial coordinates Right Ascension (R.A.) and Declination (Dec.) of both fields.

Table 1. Coordinates of the centers of the fields BUL-SC10 of OGLE and 309 of MACHO.

Catalogue	Field	R.A. (h m s)	Dec. (° ′ ″)
OGLE II	BUL-SC10	18:20:06.6	-22:23:03.0
MACHO	309	18:19:01.19	-22:23:43.5

Figure 2 shows a graph of the areas occupied by each of these fields, the center of the field is marked by a (+) (field 10 of OGLE) or a (×) (field 309 of MACHO). We see that a large part of the BUL-SC10 field is contained in field 309. The OGLE field is of the shape of a rectangle of $14'' \times 57''$ in the sky and the MACHO field is a square of $43'' \times 43''$.

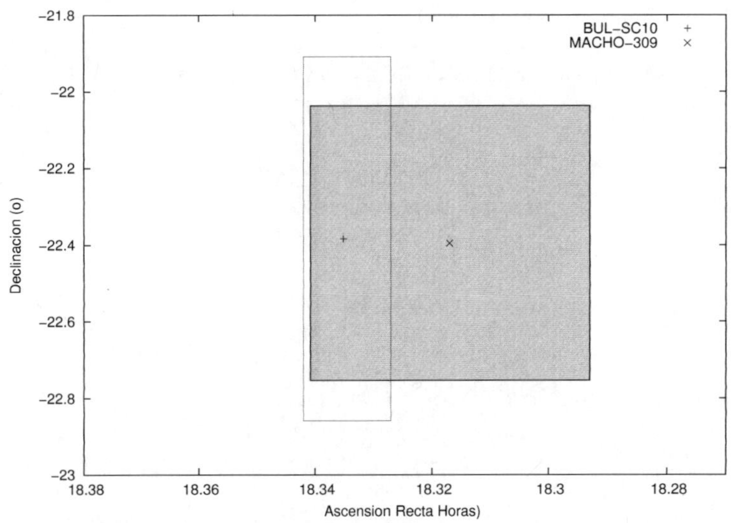

Figure 2. Intersection of the fields OGLE BUL-SC10 and MACHO-309.

2.2. We calculate the period and separate the eclipsing binaries.

We identify the periodic variable stars and separate the eclipsing binary systems from them. For each variable star of the OGLE II catalogue, we calculate the period of variation of the light curve using the String-Length method [4]. After obtaining the light curves of the OGLE variable stars that are periodic, we separate those that are eclipsing binaries, according to the shape of the light curve. For example, figures 3a and 3b show some light curves of eclipsing binaries from the BUL-SC10 field with their orbital periods (in days) and the time of their main eclipse. Of all the light curves that we have obtained for the field 10 of OGLE, we select those that correspond to eclipsing binary systems, which are the objects of study of this work. Note in these graphs that the vertical axis is called the I band and the horizontal axis is called the phase and, by convention, in eclipsing binary systems the phase is zero in the deepest eclipse; when the phase is 1 the system has given one complete orbit. The eclipse in phase zero is called the primary eclipse and the other less deep eclipse, in phase 0.5, is called the secondary eclipse. The different depth of these minima indicates that both stellar components have different surface temperatures. For example, in figure 4a the binary 1082 of field 10 of OGLE (named BUL-SC10-1082) its calculated orbital period was approximately 4.30 days.

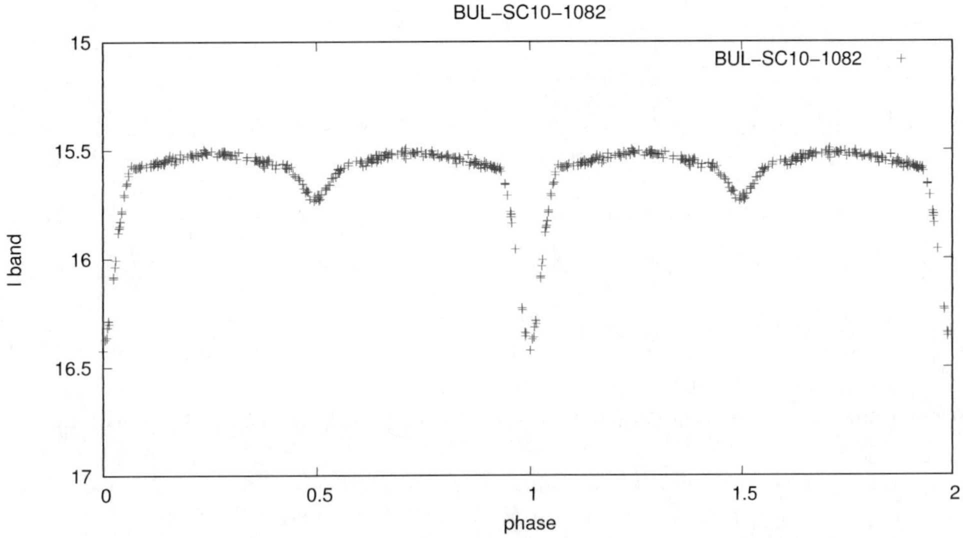

Figure 3a. Light curve of the eclipsing binary OGLE BUL-SC10-1082.

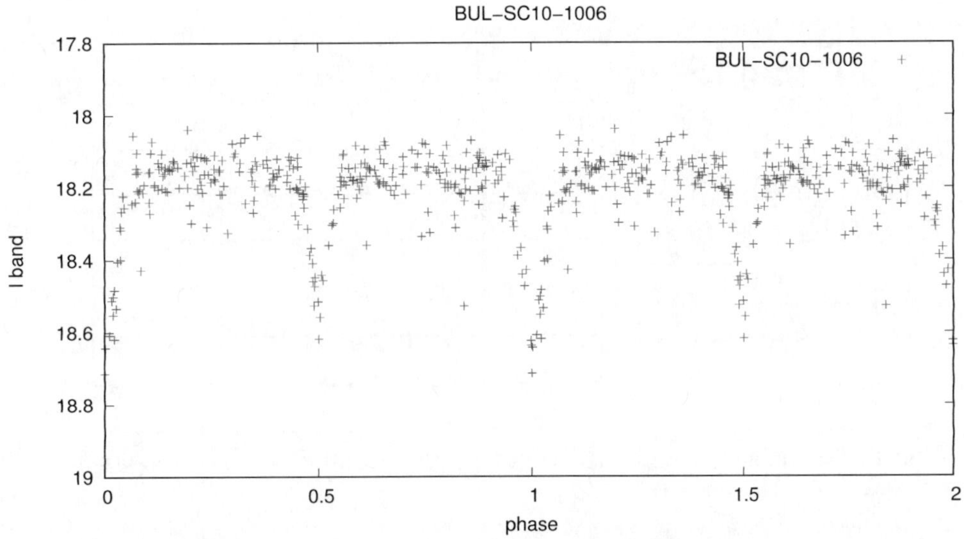

Figure 3b. Light curve of the eclipsing binary OGLE BUL-SC10-1006.

IOP Conf. Series: Journal of Physics: Conf. Series **1143** (2018) 012009 doi:10.1088/1742-6596/1143/1/012009

The light curve of the eclipsing binary OGLE BUL-SC10-1006 in the figure 3b shows that the difference between the eclipses is not so pronounced, which indicates that the temperature of the components is not very different.

2.3. For each OGLE star we look for the possible counterparts of the eclipsing binary stars in the MACHO field.

With information from the OGLE catalog of the positions of each eclipsing binary we look for the counterparts in the MACHO catalog. For example, for the binary BUL-SC10-1743 with orbital period 0.380 d its coordinates are:

R.A.: 18h19m45.55s

Dec: -22°12′44.4″

In the MACHO catalogue we search for objects with coordinates close to the OGLE binary with the condition:

$$r = \sqrt{(RA_{OGLE} - RA_{MACHO})^2 cos DEC_{OGLE}^2 + (DEC_{OGLE} - DEC_{MACHO})^2}$$

Where r is a value that indicates the proximity of one object to another in the sky and that we choose to obtain the closest counterparts to the OGLE binary. For that we write a script so that for the position of an OGLE object it looks for all the objects of the MACHO catalogue of the nearest region and only takes into account those possible counterparts with the condition $r<4''$.

As an example on the counterparts, object BUL-SC10-1743 is in field 10 of OGLE and 7 possible counterparts were found (as seen in the table 2) in field 309 of the MACHO catalog with a value of r comprised between: $4.85\times10^{-5}\,'' \leq r \leq 0.001''$.

Since the fields scanned in the MACHO catalogue can overlap, the regions in which the sky was divided and which never overlap were called TILES. Therefore, to describe the position of a star in the MACHO catalogue, 3 parameters are needed: the field, the tile and the star number in this tile [5]. Figure 4 shows how the fields and tiles of the MACHO catalogue are distributed.

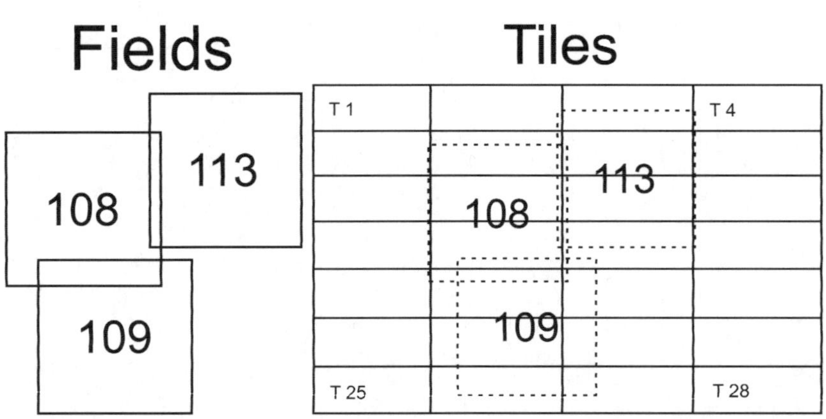

Figure 4. Distribution of FIELDS and TILES of the MACHO catalogue.

We use these 3 parameters to refer to the couterparts in the MACHO catalogue. Table 2 shows the MACHO objects, which may be possible counterparts of the eclipsing binary BUL-SC10-1743. The first column is the binary of the OGLE catalogue, the second column is the

Table 2. 7 possible counterparts of BUL-SC10-1743 found in the MACHO catalogue ordered according to their proximity in the sky.

OGLE	MACHO FIELD	TILE	STAR	$(\times 10^{-4})$
bul-sc10-1743	309	38255	169	$8.88E^{-04}$
bul-sc10-1743	309	38255	500	$4.82E^{-04}$
bul-sc10-1743	309	38255	747	$4.85E^{-05}$
bul-sc10-1743	309	38255	789	$5.36E^{-04}$
bul-sc10-1743	309	38255	1382	$6.62E^{-04}$
bul-sc10-1743	309	38255	1612	$1.00E^{-03}$
bul-sc10-1743	309	38255	1681	$7.99E^{-04}$

field number of the MACHO counterpart, the third column is the TILE or region of the observed field, the fourth column is the STAR number within the tile, the last column is the parameter r.

2.4. We identify MACHO counterparty,corresponding to the object OGLE.

The period found of the OGLE binary in step 2 will serve to identify the MACHO counterpart among the 7 possible candidates and select it. We make the light curve of each possible candidates with the period of the eclipsing OGLE. The light curve of the candidate with similar shape of the eclipsing OGLE which is the counterpart (figure 5a). In figs. 5b and 5c the light curves of the object 309-38255-747 of the MACHO catalogue are shown in filters V and R, respectively and that turned out to be the counterpart of OGLE BUL-SC10-1743. Note the similarity of light curves.

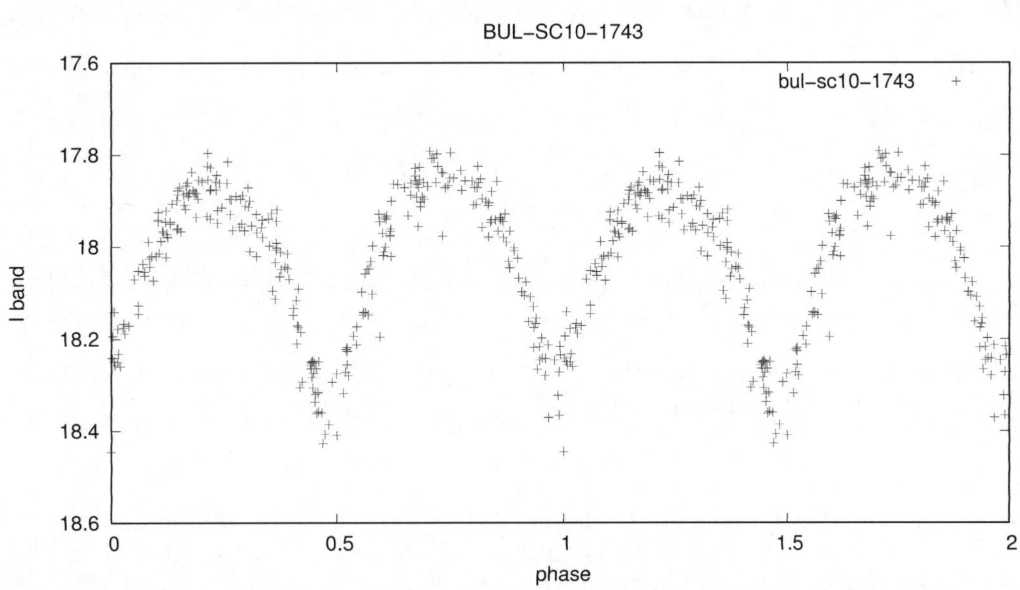

Figure 5a. Light curve OGLE BUL-SC10-1743 with period 0.380264269d.

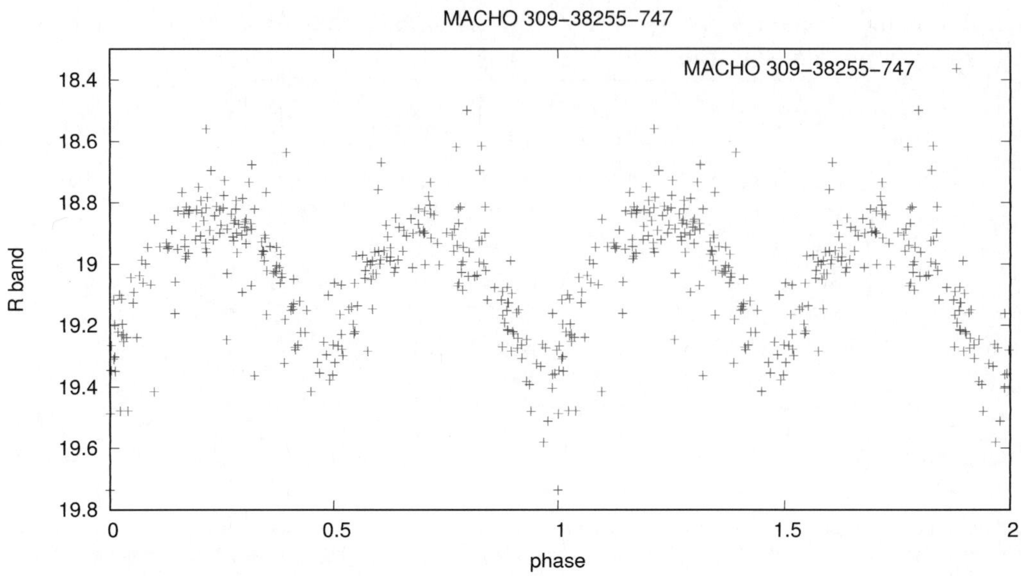

Figure 5b. Light curve of the MACHO object 309-38255-747 in filter R.

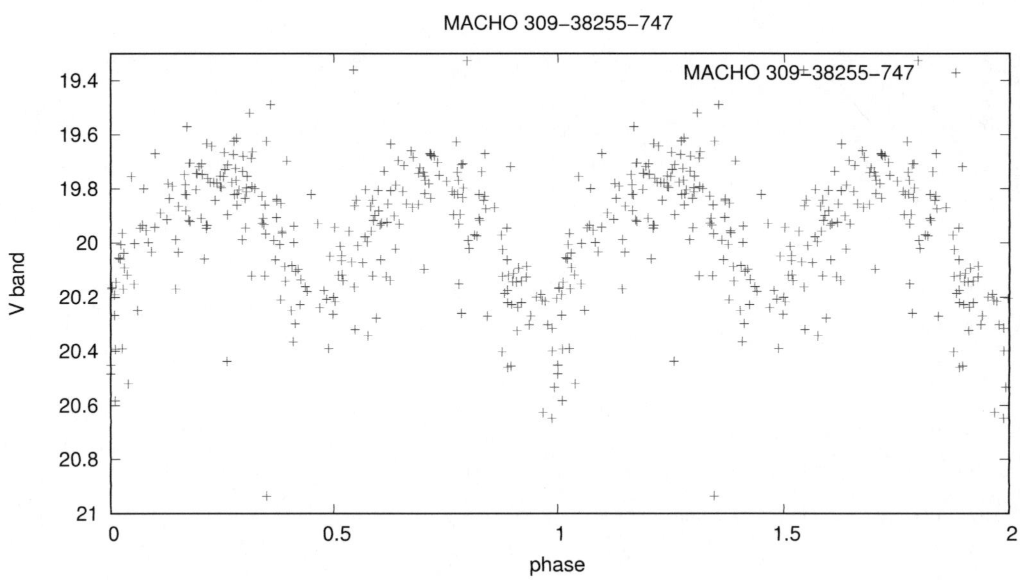

Figure 5c. Light curve of the MACHO object 309-38255-747 in filter V.

3. Results

As a result of using this method, it was found that of the 49 OGLE-II fields, only 22 fields contain counterparts in the MACHO fields. In turn, the light curves were obtained in filters I, R and V. For example, for field 10 of OGLE-II, the eclipsante 1743 has its counterpart, which is object 747 of field 309 of MACHO (figure 6).

IOP Conf. Series: Journal of Physics: Conf. Series **1143** (2018) 012009 doi:10.1088/1742-6596/1143/1/012009

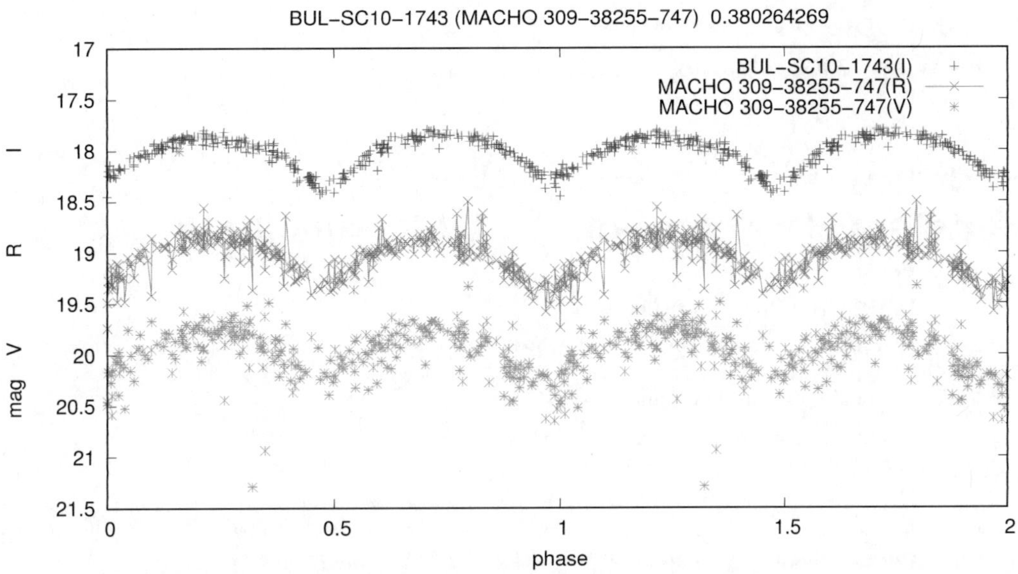

Figure 6. Shows the light curves of this object on three different filters.

4. Conclusions

Our method was successful in selecting the eclipsing binaries, obtaining the counterparts and finding the light curves in different filters. These results may be useful, in a later study, to determine some parameters of the eclipsing binaries such as the inclination angle of the orbital plane, the relative radii of the components and their mass ratio. Additionally with additional spectroscopy information it is possible to obtain parameters such as the temperatures of the system components. Finally this method can be used in other catalogs of variable stars, such as pulsating stars, to find their corresponding counterparts.

5. References

[1] Paczynski B and Stanek K Z and Udalski A and Szymanski M and Kaluzny J and Kubiak M and Mateo M and Krzeminski W and Preston G W 1995 *Proc. 187th AAS Meeting* vol 27 (USA: Bulletin of the American Astronomical Society) p 1302

[2] Wozniak P R and Udalski A and Szymanski M and Kubiak M and Pietrzynski G and Soszynski I and Zebrun K 2002 *Acta Astronomica.* **52** 129

[3] Popowski P, Griest K, Thomas C L, Cook K H, Bennett D P, Becker A C, Alves D R, Minniti D, Drake, A. J. and Alcock, C. and Allsman R A, Axelrod T S, Freeman K C, Geha M, Lehner M J, Marshall S L, Nelson C A, Peterson B A, Quinn P J, Stubbs C W, Sutherland W, and Vandehei T, Welch D and MACHO Collaboration 2005 *The Astrophysical Journal.* **631** 879

[4] Dworetsky, M. M. 1983 *Monthly Notices of the Royal Astronomical Society.* **203** 917

[5] Axelrod T S, Allsman R A, Quinn P J, Bennett D P, Freeman K C, Peterson B A, Rodgers A W, Alcock C, Cook K H, Griest K, Marshall S L, Pratt, M R, Stubbs C W and Sutherland W 1995 The MACHO Data Pipeline *Preprint* astro-ph/9508107

IOP Conf. Series: Journal of Physics: Conf. Series **1143** (2018) 012010 doi:10.1088/1742-6596/1143/1/012010

Cataclysmic variable stars: photometry of V893 Sco star and spectroscopy of HS 0220+0603 star

William Cori[1,2] **and Julio Tello**[1,2]

[1] Grupo Astronomía, Faculty of Sciences, Universidad Nacional de Ingeniería, Lima, Peru
[2] Grupo de Investigación en Astrofísica y Cosmología, Faculty of Sciences, Universidad Nacional de Ingeniería, Lima, Peru

E-mail: ecoric@uni.pe, jtellog@uni.edu.pe

Abstract. Differential photometry was applied on V893 Sco star, observed from the Astronomical Observatory of the Universidad Nacional de Ingeniería in Peru. From its light curve it was measured an orbital period of 0.076 days approximately. Moreover, spectroscopic analysis was performed on the HS 0220+0603 star, which data was acquired from a database. A fitting program was written in Python to find, out of several synthetic spectra of white dwarf and templates of red dwarf, the combination of a white dwarf and a red dwarf spectrum that fits best the HS 0220+0603 spectrum, thus we found that this spectrum is a composition of a white dwarf spectrum of 30000 K and surface gravity (log g) = 8.0, and a M5.5 type red dwarf star.

1. Introduction

Cataclysmic variable (CV) stars are binary systems composed of a white dwarf (WD) and a red dwarf (RD) orbiting each one around the center of mass of the system, where there is mass transference from the red one to the white one through the first Lagrangian (L1) point of the system, as the red dwarf has filled its Roche lobe [1]. The WD is called primary star and the RD secondary star.

If the WD has a weak magnetic field the system is called non-magnetic CV and the transferred mass forms an accretion disk around the WD, and subsequent mass flux collides with this disk producing a bright point, called "hot spot", that is sometimes the strongest source of light of the system. However, when the WD has a strong magnetic field, the flow is driven through the magnetic field lines to its poles, preventing the formation of the accretion disk, such stars are called "polar" CV stars. If the magnetic field is not so strong the disk is partially formed far from the WD, these systems are called "intermediate polar" CVs. Both polar and intermediate polar CVs are called magnetic CV.

When the orbital plane of the system is aligned to our line of sight we can observe periodic changes of the brightness caused by alternate eclipses between the stars and, eventually, the hot spot. That system is called a eclipsing system.

However, there are other characteristics related to the dynamics of the system that produce an abrupt increasing of its brightness. These increments in some cases are really big, orders of hundreds, thousands and millions in the brightening of the system have been observed. That is the reason why these systems are called cataclysmic. Such dynamical characteristic and the resulting change of brightness are used to classify to the CV stars.

IOP Conf. Series: Journal of Physics: Conf. Series **1143** (2018) 012010 doi:10.1088/1742-6596/1143/1/012010

Between non-magnetic CV stars, we have the following three types [2]: A Nova, classic or recurrent, is a CV that has presented at least one eruptive phenomenon caused by thermonuclear explosion on the WD surface, where external shells of the WD are ejected to high velocities. Conditions for thermonuclear explosions are reached by increasing in the accretion rate over the WD surface. The brightness raises in a range from six to nineteen magnitudes. A Dwarf Nova (DN) is a CV that presents quasi-periodic increments of brightness, they are not produced by thermonuclear explosions but by instabilities in the disk related to increasing in the mass flux rate through the disk. No mass is ejected. The brightness raises in a range from two to five magnitudes. A Novalike (NL) is, basically, a CV that has not presented changes of its brightness with the characteristics of a Nova or a DN. They present irregular changes in its brightness.

Moreover, many types of CVs present states where its brightness diminishes between three to five magnitudes owing to decreasing in the mass flux from the secondary through the L1 point, which lighten the density of the disk, making it colder and less bright, thus the hot spot is extincted or insignificant. This "low state" can last days, months or years, after that the brightness return to its normal high state.

In this paper we present the first observation of a CV star from the Astronomical Observatory of the Universidad Nacional de Ingeniería (OAUNI, from its Spanish initials) [3, 4], we registered images of V893 Sco dwarf nova star [5] to measure its orbital period by differential photometry technique. Furthermore, we downloaded spectroscopic data of HS 0220+0603 eclipsing Novalike SW Sex class star [6] in low state from the European Southern Observatory (ESO) archive facility web page[1] to find the characteristics of its components with our empirical fitting method, which consists of finding the best combination of a WD synthetic spectrum and a M type (RD) template that fits to a spectrum of HS 0220+0603 star. Our method is assessed here for the first time.

2. Data acquisition

2.1. Photometric data

Observations of the V893 Sco dwarf nova star were carried out at OAUNI. OAUNI is located in the Junín region at Peruvian central Andes (see table 1 for geographical coordinates). This observatory has a 0.51 m RCOS telescope, a Paramount MEII mount, and a SBIG STXL-6303 CCD. The images were taken on June 30th, 2017. 275 frames with 20 s exposure time and filter V were acquired (see table 2). Furthermore, *darks* and *flatfields* were taken.

Table 1. Geographical coordinates of OAUNI.

Longitude	Latitude	Altitude
75° 19' 14.11" W	12° 02' 28.69" S	3314 masl

Table 2. Photometric observations at OAUNI.

Date	Star	Coverage	# frames	Exp.	Filter
2017/06/30	V893 Sco	123 min	275	20 s	V

[1] http://archive.eso.org/cms.html

2.2. Spectroscopic data

We acquired data of HS 0220+0603 eclipsing SW Sex star from the ESO Science Facility web page. This star was observed on November 20th, 2004, from ESO Paranal observatory with the Very Large Telescope - Unit 1 (ESO-VLT-U1)[2], called Antu, and the second version of the instrument FOcal Reducer and low dispersion Spectrograph (FORS2), with the grism 600RI. Four draw spectra of this star with 600 s of exposure time were downloaded (see table 3), as well as *bias*, *flats*, *arcs*[3] and a spectrum of a standard star[4] (with its respective *flats* and *Arcs*).

Table 3. Spectroscopic data from ESO Science Archive Facility, observed with VLT-U1.

Date	Star	Instrument	Exp.	# frames
2004/11/20	HS 0220+0603	FORS2, GRIS_600RI	600 s	4

3. Reduction and analysis

3.1. Photometry: V893 Sco

The data (see table 2) was reduced with the software system IRAF (Image Reduction and Analysis Facility)[5]. The images were corrected by *overscan*, *dark* and *flatfield* with tasks from NOAO.IMRED.CCDRED package, then aligned with *imalign* task from IMAGES.IMMATCH package, and to calculating the magnitudes it was used the NOAO.DIGIPHOT.DAOPHOT package.

Figure 1 shows a reduced and aligned image, the V893 Sco star is pointed by the two red lines, while the two comparison stars, used for the differential photometry technique, by Comp1 and Comp2 labels, respectively.

Magnitudes of V893 Sco, Comp1 and Comp2 stars were measured, then the differences between them over each reduced and aligned image. Figure 2 shows light curves with errorbars built with Gnuplot[6] from these differences. First, on top of the figure a virtual horizontal line is shown, obtained from differences between the two comparison stars. This indicates that the two comparison stars have constant brightness, which makes them good comparison stars. The other two curves, light curves, have been built with the differences between the V893 Sco star and each comparison stars. They have similar form because, as was already said, the comparison stars have uniform brightness. For this reason, the light curves differ approximately just by a constant value.

From the light curves it is possible to infer the orbital period of the system. The figure 2 shows that the light curves have deep falls in x values of 0.516 and 0.592 approximately. These values indicate approximately the times of mid-eclipses of the primary eclipses in Julian days, i.e. the maximum eclipse of the WD by the RD. The difference between these primary eclipses is the orbital period, that is 0.076 d approximately, similar to 0.075961 d from Ritter and Kolb catalog [5].

[2] http://www.eso.org/public/teles-instr/paranal-observatory/vlt/

[3] Spectra of a combination of lamps (He, HgCd and Ar), for wavelength calibration.

[4] Spectrum of LTT 377 star, for flux calibration.

[5] IRAF is written and supported by the National Optical Astronomy Observatories (NOAO) in Tucson, Arizona. NOAO is operated by the Association of Universities for Research in Astronomy (AURA), Inc. under cooperative agreement with the National Science Foundation.

[6] http://www.gnuplot.info/

IOP Conf. Series: Journal of Physics: Conf. Series **1143** (2018) 012010 doi:10.1088/1742-6596/1143/1/012010

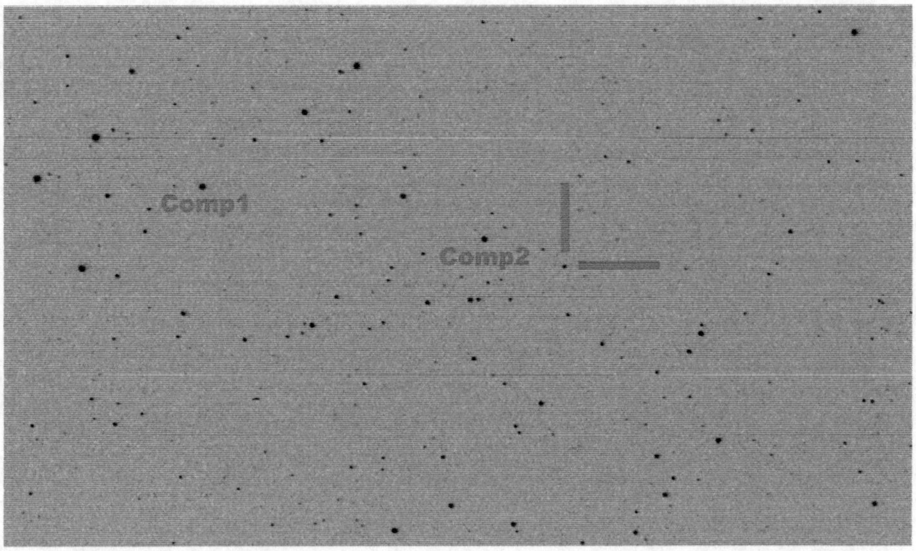

Figure 1. Field of vision of a reduced and aligned image. The V893 Sco star is in the intersection of red lines and the comparison stars are above "m" character in labels "Comp1" and "Comp2", respectively. This star was observed from OAUNI on 2017/06/30.

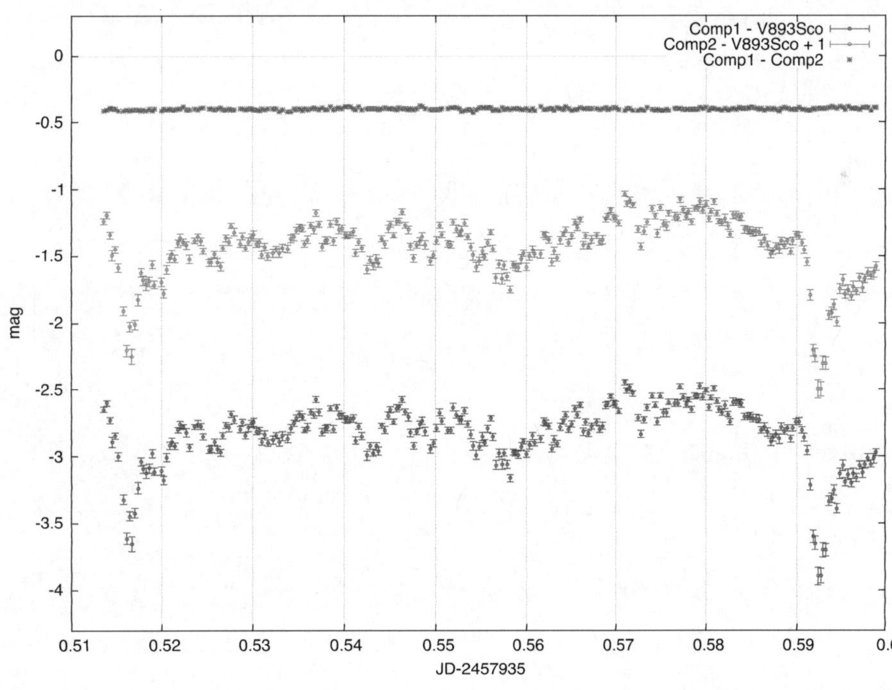

Figure 2. Light curves with errorbars of the V893 Sco star, built by differential photometry with data registered at OAUNI on 2017/06/30. Approximate orbital period of 0.076 d can be calculated from approximate positions of mid-eclipse of the primary eclipses.

IOP Conf. Series: Journal of Physics: Conf. Series **1143** (2018) 012010 doi:10.1088/1742-6596/1143/1/012010

3.2. Spectroscopy: HS 0220+0603

Spectra of HS 0220+0603 star, standard star and *Arcs* were corrected by *overscan* and *flatfield* in similar way to that of photometric data. Then spectra of the target star, calibrated on wavelength and flux, were extracted by *doslit* task from NOAO.IMRED.SPECRED, where extinction values and the directory of standard stars of Cerro Tololo Inter-American Observatory were used. Finally, the spectra were corrected by interstellar reddening with the task *deredden* from NOAO.ONEDSPEC package for $E(B-V) = 0.047$, corresponding to $A_v = 0.145$ [6]. In figure 3, an example of this procedure is shown.

It is known that a spectrum of a CV in low state has contributions of basically the WD and the RD. A fitting program in Python was written to find out the combination of a synthetic spectrum of a WD and a template of a M type star that fits best to this spectrum.

From the Spanish Virtual Observatory web page[7] was downloaded synthetic spectra of WD made by Koester [7] with temperatures from 10000 K through 90000 K with interval of 5000 K, all of them with log g = 8.0. From this web page, it was also downloaded templates of M type classified under the system defined by Kirkpatrick *et al.* [8] from M4.5 through M9.5 with interval of 0.5.

A program that combines a synthetic spectrum of a WD and a M type star template was made. Seven points between wavelengths of 6300 and 8300 Å were set for fitting. The best fitting is that composed by a synthetic spectrum of 30000 K and log g = 8.0, and a template of M5.5 type star, similar to a WD of 30000 K and log g = 8.35, and a M5.5 type star from Rodríguez *et al.*, 2015 [6].

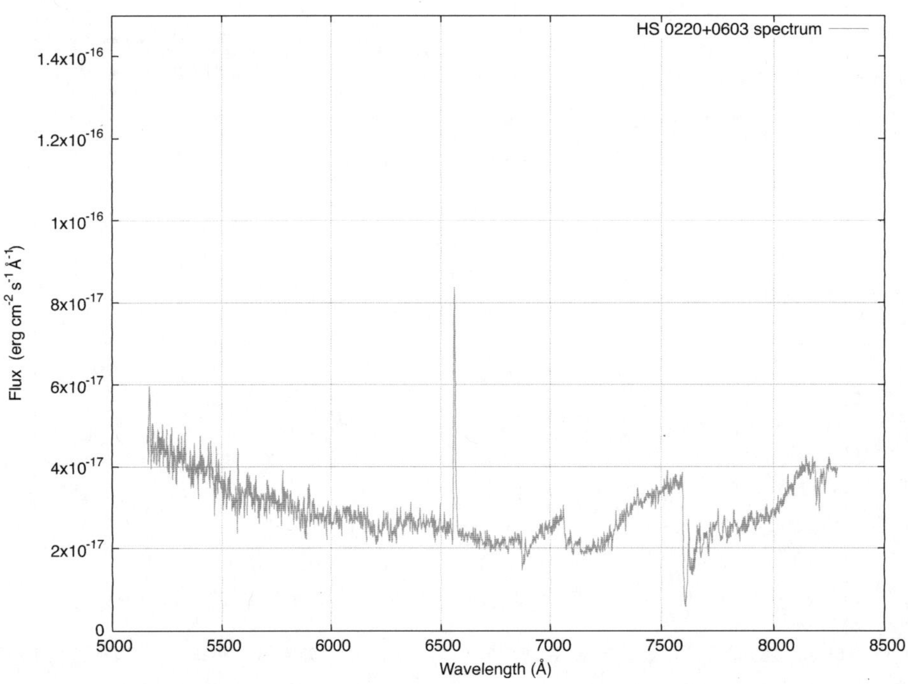

Figure 3. Spectrum of HS 0220+0603 star wavelength and flux calibrated and corrected for interstellar reddening .

[7] https://svo.cab.inta-csic.es/main/index.php

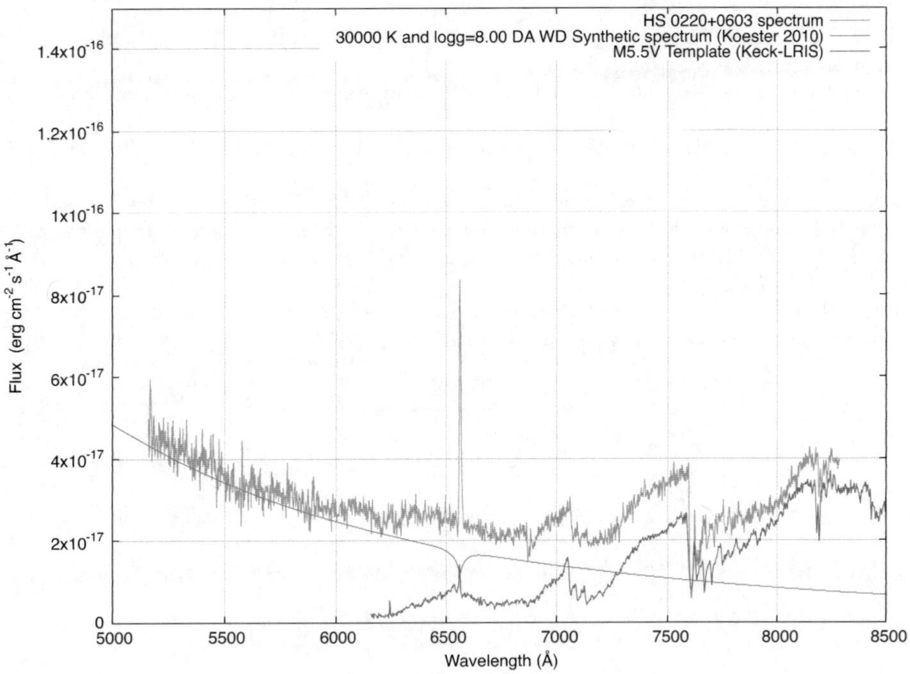

Figure 4. Observed spectrum of HS 0220+0603 (light blue), synthetic spectrum of a white dwarf of 30000 K and log g = 8.0 (green) and M5.5 type template (red). This combination is the best fit to the observed spectrum according the fitting program.

4. Conclusion

An orbital period of 0.076 d was inferred for the dwarf nova V893 Sco and it was found that the components of the novalike HS 0220+0603 star in low state are a white dwarf of 30000 K and log g = 8.0, and a red dwarf M5.5 type star. Both, orbital period inferred and components found are similar to what we can find in literature as it was said in section 3.1 and 3.2, respectively. We are confident that our program has the potential for being useful in the study of cataclysmic variable stars. However, much more tests would be necessary to confirm the efficiency of our fitting method. Moreover, in this paper it has been demonstrated the suitableness of the OAUNI for the study of cataclysmic variable stars by photometry.

Acknowledgments

This work was supported by Convenio 102-2015 Fondecyt - Cienciactiva (OAUNI project, P.I. Antonio Pereyra) and Convenio 168-2015 UNI-MINEDU-CONCYTEC (doctorate scholarship for W.C. at Universidad Nacional de Ingeniería). We thank Instituto Geofísico del Perú for lodging at Observatorio de Huancayo during photometric observations at OAUNI and Grupo Astronomía members for their help on these observations. Spectroscopic part of this work was based on observations made with ESO Telescopes at the La Silla Paranal Observatory under program ID 074.D-0657(A). This research has made use of the Spanish Virtual Observatory (http://svo.cab.inta-csic.es) supported from the Spanish MINECO/FEDER through grant AyA2014-55216. We also thank to the editors, Ener Salinas, Harold Blas and Rosendo Ochoa, for their useful suggestions and corrections.

IOP Conf. Series: Journal of Physics: Conf. Series **1143** (2018) 012010 doi:10.1088/1742-6596/1143/1/012010

References

[1] Hellier C 2008 *Cataclysmic Variable Stars: How and why they vary* Springer-Praxis Books in Astronomy and Space Sciences (Chichester: Springer-Praxis)
[2] Warner B 1995 *Cataclysmic Variable Stars* Cambridge astrophysics series (Cambridge: Cambridge university press)
[3] Pereyra A, Tello J, Meza E, Cori W, Ricra J and Zevallos M 2015 *REVCIUNI* **18** 761 (*Preprint* astro-ph/1512.03104)
[4] Cori W 2016 *Implementación del Observatorio Astronómico de la Universidad Nacional de Ingeniería* Master's thesis Universidad Nacional de Ingeniería Lima URL http://cybertesis.uni.edu.pe/handle/uni/3932
[5] Ritter H and Kolb U 2003 *Astronomy and Astrophysics* **404** 301
[6] Rodríguez-Gil P *et al.* 2015 *Mon. Not. R. Astron. Soc.* **452** 146
[7] Koester D 2010 *Memorie della Societa Astronomica Italiana* **81** 921
[8] Kirkpatrick J D, Henry T J and McCarthy(Jr) D W 1991 *ApJS* **77** 417

IOP Conf. Series: Journal of Physics: Conf. Series **1143** (2018) 012011 doi:10.1088/1742-6596/1143/1/012011

Geant4 simulation of the dual phase Time Projection Chamber TPC of argon to detect neutrinos

I. J. Huancco M.
Universidad Nacional de Ingenieria, Lima-Peru
E-mail: ivancovish.huancco@gmail.com

Abstract. The present work describes the simulation carried out in Geant4 of a Time Projection Chamber TPC of argon that uses not only a dual-phase technology (liquid and gas) but also PhotoMultiplier tubes PMT for the detection of scintillation caused by the interaction of charged particles with argon. Furthermore, the optoelectronic characteristics of the PMT are analyzed. This allows the study of particles production through the neutrino-nucleus interactions.

Keywords: Neutrinos, Time Projection Chamber, Photomultiplier tubes, ionizing particles.

1. Introduction

The study of neutrino interactions, in general, has played a tremendous role in testing the validity of several models of the electroweak weakening theory. In particular, the weak neutrino-nucleus interactions involving the associated exclusive production of stranger particles in the medium energy region have become one of the main theoretical and experimental topics in the fields of astrophysics, cosmology, particle physics and nuclear.

Neutrino experiments in general have two different simulation phases. The first simulates the line of light system that produces the neutrino flux of the primary protons and the second models the resulting energy deposits in the detector after the neutrino has traveled from its origin to the detector and interacted with a nucleus or electron objective [5]. The realms of each phase can be very different in the modeling needs of physics. For this we need a new generation of highly sensitive and high resolution detectors for neutrino-matter interaction. A novel detector is a Time Projection Chamber TPC.

2. Basic information of the dual phase Time Projection Chamber

The dual-phase TPC that uses liquid argon with a gaseous phase in the upper part as the active volume to detect scintillation and ionization signals created within the liquid Argon by incoming particles, as shown in figure 1. In the interface there is a mesh with special properties of a Large Electron Multiplier in order to amplified the signal in S2.

IOP Conf. Series: Journal of Physics: Conf. Series **1143** (2018) 012011 doi:10.1088/1742-6596/1143/1/012011

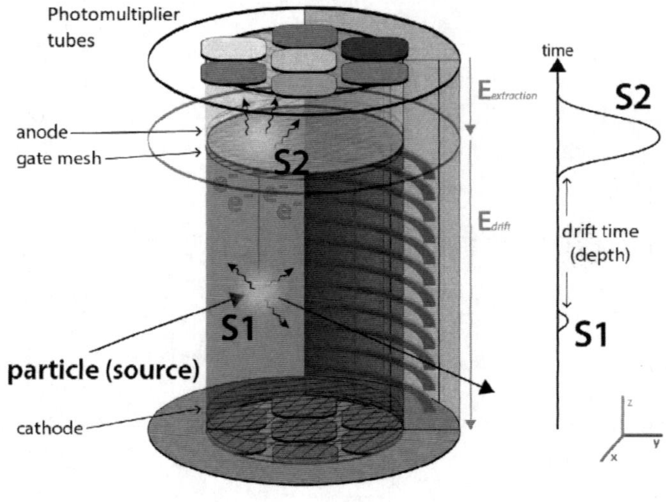

Figure 1: Sketch of the operating principle of the dual phase TPC of Argon

2.1. Large Electron Multiplier

The electron multiplier amplifies the signal that was initially sent to the detector that has upper and lower electrodes and are located at the interface between the liquid and gaseous argon and follow the following steps:

- The ionization electrons move towards the liquid surface of the argon.
- Extraction of electrons from liquid phase to gas through a grid.
- Multiplication of charge: LEM - and then load collection in the anode reading (2 orthogonal views) (without induction plane)

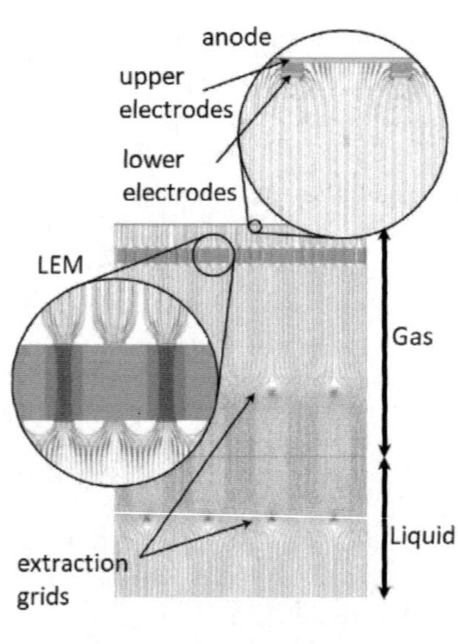

Figure 2: Extraction of electrons from liquid to gas phase through a Large Electron Multiplier LEM.[3]

IOP Conf. Series: Journal of Physics: Conf. Series **1143** (2018) 012011 doi:10.1088/1742-6596/1143/1/012011

2.2. Liquefied noble gases as neutrino target

From all noble elements, only argon and xenon are currently used as targets for dark matter detection. Neon has been suggested as a medium for low-energy neutrino detection, and could potentially be employed in the search for neutrinos. In their liquid phase, noble elements are excellent media for building large, homogeneous, compact and self-shielding detectors. Liquid xenon (LXe) and liquid argon (LAr) are excellent scintillators and good ionizers in response to the passage of radiation. The simultaneous detection of ionization and scintillation signals allows to identify the primary particle interacting in the liquid, for the ratio of the two observables depends on dE/dx [4]. Argon has been chosen for its low price and abundance in the atmosphere.

Table 1: Physical properties, volume fraction in the atmosphere and radioactive isotopes of the noble elements.[4]

Property [unit]	Xe	Ar
Atomic number	54	18
Mean atomic weight	131.3	40.0
Boiling point T_b at 1 atm [K]	165.0	87.3
Melting point T_m at 1 atm [K]	161.4	83.8
Gas density at 1 atm & 298 K [g/l]	5.40	1.63
Gas density at 1 atm & T_b [g/l]	9.99	5.77
Liquid density at T_b [g/cm^3]	2.94	1.40
Volume ratio	526	795
Dielectric constant of liquid	1.95	1.51
Volume fraction in Earths atmosphere [ppm]	0.09	9340
Radioactive isotopes	^{136}Xe,$T_{1/2} = 2.16 \times 10^{21}$ yr	^{39}Ar, $T_{1/2}$= 269 yr

2.3. Scintillation and ionization

According to [2], the argon scintillation process consists of different reactions. Briefly, the scintillation light is emitted by a state of argon dimer excited (Ar_2^*) that decomposes to the ground state. This state of arousal can be created in two different simultaneous reactions. The first reaction begins with an atom of argon excited, produced after the process of dispersion with the incoming particle, as an initial condition.

$$\begin{aligned} Ar^* + Ar + Ar &\rightarrow Ar_2^* \\ Ar_2^* &\rightarrow 2Ar + h\nu \end{aligned} \tag{1}$$

The second reaction is a collection of several reactions after an initial ionization:

$$\begin{aligned} Ar^+ + Ar &\rightarrow Ar_2^+ \\ Ar_2^+ + e^- &\rightarrow Ar**+Ar \\ Ar^{**} &\rightarrow Ar^* + heat \\ Ar^* + Ar + Ar &\rightarrow Ar_2^* + Ar \\ Ar_2^* &\rightarrow 2Ar + h\nu \end{aligned} \tag{2}$$

As can be seen in these scintillation reactions, one can delineate that argon is transparent to its own scintillation photons since the photons produced are emitted by an excited dimer state,

IOP Conf. Series: Journal of Physics: Conf. Series **1143** (2018) 012011 doi:10.1088/1742-6596/1143/1/012011

so that scintillation light is not absorbed by other atoms of argon. The average energy required to produce a scintillation photon in liquid argon is $W = 23.6^{+0.5}_{-0.3} eV$ [6].

When applying the electric field of drift, the second line in the reactions in the equation 2 is suppressed and a part of the electrons produced separates from the ions, thus being lost for the recombination process. Due to the amplification of the ionization signal by the electric field, signal S2 is always much larger than S1. With the relation of $S2/S1$, which is always lower for nuclear setbacks than for electronic setbacks, it is possible to perform a background discrimination of the electronic and nuclear recoil [2].

3. Analysis of dual-phase Time Projection Chamber

3.1. Light Collection Efficiency LCE

To extract the LCE from the simulation data for each tracked particle, an additional program is developed for the analysis, which divides the TPC into volumes and evaluates the LCE as the PMT fraction coincides with the number of fully created photons in each volume. To obtain the LCE of the TPC, 10^6 optical photons with an energy of 7 eV are simulated. These optical photons are uniformly distributed and connected to the volume of liquid argon. The absorption length of liquid argon is set at 20 cm and the reflectivity of Politetrafluoroetileno PTFE is implemented at 95 %. In addition to this, another geometric effect must be considered.

However, grids can not be implemented directly in GEANT4[1], the meshes of the grid in the simulation are added as a solid volume with properties of a grid. That means meshes have a refractive index that controls the permeability. This induces different refractive indices for meshes surrounded by different volumes. If this is not considered and an adequate refractive index is not used, the photons will no longer pass the mesh.

The absolute LCE can be obtained from the MonteCarlo MC in the following way (where (x, y, z) is the corresponding Bin):

$$LCE(x, y, z) = \frac{N_{detected}(x, y, z) \cdot QE \cdot CE \cdot QE_{increase}}{N_{generated}(x, y, z)} \quad (3)$$

- $N_{detected}(x, y, z)$: The number of photons generated at (x, y, z) and hitting a PMT photocathode (are registered by the PMTHitCollection) corrected by the PMT QE and CE (90
- QE: Quantum Efficiency which has to be applied per-PMT.
- $N_{generated}(x, y, z)$: The number of photons generated at (x,y,z).
- CE: Collection Efficiency of the PMTs (from the photocathode to the first dynode).

3.2. Calculation in the source code

The simulations are performed with 1 photon per Geant4 event, therefore we have a maximum of one PMT triggered and can use the 'pmthitID (in the code)' information for each event. This gives us directly the ID of the PMT which allows us to use the corresponding QE and check if the PMT is excluded from analysis. A PMT hit is recorded each time when a photon is registered in the PMT photocathode (100

$$LCE(x, y, z) = \frac{\sum_{eventsinsideTPC}^{n} PMThits_n \cdot QE(PMTID_n) \cdot CE \cdot QE_{increase}}{\sum_{eventsinsideTPC}^{n} photons_{generated}(x, y, z)} \quad (4)$$

IOP Conf. Series: Journal of Physics: Conf. Series **1143** (2018) 012011 doi:10.1088/1742-6596/1143/1/012011

3.3. relative Light Collection Efficiency rLCE

The rLCE is defined as the LCE relative to its mean value and can be obtained for MC and real data:

$$rLCE_{MC}(x, y, z) = \frac{LCE(x, y, z)}{LCE_{mean}} \tag{5}$$

$$rLCE_{data}(x, y, z) = \frac{ly(x, y, z)}{ly_{mean}} \tag{6}$$

3.4. Light Yield LY

TPC can also be described by a light performance. To perform this three-dimensional light performance calculation, which others can verify the small-scale symmetry of different detector volumes, the detector is divided into 1000 volumes, 10 volumes in each direction. The light yield in each of these volumes can be obtained by the following equation 7:

The LY of the MC simulation can be assumed as (for average photon yield from NEST ($W = 50ph/keV$, at 32 keV, at 150 V/cm)):

$$ly(x, y, z) = \frac{1}{W} \cdot LCE(x, y, z) \tag{7}$$

$\frac{1}{W}$: Average energy that is required to produce one scintillation photon.

4. Results

The final simulation of the dual phase TPC of argon shows in the figure 3. Where the particle incident is optical photon because is the easiest way to see the incoming particles and the signal detected by PMT. The simulation was done in Geant4 [1], which is a toolkit for the simulation of the passage of particles through matter.

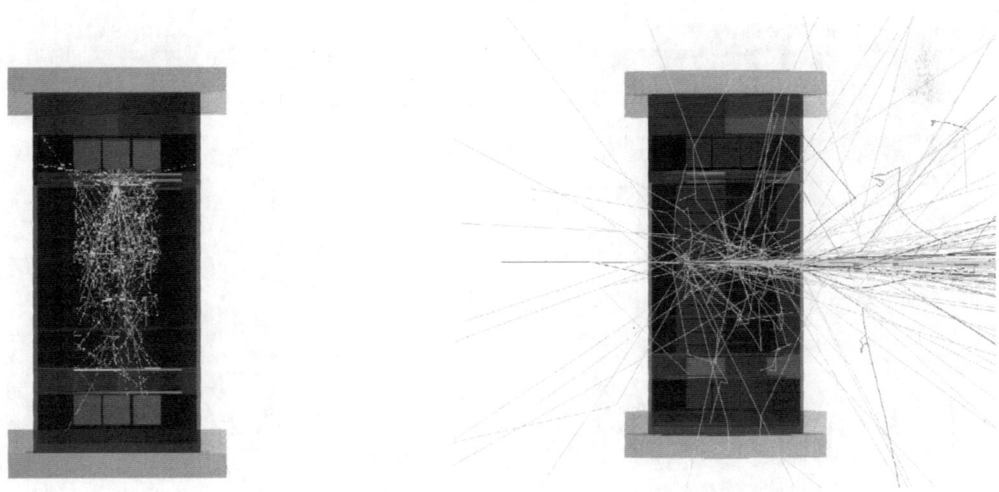

Figure 3: Left: TPC Simulated in Geant4 with the distribution of detected photons. The photons are stopped by the reflection mechanisms and the electric field. Right: 1 GeV muon beam interaction and the detector response. Primary beam (blue line), interactions (red points and lines), photons produced (green lines).

For the initial analysis of the dual phase TPC and photomultipliers, optical photon events were generated, since they interact easily with the argon leaving a trace as in the figure 3. Signal produced inside S1 and S2 are detected by PMTs (BOTTOM and TOP respectively) figure 4 which in turn are detected with photomultipliers. As has been pointed out, there is an charge amplification and more events are detected in S2 than S1. However, it will depend on beam position.

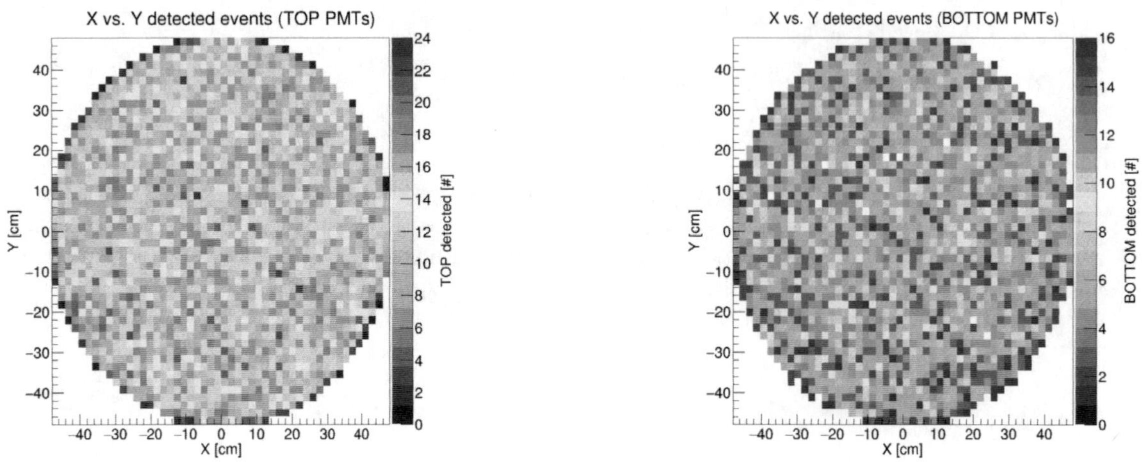

Figure 4: Events detected in the TPC (sight on the axis x e y) S2 region.

According to our source code, the events generated for the efficient collection of photons in the photomultipliers must take into account the QE and CEs. The figure 5 shows a greater interaction in the Bottom PMTs than top PMTs.

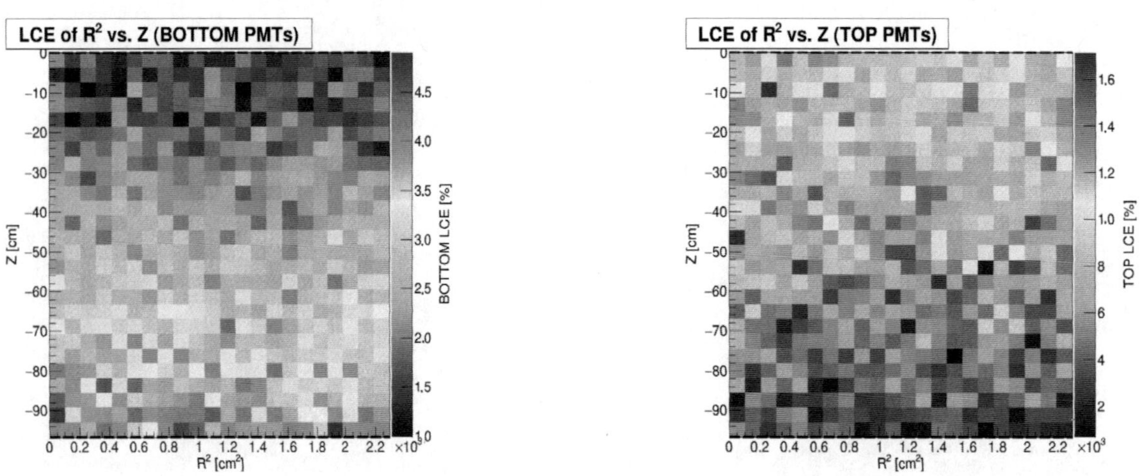

Figure 5: LCE R^2 y z (radius vs height). Left: LCE BOTTOM PMTs. right: LCE in BOTTOM PMTs.

One can point out, the bigger radius TPC has, the lowest photoelectron will be produced. Besides, the LCE is 4.5%.

On the other hand, the respective graphics for the LY 6, Bottom and top PMT in the area vs z axis

IOP Conf. Series: Journal of Physics: Conf. Series **1143** (2018) 012011 doi:10.1088/1742-6596/1143/1/012011

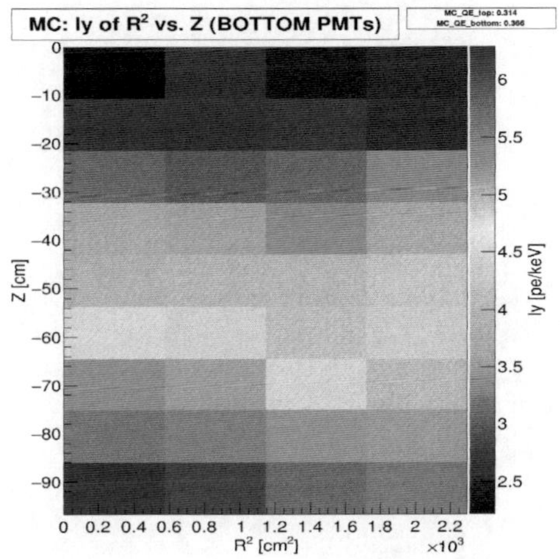

(a) Light Yield (BOTTOM PMTS).

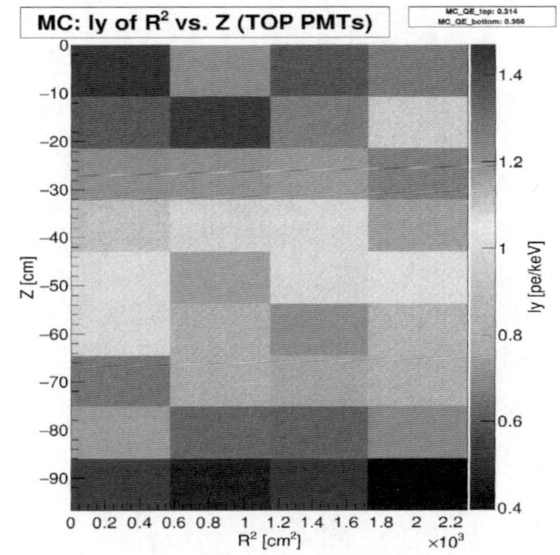

(b) Light Yield (TOP PMTS).

Figure 6: Light Yield in PMTs along z axis vs radius R^2

Finally,the dual phase TPC was proved with muon in order to obtain a signal of muon neutrino. It is well worth remembering that neutrino-argon produce muons and pions given by:

$$\nu_\mu + Ar \rightarrow \mu^- + \pi^+ + Ar$$
$$\overline{\nu}_\mu + Ar \rightarrow \mu^+ + \pi^- + Ar$$

Muons and pions are charged particle and they interact with argon. Starting from this one can reconstruct the interaction vertex in the TPC.

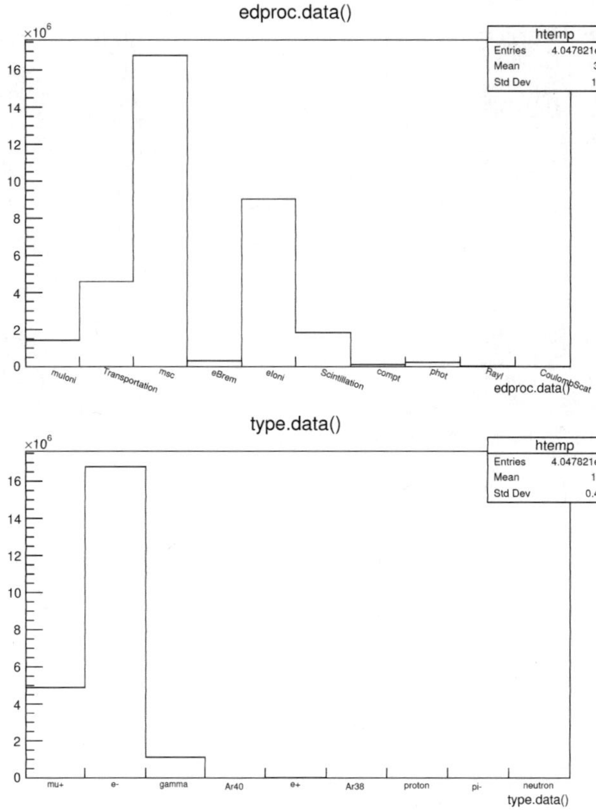

Figure 7: Generated process in muon-argon interaction

As can be seen in the figure 7, ionization and scintillation take place, these are detected by the PMTs. Also, positive muons are detected (mu+).

5. Conclusions

With this Monte Carlo-based simulation package is possible to study the effects, phenomena and parameters of the dual phase TPC of argon that are necessary for a complete understanding of the detector. After some basic tests, which verified the geometric implementation of the simulated configuration of the detector, as well as the physics implemented, the simulation package was used to characterize the TPC with calibration simulations.

In a first stage, light collection efficiency was obtained, 10^6 optical photons were simulated in the liquid argon volume of the TPC and the number of collisions detected in the PMT was determined, with LCE of 4.5% in PMTs.

References

[1] S. Agostinelli et al. "GEANT4: A Simulation toolkit". In: *Nucl. Instrum. Meth.* A506 (2003), pp. 250–303. DOI: 10.1016/S0168-9002(03)01368-8.

[2] E. Aprile et al. "The XENON100 dark matter experiment". In: *Astroparticle Physics* 35.9 (2012), pp. 573–590. ISSN: 0927-6505. DOI: https://doi.org/10.1016/j.astropartphys.2012.01.003. URL: http://www.sciencedirect.com/science/article/pii/S0927650512000059.

IOP Conf. Series: Journal of Physics: Conf. Series **1143** (2018) 012011 doi:10.1088/1742-6596/1143/1/012011

[3] A. Badertscher et al. "First operation of a double phase LAr Large Electron Multiplier Time Projection Chamber with a two-dimensional projective readout anode". In: *Nucl. Instrum. Meth.* A641 (2011), pp. 48–57. DOI: 10.1016/j.nima.2011.02.100. arXiv: 1012.0483 [physics.ins-det].

[4] Laura Baudis. "WIMP Dark Matter Direct-Detection Searches in Noble Gases". In: *Phys. Dark Univ.* 4 (2014), pp. 50–59. DOI: 10.1016/j.dark.2014.07.001. arXiv: 1408.4371 [astro-ph.IM].

[5] "MINERvA Technical Design Report". In: (2006).

[6] M. Miyajima et al. "Average energy expended per ion pair in liquid argon". In: *Phys. Rev. A* 9 (3 Mar. 1974), pp. 1438–1443. DOI: 10.1103/PhysRevA.9.1438. URL: https://link.aps.org/doi/10.1103/PhysRevA.9.1438.

6. Acknowledgements

This work was supported by Concejo Nacional de Ciencia y Tecnologia (CONCYTEC) and the faculty of sciences of the Universidad Nacional de Ingenieria. Thanks to my advisor, Dr. Javier Solano Salinas, for his corrections, suggestions and explanations that clarified the situation when analyzing the present research work.

IOP Conf. Series: Journal of Physics: Conf. Series **1143** (2018) 012012 doi:10.1088/1742-6596/1143/1/012012

Pristine graphene: functionalization, fabrication, and nanocomposite materials

Mingdi Yan

Department of Chemistry, Kennedy College of Sciences, University of Massachusetts Lowell, 1 University Ave., Lowell, MA 01854

E-mail: mingdi_yan@uml.edu

Abstract. Graphene has a wide range of unique properties including high carrier mobility, optical transparency, and exceptional electrical and thermal conductivity. The interest in graphene has been largely fueled by its potential to revolutionize a number of technologically important areas including materials science, biomedical science, and energy. Despite its high potentials, major challenges still remain. These include poor solubility, intrinsic zero band gap energy, low reactivity and the availability of well-defined pristine graphene in large quantity, which have hampered the rapid development of graphene-based functional devices. Many of these challenges can be potentially addressed through chemical functionalization of the material. Covalent functionalization can enhance graphene's properties including opening its band gap, tuning conductivity, improving solubility, and enhancing the properties of graphene-based composite materials. This paper discusses our work on the chemical functionalization of pristine graphene, and synthesis of graphene-nanoparticle composite materials and three-dimensional graphene (3DG)-TiO_2 nanocomposite photocatalyst. In photocatalytic CO_2 reduction, the 3DG-TiO_2 nanocomposite demonstrated excellent activity, about 11 times higher than TiO_2 nanoparticles.

1. Covalent functionalization of pristine graphene

The potential of graphene to revolutionize electronics, photonics, sensing, imaging and nanomaterials has been hampered by the availability and processability of graphene materials. The ability to process graphene in solution is a critical step in making new graphene materials and integrating them into high-performance devices. Covalent chemistry offers an attractive means to modify graphene with well-defined chemical functionalities, modulate solubility and enable processability.

Covalent chemistry has the potential to open a substantial band gap on the otherwise zero-gap electronic band of pristine graphene. It has been suggested that a band gap can be created by breaking the equivalence between the A and B sublattices in graphene by introducing structural perturbations.[1-2] It is also known that in semiconductors, a band gap can be altered with an applied strain. This opens up enormous opportunities for the covalent chemistry to open and fine-tune the band gap, and to control the remarkable electronic properties of graphene materials.

Graphene contains exclusively conjugated sp^2 carbons delocalized in the entire plane. This large resonance stabilization makes graphene chemically inert and poses considerable challenge for the

IOP Conf. Series: Journal of Physics: Conf. Series **1143** (2018) 012012 doi:10.1088/1742-6596/1143/1/012012

chemical modification of pristine graphene. Limited methods have been reported for the covalent functionalization of pristine graphene. Most approaches involve reactive species, for example, nitrenes[3-7] and free radicals (generated from diazonium ions,[8-11] perfluorinated alkyl iodides,[12] 2,2,6,6-tetramethylpiperidinyloxy (TEMPO),[13] or benzoyl peroxide[14]. Nitrene adds to the double bonds on graphene to give a bivalent adduct. Addition of radicals to graphene, on the other hand, converts sp^2 carbons to sp^3, and yields a single covalent bond as a result. Other reported covalent modification methods include the Diels−Alder reaction where graphene as a dienophile reacted reversibly with 2,3-dimethoxybutadiene or 9-methylanthracene.[15-16] However, a later work revealed that the reaction enthalpies were highly unfavorable for the Diels-Alder reaction to occur on the basal plane of graphene. The authors concluded that the experimentally observed reaction was mostly on the graphene edges that have higher reactivity due to defects, or from physical absorption of the reagents on the basal plane of graphene.[17]

2. Functionalization of pristine graphene with perfluorophenyl azides (PFPAs)

We have developed an efficient chemistry to covalently functionalize pristine graphene using PFPAs.[4, 18-21] Activation of PFPA by heat, microwave, or UV generates the highly reactive perfluorophenyl singlet nitrene, which reacts with graphene to give the covalent adduct. By changing the functional groups on PFPAs, we could make the graphene soluble in common organic solvents as well as in water.[5] This coupling chemistry, which was highlighted by *Nature*,[22] represents a powerful method for the covalent functionalization of graphene, especially on the basal planes which are much less reactive than the edges and are difficult to modify.

Several groups have conducted computational studies on the reaction of PFFA with graphene.[23-25] DFT calculations concluded that the aziridine product from the (2+1) cycloaddition of perfluoroaryl nitrene with graphene (**Figure 1A**) would induce bond cleavage between adjacent carbons in the basal plane to give the azepine structure (**Figure 1B**). One study found that unlike the single covalent bond formed between a radical and graphene where the adduct acts as a strong scattering center, the divalent azepine adduct has a very small effect on the conductance near the Fermi level.[23] This opens up the possibility of introducing chemical functionalities while at the same time preserving the conductance of graphene. In another work, the computation as well as experimental results showed that after functionalization with PFPA, the conductivity of the pristine graphene degraded only about 50%.[25] The authors attributed this to the interaction of N with graphene's p orbitals, i.e., σ–π interaction, which causes rehybridization of the C atoms from the sp^2 to $sp^{2+\eta}$ state (η is defined as the degree of additional hybridization).

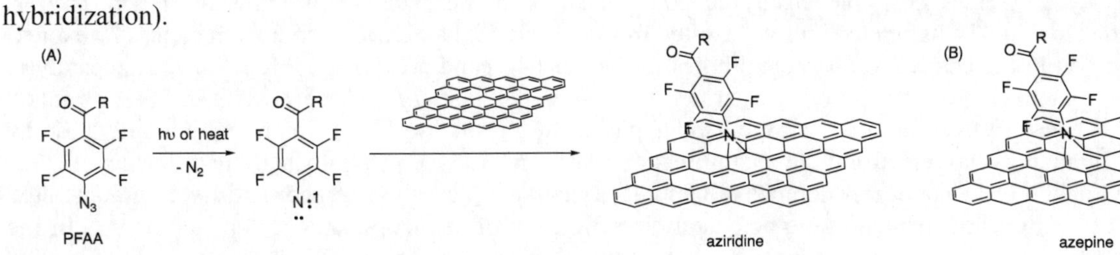

Figure 1. (A) Functionalization of graphene with PFPA to give the aziridine structure. (B) The azepine structure.

3. Synthesis of nanoparticle-graphene nanocomposites

To covalently conjugate nanoparticles to pristine graphene, nanoparticles were first functionalized with PFPA, and were then covalently attached to graphene by UV irradiation (**Figure 2a**). We have successfully conjugated gold nanoparticles (AuNPs), silica nanoparticles (SNPs), iron oxide nanoparticles and quantum dots to CVD graphene, mechanically exfoliated graphene, and few layer

IOP Conf. Series: Journal of Physics: Conf. Series **1143** (2018) 012012 doi:10.1088/1742-6596/1143/1/012012

graphene (FLG) flakes.[21] **Figure 2b-d** show that the covalently attached gold and silica nanoparticles on FLG.

(a)

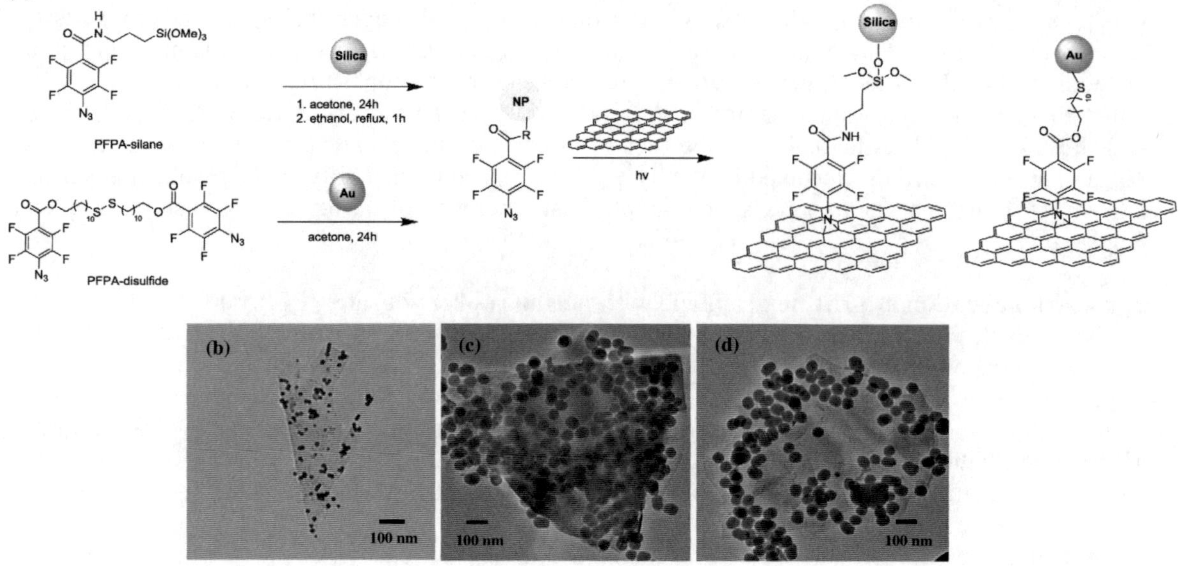

Figure 2. a) Synthesis of graphene-nanoparticle conjugates. TEM images of (b) AuNPs and (c, d) SNPs conjugated on FLG. Samples in c and d were prepared from FLG and PFPA-SNPs at 1:1 and 1:0.1 mass ratio, respectively.

4. TiO_2 nanoparticle-graphene composite

A heterogeneous photocatalyst is often a semiconductor that is capable of catalyzing photochemical reactions. There are various materials that show photocatalytic properties, among which, TiO_2 is one of the most widely used heterogeneous photocatalysts.[26] It exhibits excellent physical and chemical stability, possesses desirable electronic and optical properties, and is inexpensive. When TiO_2 is irradiated with UV light, electrons are excited from the valence band to the conduction band. The excited electrons in the conduction band and holes in the valence band are responsible for the photocatalytic activity, carrying out reduction and oxidation with reagents at the surface of TiO_2. Albeit its many advantages, TiO_2 suffers from a large band gap (3.2 eV for anatase TiO_2) which requires activation by UV light, and a fast electron-hole recombination rate which reduce its photocatalytic efficiency.[26-27]

Graphene-semiconductor composite nanomaterials have recently emerged, and have demonstrated high potential in enhancing the photocatalytic efficiency of semiconductor nanomaterials.[28-35] In the composite nanomaterial, graphene acts as an electron acceptor to ensure fast charge transfer. Effective charge separation can therefore be achieved to slow down the electron-hole recombination, resulting in enhanced photocatalytic activity of TiO_2. Moreover, because graphene can absorb light in the visible region, the photo-excitation wavelength of graphene-TiO_2 hybrid nanomaterials can be extended to visible (400-700 nm). Consequently, the photocatalytic reactions can be carried out using the visible light instead of UV. The large surface area of graphene can additionally increase the concentration of adsorbed reagents as well as the surface interactions.[28] Because the photocatalysis occurs at the surface of TiO_2, the increased adsorption and interactions with the adsorbed reagents lead to enhanced photocatalytic activities of graphene-TiO_2 hybrid nanomaterials.[29-32]

Studies have shown that graphene-based composite materials exhibit better performance when the components are covalently linked, owing to the higher stability and stronger interactions between the individual components than the non-covalent composites.[36-37] Graphene is relatively inert and has limited dispersibility in solvents, which poses considerable challenges in graphene functionalization. The most common covalent functionalization uses graphene oxide (GO) owing to its abundant functional groups. However, the oxidation process produces various oxygen-containing species, and the structure and density of which are difficult to control. Even after reduction to reduced GO (rGO), the properties cannot be completely restored to those of the pristine graphene.[38]

Graphene-TiO_2 composite nanomaterials are mostly prepared by hydrothermal, solvothermal and hydrolysis reactions.[29-30, 39-41] In these cases, GO instead of pristine graphene was used. The hybrid materials were prepared by reducing GO in the presence of a TiO_2 precursor (titanium(IV) isopropoxide, titanium(IV) fluoride, titanium(IV) butoxide), resulting in the deposition of TiO_2 particles on the surface of rGO. Because rGO was used, oxygen-containing species remained and could not be removed completely even after reduction. These defects disrupt the conjugated structure and the electron conduction of graphene.[42]

We synthesized graphene-TiO_2 nanocomposite photocatalyst using pristine graphene and pre-made TiO_2 nanoparticles by conjugating PFPA-functionalized TiO_2 nanoparticles on few-layer graphene (FLG) flakes. Following the protocol shown in **Figure 2a**, TiO_2 nanoparticles were first functionalized with PFPA. The PFPA-functionalized TiO_2 nanoparticles were then treated with FLG under UV irradiation. When the FLG-TiO_2 samples were examined with TEM, agglomerated TiO_2 nanoparticles were observed (**Figure 3**). Agglomeration can reduce the catalytic activity of TiO_2 nanoparticles through recombination of excited electrons and holes in TiO_2.

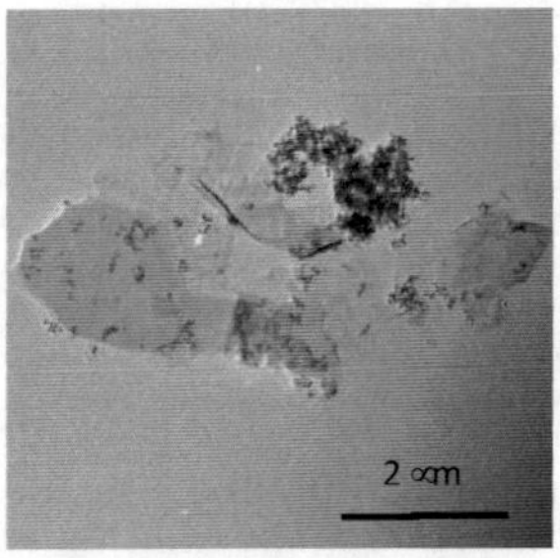

Figure 3. TEM image of TiO_2 nanoparticles conjugated on FLG flakes.

5. 3-Dimensional graphene-TiO_2 photocatalyst composite

5.1. 3-Dimensional graphene

The recent development of three-dimensional graphene (3DG) represents a significant advance in the availability of pristine graphene materials.[43-45] Pristine 3DG can be prepared by chemical vapor deposition (CVD) using a macroporous Ni or Cu foam during graphene synthesis.[46] Ni or Cu foam serves two purposes: (1) to catalyze the dissolution/deposition of graphene during pyrolysis of carbon precursors; (2) to serve as a physical template onto which graphene is deposited. When the template is removed by acid etching, a free-standing three-dimensional network of pristine graphene is obtained (**Figure 4a**). The macroporous structure of the template is maintained (**Figure 4b**). Because 3DG is fabricated on a continuous interconnected structure, it has minimal edge defects and more uniform surface (**Figure 4c**). Depending on the substrate, the 3DG produced can range from single to several layers.

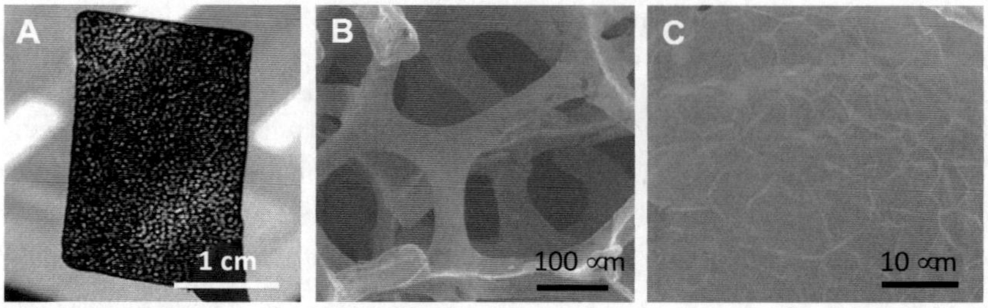

Figure 4. (a) Optical image of a CVD 3DG sample (2 × 2 × 0.1 cm^3). (b) SEM of 3DG showing the macroporous interior of 3DG. (c) SEM of 3DG surface.

3DG possesses many unique properties in addition to those of the typical graphene nano-sheets.[43-45] It has high specific surface area resulting from the interconnected graphene network. While 2D graphene sheets have conducting pathways restricted to the planar direction, 3DG provides multiple channels for fast electron transport. Its excellent electrical conductivity is also due to the absence of inter-sheet contact resistance. Moreover, 3DG reduces the agglomeration of 2D graphene sheets caused by the strong inter-sheet π-π stacking. While 2D graphene sheets are fragile and difficult to handle, 3DG is a freestanding material (**Figure 4a**). Practically, 3DG can be made into different sizes and shapes. The surface area and porosity can be controlled by the template. High purity Ni and Cu foams are commercially available and can be purchased as large pieces, e.g., 1 m^2, or cut into desired size and shape.

5.2. 3-Dimensional graphene-TiO_2 photocatalyst composite

We covalently conjugated TiO_2 nanoparticles on a pristine 3D graphene.[47] The 3D graphene was prepared by CVD on a macroporous Ni foam substrate.[46] Because the microporous structures of the Ni foam template are interconnected, the 3D graphene synthesized on this template would consist of a continuous piece of pristine graphene free of edge defects (**Figure 4**). To conjugate TiO_2 nanoparticles, 3D graphene was first functionalized with PFPA-COOH (**Figure 5**). The TiO_2 nanoparticles were conjugated by a solvothermal treatment through binding with the carboxyl groups on the graphene surface.

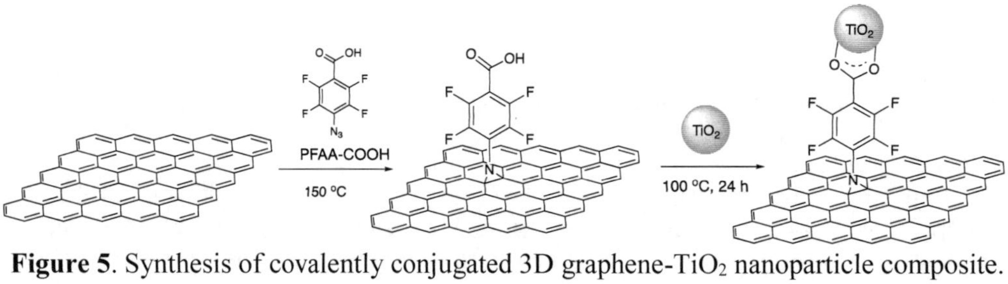

Figure 5. Synthesis of covalently conjugated 3D graphene-TiO_2 nanoparticle composite.

Densely packed and uniformly distributed TiO_2 particles were observed on 3D graphene (**Figure 6A**).[47] This is in sharp contrast with the product prepared on FLG, where the conjugated TiO_2 nanoparticles were sparse and were substantially agglomerated (**Figure 3**). To test the photocatalytic performance of the composite catalyst, CO_2 reduction was carried out in the presence of TiO_2-3D graphene by irradiating the sample with a 200 W mercury lamp. As a control, TiO_2 nanoparticles of TiO_2-3D graphene for comparison. The result shows that the amount of CO produced by TiO_2-3D graphene after 4 h was 11-fold higher than that by bare TiO_2 (**Figure 6B**). This result can be attributed to a combination of 3D graphene substrate, high density of TiO_2 nanoparticles and excellent particle dispersity.

IOP Conf. Series: Journal of Physics: Conf. Series **1143** (2018) 012012 doi:10.1088/1742-6596/1143/1/012012

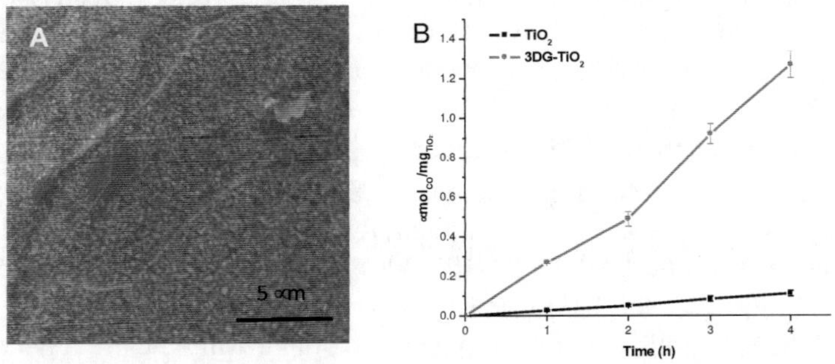

Figure 6. (A) SEM image of TiO_2 nanoparticles conjugated on 3D graphene. (B) The production of CO in photocatalytic CO_2 reduction on 3D graphene-TiO_2. CO was observed as the only gaseous product under this condition (triethylamine as the sacrificial electron donor, irradiated with a 200 W mercury lamp (140 mW/cm^2)).

Conclusions

Despite its high potential, the utility of graphene in practical applications has been hindered by the availability of high quality pristine graphene, and the integration with other technologically important materials. The ability to covalently functionalize graphene can enhance graphene's properties as well as produce high-performance composite materials. We have developed an efficient method to covalently functionalize pristine graphene. Covalently conjugated nanoparticle-graphene composite materials showed high density of nanoparticles on graphene basal planes. Finally, a 3D graphene-TiO_2 nanocomposite demonstrated excellent photocatalytic activity, about 11 times higher than TiO_2 nanoparticles in reducing CO_2.

References

[1] Nakada, K.; Fujita, M.; Dresselhaus, G.; Dresselhaus, M. S., Edge state in graphene ribbons: Nanometer size effect and edge shape dependence. *Phys. Rev. B* **1996,** *54* (24), 17954-17961.

[2] Brey, L.; Fertig, H. A., Electronic states of graphene nanoribbons studied with the Dirac equation. *Phys. Rev. B* **2006,** *73* (23), 235411.

[3] Choi, J.; Kim, K. J.; Kim, B.; Lee, H.; Kim, S., Covalent Functionalization of Epitaxial Graphene by Azidotrimethylsilane. *J. Phys. Chem. C* **2009,** *113* (22), 9433-9435.

[4] Liu, L.-H.; Yan, M., Simple Method for the Covalent Immobilization of Graphene. *Nano Letters* **2009,** *9* (9), 3375-3378.

[5] Liu, L.-H.; Zorn, G.; Castner, D. G.; Solanki, R.; Lerner, M. M.; Yan, M., A simple and scalable route to wafer-size patterned graphene. *J. Mater. Chem.* **2010,** *20*, 5041-5046.

[6] Liu, L.-H.; Lerner, M. M.; Yan, M., Derivitization of pristine graphene with well-defined chemical functionalities. *Nano Lett.* **2010,** *10*, 3754-3756.

[7] Strom, T. A.; Dillon, E. P.; Hamilton, C. E.; Barron, A. R., Nitrene addition to exfoliated graphene: a one-step route to highly functionalized graphene. *Chem. Commun.* **2010,** *46* (23), 4097-4099.

[8] Koehler, F. M.; Luechinger, N. A.; Ziegler, D.; Athanassiou, E. K.; Grass, R. N.; Rossi, A.; Hierold, C.; Stemmer, A.; Stark, W. J., Permanent Pattern-Resolved Adjustment of the Surface Potential of Graphene-Like Carbon through Chemical Functionalization. *Angew. Chem. Int. Ed.* **2009,** *48* (1), 224-227.

[9] Koehler, F. M.; Jacobsen, A.; Ensslin, K.; Stampfer, C.; Stark, W. J., Selective Chemical Modification of Graphene Surfaces: Distinction Between Single- and Bilayer Graphene. *Small* **2010,** *6* (10), 1125-1130.

[10] Sun, Z.; Kohama, S.-i.; Zhang, Z.; Lomeda, J.; Tour, J., Soluble graphene through edge-selective functionalization. *Nano Res.* **2010,** *3* (2), 117-125.

[11] Niyogi, S.; Bekyarova, E.; Itkis, M. E.; Zhang, H.; Shepperd, K.; Hicks, J.; Sprinkle, M.; Berger, C.; Lau, C. N.; deHeer, W. A.; Conrad, E. H.; Haddon, R. C., Spectroscopy of Covalently Functionalized Graphene. *Nano Lett.* **2010,** *10* (10), 4061-4066.

[12] Hamilton, C. E.; Lomeda, J. R.; Sun, Z. Z.; Tour, J. M.; Barron, A. R., Radical addition of perfluorinated alkyl iodides to multi-layered graphene and single-walled carbon nanotubes. *Nano Res.* **2010,** *3* (2), 138-145.

[13] Choi, J.; Lee, H.; Kim, K. J.; Kim, B.; Kim, S., Chemical Doping of Epitaxial Graphene by Organic Free Radicals. *J. Phys. Chem. Lett.* **2010,** *1* (2), 505-509.

[14] Liu, H.; Ryu, S.; Chen, Z.; Steigerwald, M. L.; Nuckolls, C.; Brus, L. E., Photochemical Reactivity of Graphene. *J. Am. Chem. Soc.* **2009,** *131* (47), 17099-17101.

[15] Sarkar, S.; Bekyarova, E.; Niyogi, S.; Haddon, R. C., Diels−Alder Chemistry of Graphite and Graphene: Graphene as Diene and Dienophile. *J. Am. Chem. Soc.* **2011,** *133* (10), 3324-3327.

[16] Sarkar, S.; Bekyarova, E.; Haddon, R. C., Chemistry at the Dirac Point: Diels–Alder Reactivity of Graphene. *Acc. Chem. Res.* **2012,** *45* (4), 673-682.

[17] Cao, Y.; Osuna, S.; Liang, Y.; Haddon, R. C.; Houk, K. N., Diels–Alder Reactions of Graphene: Computational Predictions of Products and Sites of Reaction. *J. Am. Chem. Soc.* **2013,** *135* (46), 17643-17649.

[18] Liu, L.-H.; Yan, M., Functionalization of pristine graphene with perfluorophenyl azides. *Journal of Materials Chemistry* **2011,** *21*, 3273-3276.

[19] Park, J.; Yan, M., Covalent functionalization of graphene with reactive intermediates. *Accounts of Chemical Research* **2013,** *46* (1), 181-189.

[20] Liu, L.-H.; Yan, M., Simple Method for the Covalent Immobilization of Graphene. *Nano Lett.* **2009,** *9* (9), 3375-3378.

[21] Park, J.; Jayawardena, H. S. N.; Chen, X.; Jayawardana, K. W.; Sundhoro, M.; Ada, E.; Yan, M., A general method for the fabrication of graphene-nanoparticle hybrid material. *Chem. Commun.* **2015,** *51* (14), 2882-2885.

[22] *Nature* **2010,** *466*, 904.

[23] Suggs, K.; Reuven, D.; Wang, X.-Q., Electronic Properties of Cycloaddition-Functionalized Graphene. *The Journal of Physical Chemistry C* **2011,** *115* (8), 3313-3317.

[24] Denis, P. A.; Iribarne, F., Monolayer and Bilayer Graphene Functionalized with Nitrene Radicals. *The Journal of Physical Chemistry C* **2011,** *115* (1), 195-203.

[25] Plachinda, P.; Evans, D.; Solanki, R., Electrical conductivity of PFPA functionalized graphene. *Solid-State Electronics* **2013,** *79* (Supplement C), 262-267.

[26] Linsebigler, A. L.; Lu, G.; Yates, J. T., Photocatalysis on TiO_2 Surfaces: Principles, Mechanisms, and Selected Results. *Chem. Rev.* **1995,** *95* (3), 735-758.

[27] Kazuhito, H.; Hiroshi, I.; Akira, F., TiO_2 Photocatalysis: A Historical Overview and Future Prospects. *Jpn. J. Appl. Phys.* **2005,** *44* (12R), 8269.

[28] Kamat, P. V., Graphene-Based Nanoassemblies for Energy Conversion. *The Journal of Physical Chemistry Letters* **2011,** *2* (3), 242-251.

[29] Zhang, H.; Lv, X.; Li, Y.; Wang, Y.; Li, J., P25-Graphene Composite as a High Performance Photocatalyst. *ACS Nano* **2009,** *4* (1), 380-386.

[30] Zhang, Y.; Tang, Z.-R.; Fu, X.; Xu, Y.-J., Engineering the Unique 2D Mat of Graphene to Achieve Graphene-TiO_2 Nanocomposite for Photocatalytic Selective Transformation: What

Advantage does Graphene Have over Its Forebear Carbon Nanotube? *ACS Nano* **2011,** *5* (9), 7426-7435.

[31] Long, R.; English, N. J.; Prezhdo, O. V., Photo-induced Charge Separation across the Graphene–TiO_2 Interface Is Faster than Energy Losses: A Time-Domain ab Initio Analysis. *J. Am. Chem. Soc.* **2012,** *134* (34), 14238-14248.

[32] Perera, S. D.; Mariano, R. G.; Vu, K.; Nour, N.; Seitz, O.; Chabal, Y.; Balkus, K. J., Hydrothermal Synthesis of Graphene-TiO_2 Nanotube Composites with Enhanced Photocatalytic Activity. *ACS Catalysis* **2012,** *2* (6), 949-956.

[33] Qiu, B.; Xing, M.; Zhang, J., Mesoporous TiO_2 Nanocrystals Grown in Situ on Graphene Aerogels for High Photocatalysis and Lithium-Ion Batteries. *Journal of the American Chemical Society* **2014,** *136* (16), 5852-5855.

[34] Mo, R.; Lei, Z.; Sun, K.; Rooney, D., Facile Synthesis of Anatase TiO_2 Quantum-Dot/Graphene-Nanosheet Composites with Enhanced Electrochemical Performance for Lithium-Ion Batteries. *Adv. Mater.* **2014,** *26* (13), 2084-2088.

[35] Lightcap, I. V.; Kamat, P. V., Fortification of CdSe Quantum Dots with Graphene Oxide. Excited State Interactions and Light Energy Conversion. *J. Am. Chem. Soc.* **2012,** *134* (16), 7109-7116.

[36] Xiang, Q.; Cheng, B.; Yu, J., Graphene-Based Photocatalysts for Solar-Fuel Generation. *Angewandte Chemie International Edition* **2015,** *54* (39), 11350-11366.

[37] Zhang, N.; Yang, M.-Q.; Liu, S.; Sun, Y.; Xu, Y.-J., Waltzing with the Versatile Platform of Graphene to Synthesize Composite Photocatalysts. *Chemical Reviews* **2015,** *115* (18), 10307-10377.

[38] Compton, O. C.; Nguyen, S. T., Graphene Oxide, Highly Reduced Graphene Oxide, and Graphene: Versatile Building Blocks for Carbon-Based Materials. *Small* **2010,** *6* (6), 711-723.

[39] Nethravathi, C.; Rajamathi, M., Chemically modified graphene sheets produced by the solvothermal reduction of colloidal dispersions of graphite oxide. *Carbon* **2008,** *46* (14), 1994-1998.

[40] Lee, J. S.; You, K. H.; Park, C. B., Highly Photoactive, Low Bandgap TiO_2 Nanoparticles Wrapped by Graphene. *Adv. Mater.* **2012,** *24* (8), 1084-1088.

[41] Tang, Y.-B.; Lee, C.-S.; Xu, J.; Liu, Z.-T.; Chen, Z.-H.; He, Z.; Cao, Y.-L.; Yuan, G.; Song, H.; Chen, L.; Luo, L.; Cheng, H.-M.; Zhang, W.-J.; Bello, I.; Lee, S.-T., Incorporation of Graphenes in Nanostructured TiO_2 Films via Molecular Grafting for Dye-Sensitized Solar Cell Application. *ACS Nano* **2010,** *4* (6), 3482-3488.

[42] Gómez-Navarro, C.; Weitz, R. T.; Bittner, A. M.; Scolari, M.; Mews, A.; Burghard, M.; Kern, K., Electronic Transport Properties of Individual Chemically Reduced Graphene Oxide Sheets. *Nano Letters* **2007,** *7* (11), 3499-3503.

[43] Patil, U.; Lee, S. C.; Kulkarni, S.; Sohn, J. S.; Nam, M. S.; Han, S.; Jun, S. C., Nanostructured pseudocapacitive materials decorated 3D graphene foam electrodes for next generation supercapacitors. *Nanoscale* **2015,** *7* (16), 6999-7021.

[44] Nardecchia, S.; Carriazo, D.; Ferrer, M. L.; Gutierrez, M. C.; del Monte, F., Three dimensional macroporous architectures and aerogels built of carbon nanotubes and/or graphene: synthesis and applications. *Chem. Soc. Rev.* **2013,** *42* (2), 794-830.

[45] Park, J.; Yan, M., Three-dimensional graphene and TiO2 hybrid nanomaterial for high efficient photocatalysis. *Nanotechn. Rev.* **2016.**

[46] Chen, Z.; Ren, W.; Gao, L.; Liu, B.; Pei, S.; Cheng, H.-M., Three-dimensional flexible and conductive interconnected graphene networks grown by chemical vapour deposition. *Nature Materials* **2011,** *10* (6), 424-428.

[47] Park, J.; Jin, T.; Liu, C.; Li, G.; Yan, M., Three-Dimensional Graphene-TiO2 Nanocomposite Photocatalyst Synthesized by Covalent Attachment. *ACS Omega* **2016,** *1*, 351-356.

IOP Conf. Series: Journal of Physics: Conf. Series **1143** (2018) 012013 doi:10.1088/1742-6596/1143/1/012013

Instrumentation-level improvements in shear-force near-field acoustic microscopy

J Bai, P Devulapalli, T Brockman and Andres H La Rosa
Portland State University, Portland, Oregon 97201, USA

E-mail: bai@pdx.edu

Abstract. The recently introduced Shear-force Near-field Acoustic Microscopy (SANM) brings a new sensing mechanism to the scanning probe microscopy family. SANM's ability to simultaneously monitor, in real time, several physical sample's responses presents some challenges for ensuring optimal operation; namely, avoid "cross-talk" among the multiple signals, address the compensation of thermal drift to ensure reproducibility of the measurements, and measuring the typical low-level signals obtained from nanometer-sized tested regions. Here, several improvements relevant to SANM, but valid for SPM in general, are addressed. *i*) The probe's coarse approach is performed via stepper motors, which are controlled either by a computer software interface, or simply by a user-friendly RF remote control. *ii*) The inherent mechanical drift of the stage is evaluated first (by monitoring the feedback voltage that acts on the sample to maintain the probe-sample distance constant), and then automatically compensated (via linear interpolation) in the immediate subsequent probe's approach/retraction measurements. *iii*) To determine the absolute position of the substrate relative to the probe, the probe-substrate contact current is measured with circuitry properly shielded from eventual electrical ground loops. Buffer amplifiers are used to drive the quartz tuning fork (that holds the probe) and to measure the AC and DC tunneling currents between the tip and sample. *iv*) To improve the signal to noise ratio, a Kalman filter is implemented into the SANM's field-programmable gate array board, which processes the signals in real time. Finally, *v*) to simplify the operation of the microscope, an intuitive LabVIEW host program is developed to control the whole system and offer the user a visualization of the data in real time.

1. Introduction

Shear-force Near-field Acoustic Microscopy (SANM) has recently been introduced as a supportive analytical tool to investigate the viscoelastic properties of mesoscopic fluids trapped between two solid substrates under relative shear motion [1]. SANM joins the family of scanning probe microscopy by bringing near-field acoustic sensing to interrogate such challenging fluid systems. New sensing mechanisms are indeed pertinent in this field since, as it turns out, understanding the striking properties that mesoscopic fluids display when placed under confinement remain as major challenges in condensed matter physics [2], [3]. Some of these properties, quite different from the bulk, include enhanced shear viscosity, prolonged relaxation time, ice formation at room temperature, and hydrophobic interaction range well beyond

IOP Conf. Series: Journal of Physics: Conf. Series **1143** (2018) 012013 doi:10.1088/1742-6596/1143/1/012013

commonly accepted molecular interactions ranges. In addition to their fundamental physics interest, an understanding of these properties is also relevant to technological areas like adhesion, wetting processes, and nanotribology [4], [5] (including finding ways to reduce corrosion and wear [6], [7].) It will also have an impact on attaining correct interpretation of high-resolution images acquired by scanning probe microscopy (SPM). The reason lies in the fact that, in SPM, the topography of the surface is inferred from the assumed shear-force interactions [8], [9], the latter comprising the participation of *i*) the SPM probe, *ii*) the interrogated substrate, and *iii*) the mesoscopic fluid trapped in between. It is possible that, in some cases, the features captured in an SPM image are not actual topographic signatures of the substrate, but rather artifacts produced by the different local wetting properties (adsorption, hydrophobicity) of the substrate's surface. In the spirit, then, of providing a reliable metrology tool to characterize solid-fluid interfaces, here we describe few improvements recently incorporated into the SANM, which can be extrapolated to any SPM in general.

2. SANM experimental setup

Figure 1a shows schematically a water meniscus that forms when a solid probe (laterally oscillating at ~ 32 kHz with an amplitude of few nanometers) and a solid substrate are brought in close proximity (~ 20 nm) at ambient conditions. The mesoscopic fluid *i*) affects the lateral motion of the probe, and *ii*) engenders an acoustic wave. The SANM is used to investigate correlations between these two signals and extract from them the elastic and damping properties (k, γ) of the fluid.

Figure 1b shows three distinct sections of the experimental arrangement: *i*) The central part displays a tapered probe attached to one of the tines of a quartz tuning fork (QTF), and a sample (referred here as a metal substrate with its naturally adsorbed—and few monolayer thick—fluid, at ambient conditions. *ii*) The upper part shows a synchronous detection for monitoring the QTF signal, which indicates the probe's lateral oscillation amplitude. *iii*) The bottom side shows an acoustic transducer in intimate contact with the substrate (typically a 2 mm thick piece of glass or silicon wafer), which synchronously senses the near-field acoustic waves engendered at the trapped fluid. At 32 kHz operating frequency, the wavelength of sound in bulk water is ~ 5 cm. For comparison, this acoustic wavelength is much greater than the distance at which the acoustic transducer is placed relative to the mesoscopic fluid in our case; that is, we are detecting the acoustic excitation in the *near field* region from the source.

Figure 1c shows a typical experimental result obtained in the SANM. The probe is initially positioned far away from the substrate and driven at its resonance frequency (~ 32 kHz) and 15 nm initial oscillation amplitude. As the probe approaches the surface (at a rate of 1.5 nm/s) the QTF-signal initially does not change. But at a given distance (which varies stochastically from trial to trial) a water bridge forms between the probe's apex and the substrate, causing the QTF signal to gradually decrease. By about the same time, an acoustic wave is also engendered at the coalesced fluid meniscus, which couples to the substrate and propagates downwards towards the acoustic transducer (SE32-Q, Score Atlanta Inc.) The transducer's signal is synchronously detected with the lock-in amplifier #2; the output of this second lock-in is referred to as the "*acoustic-signal*". By the end of the approach the acoustic signal decreases, which may be the result (among other possibilities that will be addresses elsewhere) that the trapped fluid is quizzed out; that is, the lower the fluid volume, the lower the acoustic signal.

The stochastic nature of the meniscus formation makes each approach/retraction measurement different. Still, some variations may be due to unwanted perturbation from the user (if the measurements are not automated), or by the natural thermal drift (due to the geometrical asymmetries of different parts comprising the microscope stage), or because the noise level is in excess to discriminate the potentially different dynamic response from the tiny volumes (10^{-9} nano-liters) of fluid. These aspects are addressed below, leading to an improvement in the detection of the signals in the SANM.

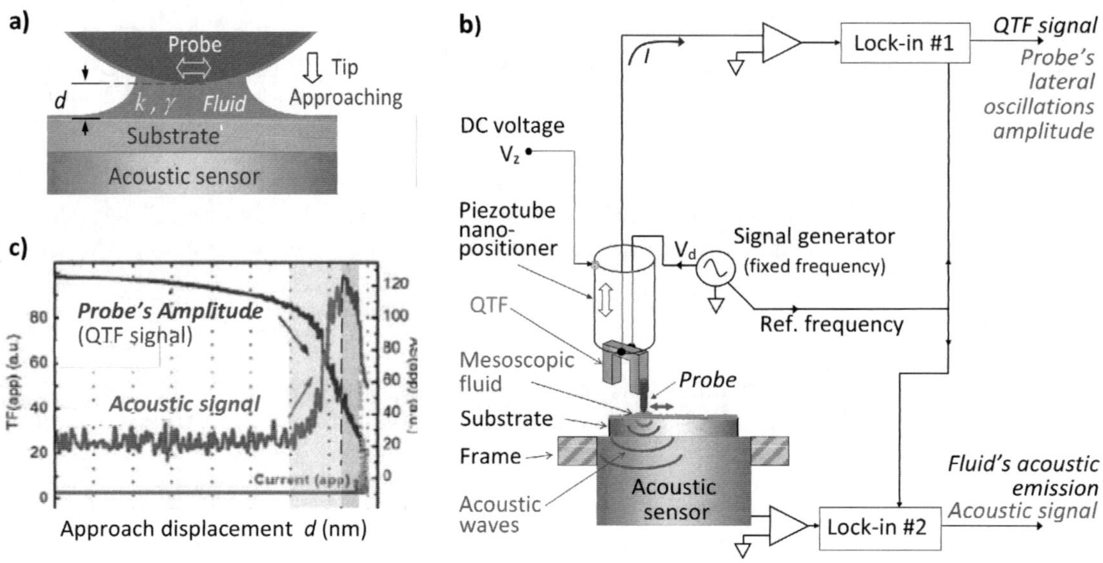

Figure 1. a) Illustration of the mesoscopic fluid (meniscus), confined in the vertical direction by two solid surfaces under shear motion, studied with the SANM. **b**) Schematic experimental setup of the SANM. The probe's lateral oscillations amplitude and the acoustic emission form the mesoscopic fluid are synchronously and simultaneously detected. **c**) Typical variation of the probe's amplitude (blue trace) and acoustic (green trace) signals as the probe approaches the sample. For cases where the probe and the substrate are electrically conductive, the red trace tracks the electrical current that onsets when the probe's apex gets in contact with the substrate. The ability to detect acoustic waves even when the probe is not in mechanical contact with the substrate (*i.e.* sound is engendered at the mesoscopic fluid) constitutes one of the main features of the SANM.

3. The mechanical motion controller of the head stage

During the SANM operation it is typically observed that the probe-sample distance at which a water bridge forms is influenced by the environmental humidity, and that the experimental results are more reproducible at relative humidity greater than 50%. Hence a humidity chamber able to maintain a stable humidity becomes necessary (see Figure 2). But the enclosure obviously imposes access limitations to manually operate the coarse approach of the stage, which is typically implemented by gradually turning three supporting fine screws. To overcome this obstacle, we have implemented a remote control system that operates the screws by activating their corresponding stepper motors. Approaching speed and steps of different approach resolution are available to the user. The implementation is detailed next.

IOP Conf. Series: Journal of Physics: Conf. Series **1143** (2018) 012013 doi:10.1088/1742-6596/1143/1/012013

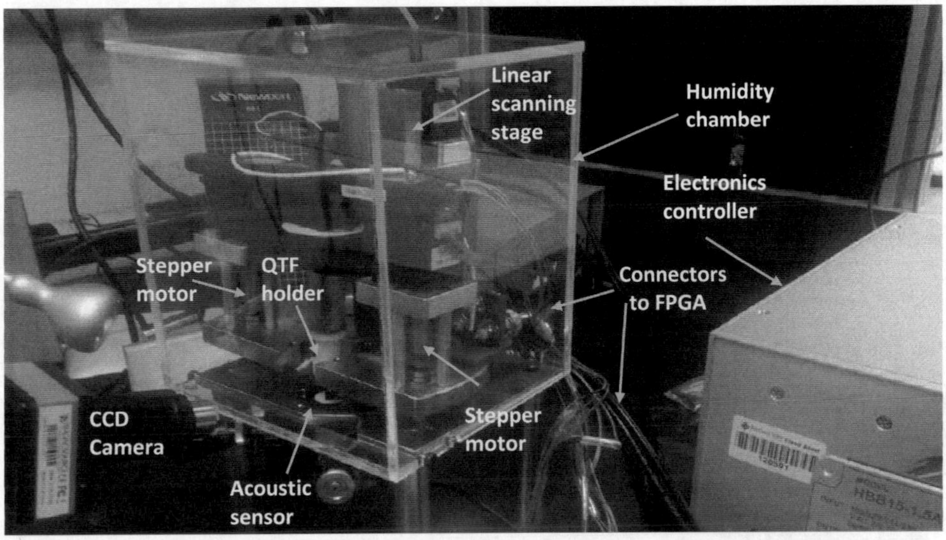

Figure 2. Front side view of the shear-force acoustic near-field microscope.

3.1 Hardware of the stepper motor controller.
A low-cost alternative comprises building the control circuit on a protoboard with three A4988 motor driver boards, and one 315 MHz RF module for portability and accessibility purposses (Figure 3.) Coupling and protection circuits are also built to ensure that the motors run smoothly, quietly, and without skidding or lossing steps. Finally, the motor drive circuit is mounted on an Arduino microcontroller board.

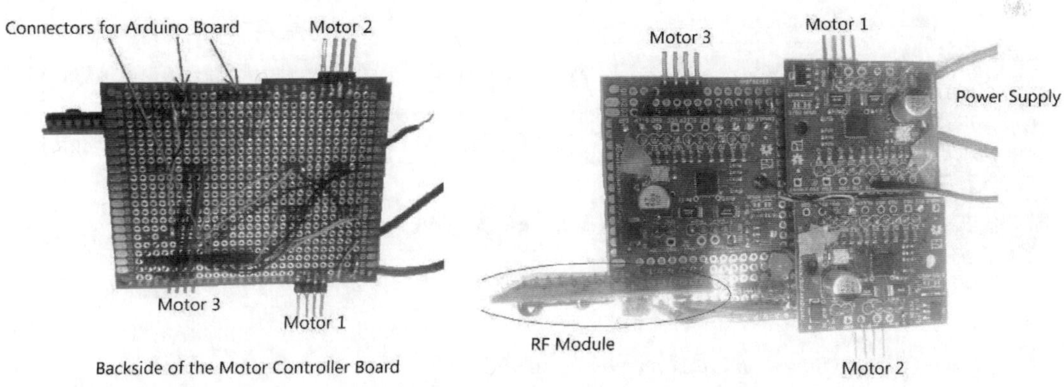

Figure 3. Hardware circuit for controlling the stepper motors.

The inputs of the A4988 driver boards are connected to pins 2 to 13 of the Arduino board. The RF module is connected to pins A0 to A3 of the Arduino board. Pins A4 and A5 of the Arduino board are connected to 2 LEDs for display purposes (Figure 4). More details can be found in the spec of the A4988 chip [10]. The RF module is paired with a four-button remote controller unit. Usually, when one button is

IOP Conf. Series: Journal of Physics: Conf. Series **1143** (2018) 012013 doi:10.1088/1742-6596/1143/1/012013

pressed, one line of the RF module will be pulled to High. The mapping between the controller's buttons, the RF data lines, and the Arduino pins should be carefully verified. Figure 5 shows the mapping selected for the current implementation.

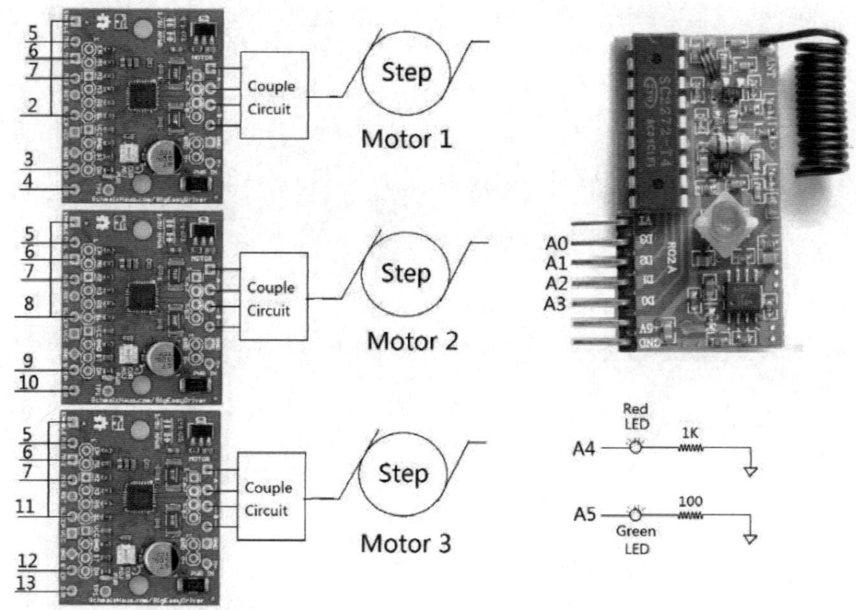

Figure 4. Electric connections of the modules. The left side shows the three A4988 driver boards; the right side shows the RF module.

Controller	A	B	C	D
RF Module	D2	D0	D3	D1
Arduino	A1	A3	A0	A2

Figure 5. The 4-button remote controller and mapping of its buttons to the Arduino pins.

3.2 The logic structure of the mechanical motion controller

A 3-layer model would be sufficient to model this motor controller. But, for future integration purposes involving a larger project, we have built the system with a 4-layer model. Accordingly, the test applications are on the fourth layer, while the stage motion functions are on the third layer. The instructions to control the specific motion of each motor are located on the second layer. The first layer is the relevant hardware. The final structure of the controller is shown in table 1. The benefits of a 4-layer model include better

IOP Conf. Series: Journal of Physics: Conf. Series **1143** (2018) 012013 doi:10.1088/1742-6596/1143/1/012013

hardware compatibility, easier programming on each abstract level, and flexibility in software updating and maintenance [11], [12].

Table 1. 4-layer model for the mechanical motion controller

Abstract Level	Description	Targets
Layer 4	Apication Layer	Running on PC
Layer 3	Functional Block	Running on PC
Layer 2	Instructions	Arduino
Layer 1	Hardware	Arduino, Analog and Digital Circuits

3.3 Layer 2 description of the mechanical controller.
Table 2 shows the detailed components of the mechanical controller on each layer. Layer 2 of the system is implemented in an Arduino UNO microcontroller board. The *RF Signal Decoding Module* builds a transparent bridge between button actions of the RF remote and the Arduino instructions. The *RF Signal Decoding Module* will interpret the button pushing actions as different commands sending to the Arduino board. Human behaviors and reactions are integrated into the module already. This makes the RF control more intuitive.

Table 2. Modules of the mechanical controller.

Abstract Level	Hardware and Software Resources
Layer 4	Applications
Layer 3	LabVIEW FSM Controlling Stage Motions, Human Controlling the RF Remote
Layer 2	RF Signal Decoding Module, A4988 Encoding Module, Stage Action Module, Communication FSM
Layer 1	A4988 Motor Driver Boards, RF Receiver Module, LEDs, Arduino Board, 4 Button RF Remote

The *A4988 Module* masks the detailed signal waveforms of the motor controllers. It receives commands from the *Stage Action Module* and controls the actual movement of the motors. The *Stage Action Module* controls the head stage to move down or up, fast or slow, as well as continuously or by one step at a time.

The *Communication Finite State Machine* (FSM) is the interface between the Arduino board and the outside world. The FSM is continuously running in a loop, taking commands from the third layer of the system and passing them to the *Stage Action Module*. Figure 8 shows the state diagram of the FSM. The default mode of the FSM is *Computer Control*, where the FSM reads commands from the third layer program running on a computer. In RF mode, the FSM receives signals from an RF remote and does various actions depending on the RF signal sequences. See figure 6.

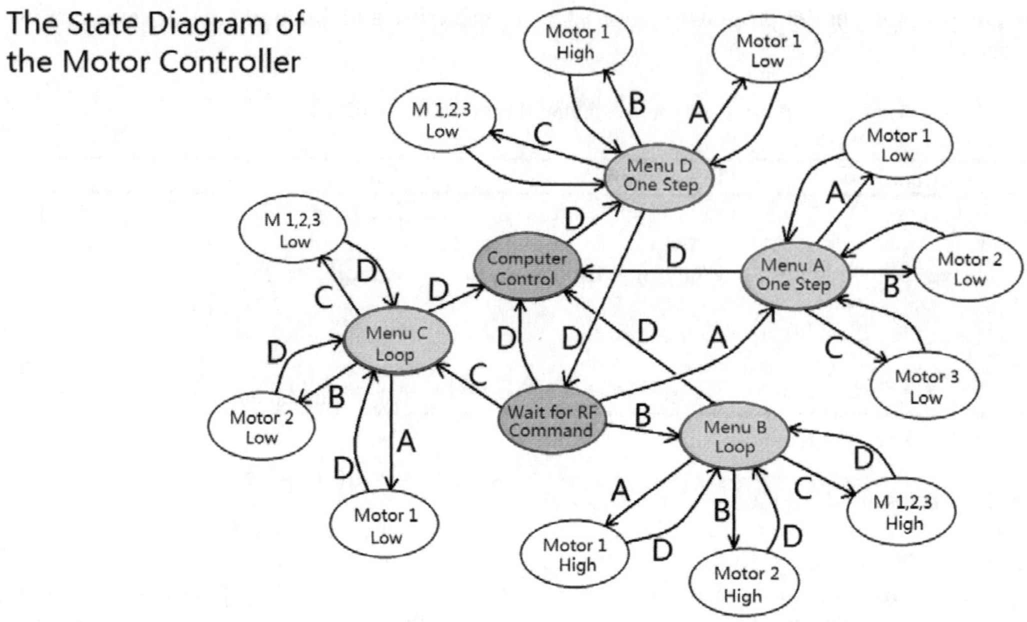

Figure 6. State diagram of the *Communication Finite State Machine* (FSM) running on an Arduino microcontroller.

The FSM is the interface between the Arduino microcontroller, the third layer program, and the RF remote. For simplicity, the LED display actions are not shown in Figure 6. The default state of the FSM is *Computer Control. i)* When button *D* is pressed, the controller enters into *Menu D*, which is the default RF state. Pressing buttons A, *B* or *C* will trigger 3 actions (see the state diagram). When button *D* is pressed again, the controller will leave the default RF state and enter into the menu selection state, *Wait for RF Command.* Then, pressing buttons A, *B* or *C* will drive the controller into *Menu A, Menu B*, or *Menu C,* respectively. *ii)* Menu A is a one-step function menu, where each function will only execute once upon receiving an RF signal. *iii) Menu B* and *Menu C* are continuous function menus, where the function will execute continuously until a button *D* signal arrives. *iv)* While in *Menu A, B, C* or *D*, one more push of button *D* will drive the controller back to *Computer Control* state. During all the transitions, two LEDs will indicate the state of the controller in real time.

The FSM also controls two 5-state LEDs to indicate the current state of the FSM. The 5 states of the LEDs are *ON, OFF, BLINKING, FLASHING*, and *DIM* (see table 3). Combining the two LEDs, one can obtain 10 to 25 state capacity. This capability would be very convenient in cases where the stage is in, for example, a vacuum chamber or other enclosed environment, and therefore the operator cannot see the movement of the stage. In those cases, the LED state indicators will show the users the hidden motion of the stage.

When the FSM is in the *Computer Control* state, it uses a 3-byte array to communicate with the computer. Every time the FSM reads 3 bytes of string from the computer, it will verify the first 2 bytes. If the first 2 bytes are character "L" and "C", the third byte will be decoded. Based on the decoding results, various functions will be called to carry out the corresponding operations. Communication via formatted strings helps reduce the error during data transfer. Strategies to format the control commands can be found in the references [11].

Table 3. The 5 state display scheme of the LED indicators.

LED States	FSM States
Green ON	Computer Control
Red ON	Menu D
OFF	Menu Select
Dim	Functions Are Running
Blink	Menu A, B, or C
Flash	Stop, Quit

3.4 Layer 3 description of the mechanical controller.

The layer 3 program runs on the host computer (PC). An example of layer 3 is shown in figure 7. Six latch buttons control the vertical motion of the stage. When they are pressed (released) the stage will move (stop) accordingly. M1, M2, and M3 are used to control the 3 motors to move with one step, during the final approach. The one-step mechanical motion of the stage has a resolution of about 650 nm.

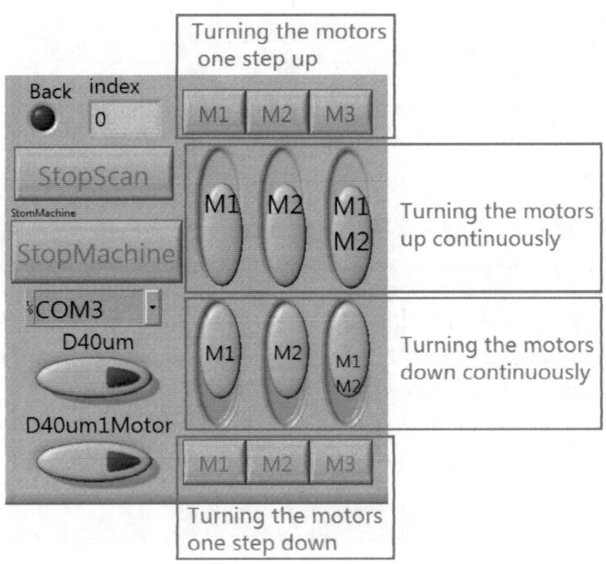

Figure 7. User-interface for controlling the motion of the SANM head-stage.

The Layer 3 program is also coded as an FSM (a section of the LabVIEW code is shown in figure 8.) An attractive feature of this programming method is that Layer 3 and Layer 2 can work at their own speed, and interact with each other asynchronously, which greatly simplifies the programming efforts and enables the

IOP Conf. Series: Journal of Physics: Conf. Series **1143** (2018) 012013 doi:10.1088/1742-6596/1143/1/012013

whole system to be more stable and reliable. The methods about how to coordinate the functions running on layer 2 and layer 3 are covered in the references [11] and [13].

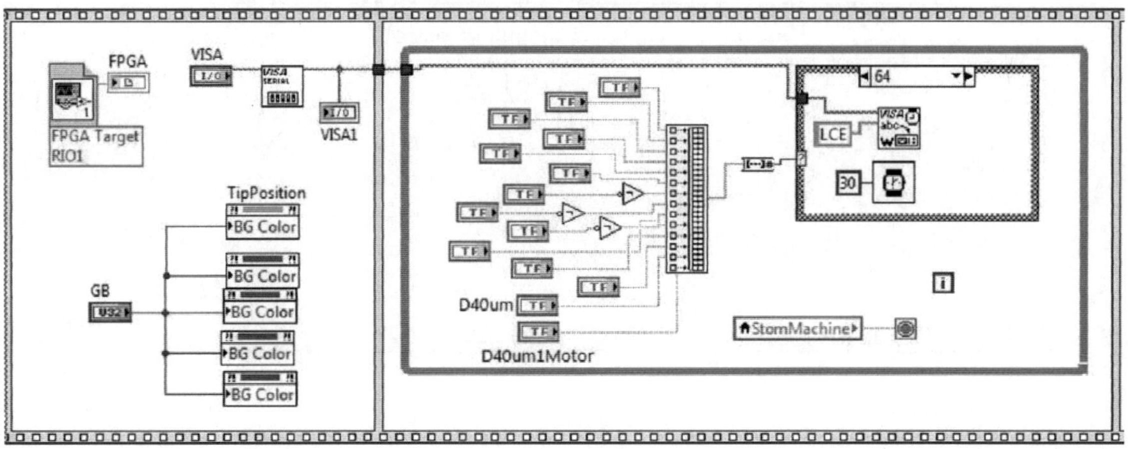

Figure 8. A section of the LabVIEW block diagram for controlling the head stage motion (running on layer 3 of the mechanical controller), coded according to the FSM displayed in figure 6. Motion actions are represented with Boolean buttons, which are built into a vector to control the execution of the FSM. States of the FSM are coded with LabVIEW case structures.

4. Compensation of the thermal drift

Mechanical drift is a problem in scanning acoustic microscopy. Due to ambient temperature variations, mechanical structures are subjected to expansions or contractions. The size of different sections of the microscope's head-stage, in particular, may vary differently (due to their asymmetric geometries), which introduces uncertainty to the actual location of the analyzed sample surface. This is referred to as the thermal drift to the system. In order to have an accurate measurement of the actual (and no juts relative) distance between the tip and the sample, this drift must be compensated. Our approach is outlined in figures 9 and 10. Figure 9 outlines the system used to perform the approach of the tip to the sample: *i*) a coarse approach, using screws controlled by stepper motors (as described in the previous section), followed by *ii*) a fine approach, using a high-voltage controlled piezoelectric linear stage (Mad City Labs, OP65; 60 microns travel range.) Then the probe is engaged into feedback with the sample, which maintains the probe-sample distance constant.

IOP Conf. Series: Journal of Physics: Conf. Series **1143** (2018) 012013 doi:10.1088/1742-6596/1143/1/012013

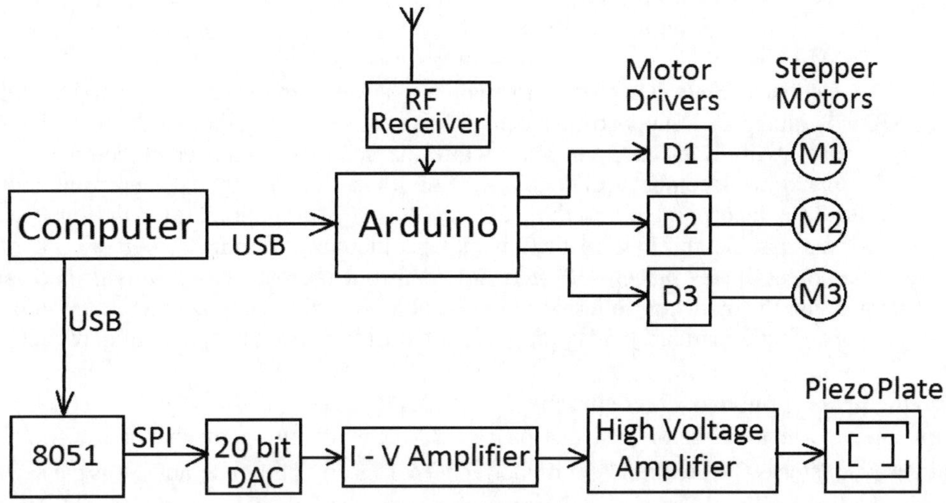

Figure 9. Signal flow graph of the coarse (top) and fine (bottom) motion control mechanism.

Figure 10 outlines the thermal drift compensation scheme. While the probe and sample are engaged in feedback, the thermal drift is evaluated by monitoring the output feedback voltage needed to keep the probe-sample distance constant. In our system, the probe is held by the piezoelectric linear stage, and the sample rests on a nominally fixed position stage. The sensing signal is provided by the QTF response (the probe's lateral oscillations amplitude) and the corresponding corrected output voltage is applied to the piezoelectric stage.

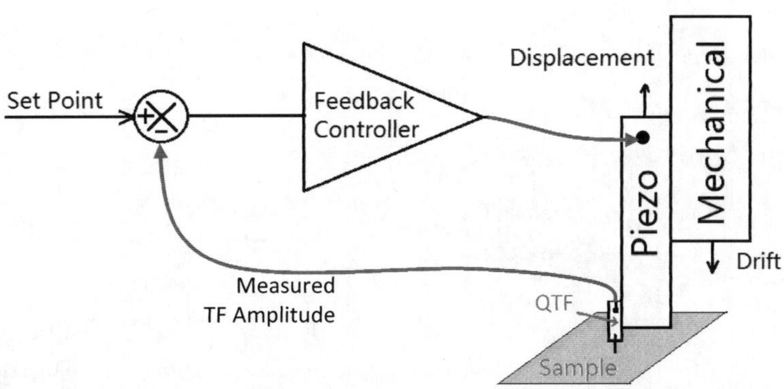

Figure 10. Illustration of the drift measurement scheme. A PID feedback controller is used to first measure the mechanical drift of the sample stage relative to the head stage. The QTF signal (that monitors the probe's amplitude of oscillations) is selected as the control variable of the feedback controller. After the tip is engaged to the sample surface, the voltage feedback control allows to measure the drift. This information is then read by the microscope controller to take it into account in the subsequent approach/retraction measurement. All this process is automated.

IOP Conf. Series: Journal of Physics: Conf. Series **1143** (2018) 012013 doi:10.1088/1742-6596/1143/1/012013

Recapitulating: *i*) One starts lowering the scanning stage at coarse steps, via the activation of stepper-motors, followed by a piezoelectric fine approach. *ii*) Subsequently, the tip is engaged in feedback with the sample surface. To avoid probe's damage, a conservative set point of 80% of the free run TF amplitude is selected; then the feedback controller strives to maintain the TF amplitude constant by continuously applying a correction voltage. *iii*) Suppose the mechanical structure holding the sample were drifting down; then a voltage feedback will drive the top stage down to maintain a constant probe-sample distance (see figure 10). *iv*) The piezo displacements (drift compensation data) are recorded, linearly interpolated; this added drift information is incorporated into the piezo driver that is programmed for the subsequent probe approach process. We assume the system drift would be the same during a short period of time an approach/retraction process takes place (~30 seconds). *v*) In our current system, this drift compensation scheme is integrated into the main controller. Every time, before an approaching and retraction process, the system measures the drift automatically and compensate the drift for the next approaching/ retraction process.

5. Monitoring probe-sample contact current

The strategy followed in the SANM—using a probe that successively approaches and retracts from the substrate as a way to interrogate the liquid-solid interface—places on a preponderant status the need to locate with precision the location of the surface relative to the probe. The drift compensation addressed above, ensuring an accurate probe-sample distance control, is a good step. But to obtain the absolute distance between the tip and the sample surface it is necessary to locate the point at which the probe encounters the substrate. Here we establish that point by monitoring the contact current. We apply a bias voltage to the tip and measure the ac-component of the tunneling current synchronously with a lock-in amplifier and the dc-component with a buffer amplifier (Figure 11). When a metal tip touches the metallic substrate, the current passing through the substrate should grow exponentially.

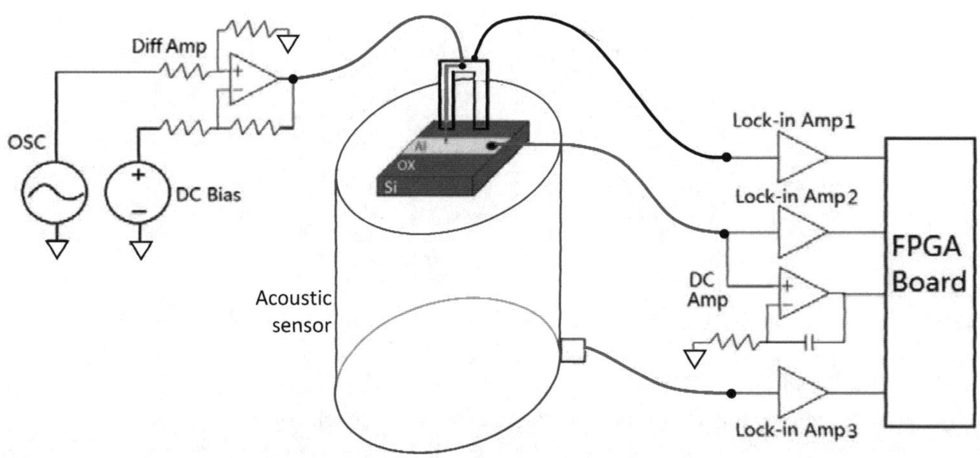

Figure 11. Experimental setup for measuring the probe-sample contact current, and avoid electrical interference with the simultaneously measured QTF signal (with lock-in #1) and the acoustic signal (lock-in #3).

The multiple signals monitored simultaneously in the SANM makes also imperative to put electrical guards against potential "crosstalk" among the signals. Figure 11 shows the experimental setting aiming at fulfilling that objective; Figure 12 shows the corresponding circuit diagram. The left side of Figure 11 shows how we use a differential amplifier to drive the QTF. An AC voltage (~ mV amplitude) is connected to the positive input of the OpAmp. A DC bias voltage is connected to the negative input of the OpAmp. Because the OpAmp has a very large input impedance (~ 10^6 Ω), the AC and DC signal sources are practically isolated. This way the potential difference between the tip and the sample can be safely be controlled with the DC signal. At the right side, three lock-in amplifiers are used to independently measure the QTF, acoustic SANM, and AC tunneling current signals. Notice the additional buffer amplifier to measure the DC tunneling current. The buffer is configured as an integrator; by choosing a proper value of the capacitor the ac-components of the tunneling current signal is filtered by the integrator, and only its DC component is picked up by the buffer amplifier. It is worth highlighting also that the buffer amplifier establishes a virtual ground to the sample. This virtual ground at the sample (introduced by the buffer amplifier) and the input bias voltage (provided by the differential amplifier) stablishes the bias between the tip and sample. This biasing mechanism adds confidence to the control and monitoring of the signals. Because the OpAmp has a very large input impedance, the buffer amplifier will not leak the high-frequency tunneling signal away. Then both AC and DC tunneling signals can be measured accurately. Figure 13 shows typically measured curves of the QTF, acoustic and tunneling current during an approaching process.

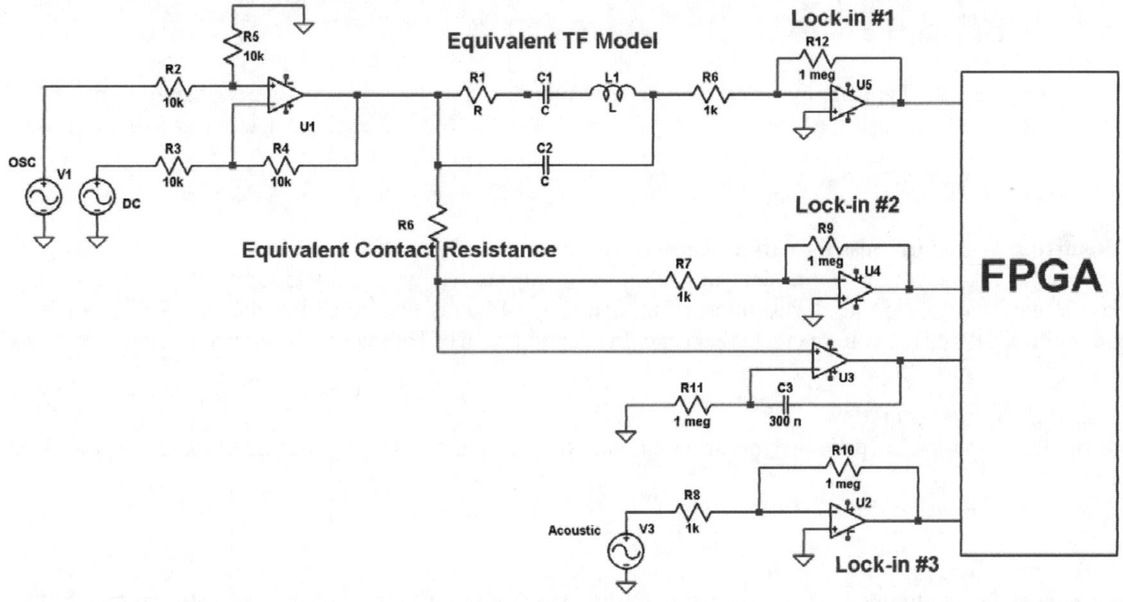

Figure 12. Circuit diagram for monitoring the probe's lateral oscillations amplitude (QTF signal, top lock-in #1), fluid's acoustic emission signal (bottom lock-in #3), and the probe-sample ac/dc contact current (lock-in #2 and buffer circuit) in the SANM. R6 is the equivalent resistance between the tip and the sample. Lock-in #1, Lock-in #2 and Lock-in #3 are the equivalent input circuits of the lock-in amplifiers when they are configured as current input with a 1kΩ virtual impedance to ground. R11 can

IOP Conf. Series: Journal of Physics: Conf. Series **1143** (2018) 012013 doi:10.1088/1742-6596/1143/1/012013

be selected from 1kΩ to 1MΩ, since the attenuation of the 32 kHz signal varies from -35dB to -95dB. If we were to set the high frequency attenuation to be -40dB, then R11 should be 2 k Ω.

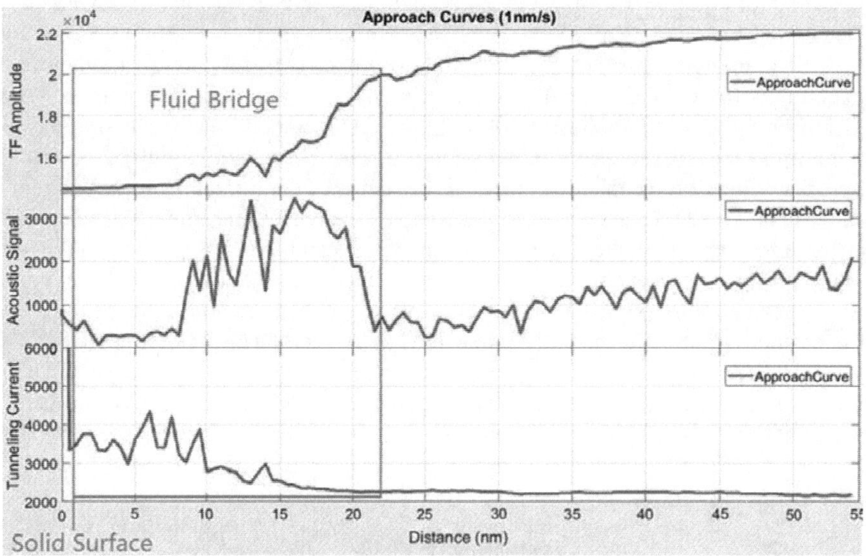

Figure 13. Typical approach curves measured in the SANM. Notice, some current is measured through the fluid bridge prior to the state where the probe encounters the solid substrate. The latter is identified when the current drastically increases.

6. Real-time signal processing with a Kalman filter

At this stage, the detection of the acoustic SANM signal has not been fully optimized yet. Due to the stochastic nature of the tip and fluid interaction, a Kalman filter is implemented into the FPGA board to obtain optimal filtered measurements in real time. Implementation of Kalman filter requires a system process model and a measurement model of the system being masured [14].

6.1 The system process model

Since the QTF and the acoustic are the major signals used in the SANM, we define the state vectors x as follows,

$$x = \begin{bmatrix} A_{TF} \\ A_{AC} \end{bmatrix} \tag{1}$$

where A_{TF} is the amplitude of the QTF signal, and A_{AC} is the amplitude of the acoustic signal. Since, experimentally, the QTF is excited by an external voltage, the tuning fork signal A_{TF} will be modeled as a random walk. Prior experiments indicates a correlation between the acoustic response and the QTF oscillations amplitude; hence we model the acoustic signal A_{AC} also as a random walk.

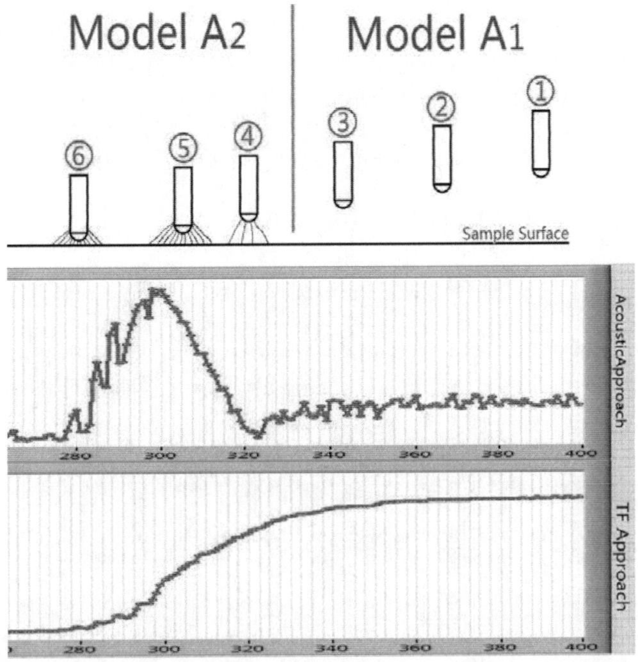

Figure 14. The QTF and acoustic signal profiles are used to build the process models. In Model A1, the tip is far away from the sample surface; hence there is no interaction between them. Both signals are modeled with a random walk. Model A2 incorporates information that a fluid bridge builds up between the tip and the sample surface at a stochastic probe-sample distance during the approach. Then the vibration of the tip is propagating through the bridge to the sample. The acoustic signal is modeled as a random walk with a small linear influence from the QTF. The figures show the results of a probe to sample approach; the filtering process was applied in real-time.

Let's start with the tip at position 1 (see Figure 14). Then the process model of the system is:

$$x_{n+1} = A_1 x_n + w_n \tag{2}$$

$$A_1 = \begin{bmatrix} 0.98 & 0 \\ 0 & 0.98 \end{bmatrix} \tag{3}$$

where A_1 embodies the random walk model; x_n is the state at time-step n, and w_n is the process noise at the time-step n.

After the tip makes a contact with the fluid layer, we model the acoustic signal with an inertial function lumped by a linear function of the QTF signal. This springs from the fact that in previous experiments it has been verified that the acoustic signal is proportional to the QTF amplitude [15]. On a first-order approximation, the acoustic signal is modeled as, ${A_{AC}}_{n+1} = 2\alpha\, {A_{TF}}_n + {A_{AC}}_n$, where the coefficient α is a relative small number. As a first order approximation, α represents the linear corelation between the QTF oscillation and the acoustic signal strength. For different lab setups, this corelation may vary. So after many lab trials, α needs to be set very carefully to have the best fitlter effects. After hundereds of trials and comparisons, α will be set automatically by an algorithm. Then the process model becomes:

$$x_{n+1} = A_2 x_n + w_n \tag{4}$$

$$A_2 = \begin{bmatrix} 0.98 & 0 \\ 2\alpha & 0.98 \end{bmatrix} \tag{5}$$

where A_2 is the process model after a fluid bridge is built between the tip and the sample surface; x_n and w_n are the state and the process noise at time step n is respectively. The time at which model-2 kicks from model-1 is described below.

6.2 The measurement model

Since the QTF and acoustic signals are the ones we want to measure, the following measurement model is used:

$$y_n = C x_n + v_n \tag{6}$$

$$C = \begin{bmatrix} 1 & 0 \\ 0 & 1 \end{bmatrix} \tag{7}$$

where y_n is the measured vector at time step n; C is the measurement matrix, v_n is the measurement noise at time step n.

The Kalman filter is initialized with a direct measurement of x_0, at time step 0. At time step n, the Kalman filter is updated with the sequence shown below [14].

$$\hat{x}_n^- = A_1 \hat{x}_{n-1} \tag{8}$$

$$\hat{x}_n^- = A_2 \hat{x}_{n-1} \tag{8b}$$

$$P_n^- = A_1 P_{n-1} A_1^T + Q_{n-1} \tag{9}$$

$$P_n^- = A_2 P_{n-1} A_2^T + Q_{n-1} \tag{9b}$$

$$K_n = \frac{P_n^- C^T}{C P_n^- C^T + R_n} \tag{10}$$

$$e_n = (y_n - C\hat{x}_n^-) \tag{11}$$

$$\hat{x}_n = \hat{x}_n^- + K_n e_n \tag{12}$$

$$P_n = (I - K_n C) P_n^- \tag{13}$$

where $\hat{x}_n^-$ is the estimated state before a measurement is taken at the time step n; P_n^- is the state covariance before a measurement is taken at the time step n; K_n is the Kalman gain at the time step n; $\hat{x}_n$ is the state estimate after a measurement is taken at the time step *n*; P_n is the state covariance after a measurement is taken at the time step *n*.

A threshold was set up to monitor the values of e_n in 5 steps. According to different value history of e_n , the system may switch between models A_1 and A_2. In model A_2, equations (8b) and (9b) would be used to update the system. The Kalman filter was implemented in the FPGA board, parameters Q_n and R_n are the process covariance and measurement covariance, respectively. For different setups, Q_n and R_n should be carefully chosen and tuned. Figure 15 shows acoustic signals acquired with and without a filtering process.

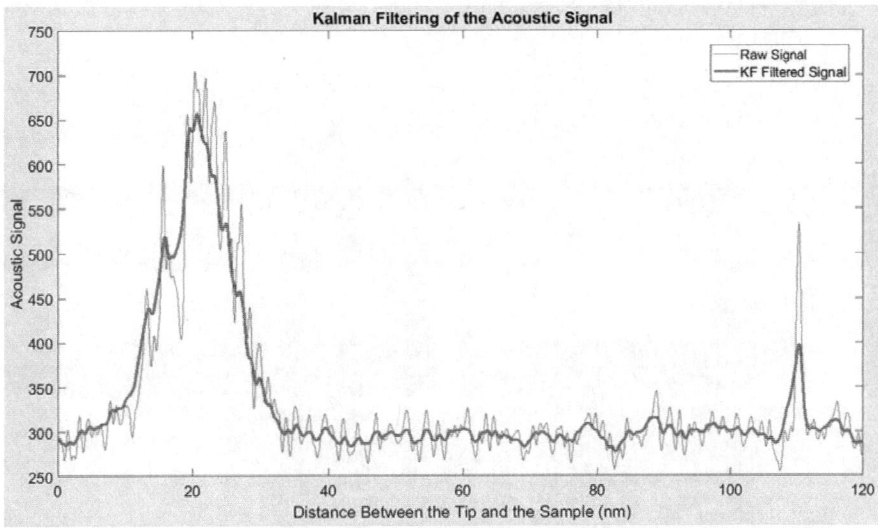

Figure 15. A comparison between acoustic traces logged without the filter in place (blue trace) and with the real-time filter in place (red trace).

The information of the SANM from previous measurements makes the user know more about the experiments and the instrument; wqith this information, one generates a simple linear model. As shown in Fig. 5, a simple model improves the signal quality significantly already. With a more accurate model better results can be achieved, by using, for example an Unscented Kalman filter (UKF).

Since the Kalman filter offers the optimal estimates of the measured signals, in the future we plan to use whispering-galley acoustic sensing [16] as an on-line sense signal for a feedback controller. Then an online-filtered acoustic SPM image of the sample can be obtained.

7. The integrated host program

In order to simplify the operation of the SANM microscope, an intuitive host program was developed to control the system resources and display the measured signals. Figure 16 shows the interface of the scanning acoustic microscope host program coded with LabVIEW. At the beginning, the operator clicks the various motor control buttons to lower the stage at coarse steps. The minimum mechanical movement of the stage is about 650 nm. With the aid of a microscope lens attached to a CCD camera (see figure 2), the user is able to monitor the motion of the probe. When the probe is in the proximity of the sample surface, one switches to finer step approaches with three piezo controlled slides, as shown in figure 16 (the three slides activate the same piezoelectric linear stage.) The piezo is controlled by a 20 bit DAC chip (see figure 9). The total range of the piezo stage is 65 µm. So for a 20 bit DAC, the nominal theoretical resolution is 0.06 nm. Here. the left slide is designed to have 200 nm resolution, the middle slide has 20 nm resolution steps. The right slide has a nominal 0.2 nm resolution, but due to the noise level of the piezo controller and thermal fluctuation of the whole system, the probe may not have a real resolution as fine as indicated on the front panel. For different hardware setup, calibration and testing would be needed. The real-time probe position indicator in the panel shows the relative probe position being driven by the piezo. When the probe is driven by the piezo either manually or automatically, this window will show the trace of the probe.

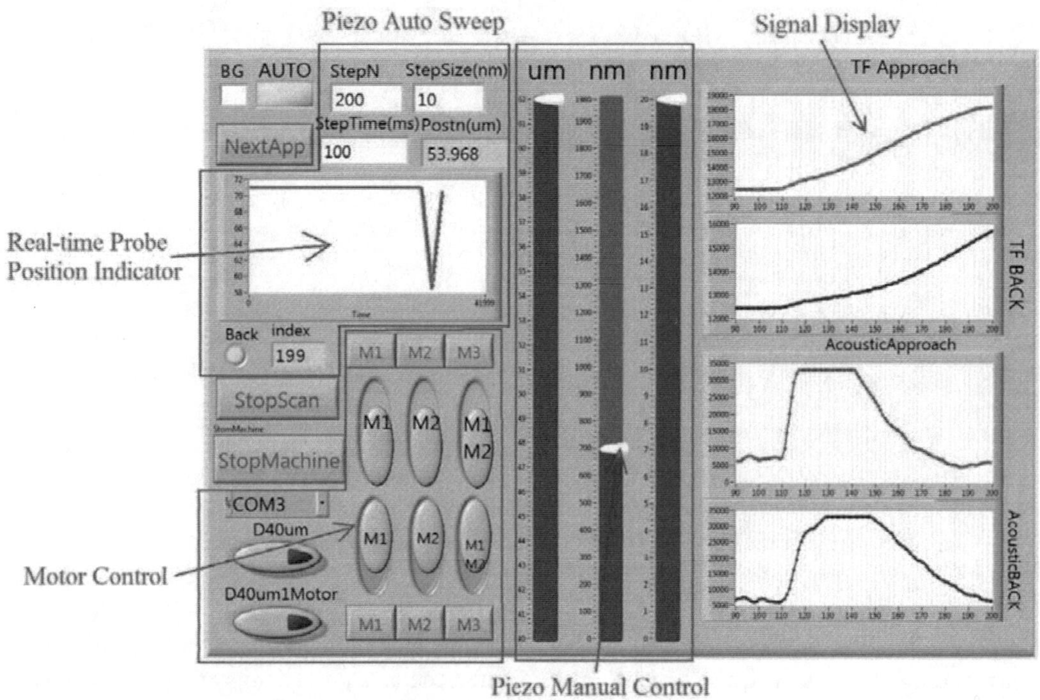

Figure 16. The front panel of the controller.

When the *AUTO* button is hit, the probe will perform an automatic approach and retraction along the Z-axis. *StepN* controls the step number of the approach and retraction curves. *StepTime* controls the time interval of each step. During an automatic approach and retraction, the measured signals will be logged by the FPGA board and showed on the signal display windows.

The *BG* button is used to control the background color. For computer monitoring, one may want a black background; while for printing, one may want a white background.

8. Conclusions

Several instrumentation-level improvements for applications in scanning probe microscopy, in general, and for shear-force near-field acoustic microscopy, in particular, were addressed. The automated coarse (micrometer-sized steps) and fine (nanometer steps) approach allow minimal interference of the operator on the functioning of the microscope. A friendly graphic user interface allows the operation of the microscope remotely. This feature opens the opportunity to implement, in the near future, operation of the microscope via an internet connection, which would be valuable to establish international collaborations since the microscope could be operated from any place in the world. Another important feature of the new SANM is to be immune to electrical ground loops when monitoring the probe-sample contact current (in addition to the probe's amplitude of lateral oscillations and the acoustic signals.) The circuits that enable this capability were shown explicitly. Finally, a real-time Kalman filtering of the signals was implemented for the first time in the SANM. The effectiveness of the Kalman filter was demonstrated even with a relatively simple linear model. In the near future, real-time Kalman filter implemented with whispering-gallery acoustic sensing signal will be used to feedback-control the tip's movement, which will allow to acquire on-line filtered SPM images can be obtained. All the instrumentation-level improvements introduced in this paper will make the SANM experiments easier to perform, accurate, and more reproducible.

Bibliography

[1] La Rosa A H, R. Nordstrom, Cui X, McCollum J and Li N 2005 The Ultrasonic/Shear-Force Microscope: Integrating Ultrasonic Sensing into NSOM *Rev. Sci. Instrum.* **76** 093707 http://pdxscholar.library.pdx.edu/cgi/viewcontent.cgi?article=1009&context=phy_fac

[2] Kim B, Kwon S, Mun H, An S, Jhe W 2014 Energy dissipation of nanoconfined hydration layer: Long-range hydration on the hydrophilic solid surface *Scientific Reports* **4** 6499

[3] Granick S, Motions and Relaxations of Confined Liquids *Science* **253** 1374

[4] Urbakh, Klafter J M, Gourdon D and Israelachvili J 2004 The nonlinear nature of friction *Nature* 430, 525

[5] Hammer M U, Anderson T H, Chaimovich A, M. Shell S and Israelachvili J 2010 The search for the hydrophobic force law *Faraday Discuss.* **146** 299

[6] N. J. Persson 2000 Sliding Friction (Springer)

[7] Kim S H, Asay D B and Dugger M T 2007 Nanotribology and MEMS *Nano Today* **2** 22

[8] Daniel Courjon 2003 Near-Field Microscopy and Near-Field Optics (Imperial College Press)

[9] Zhang Z, Ahn P, Dong B, Balogu O, Sun C 2013 Quantitative Imaging of Rapidly Decaying Evanescent Fields Using Plasmonic Near-Field Scanning Optical Microscopy *Scientific Reports* **3**, Article number: 2803

[10] Allegro, A4988 DMOS Microstepping Driver with Translator and Overcurrent Protection, http://www.allegromicro.com/en/Products/Motor-Driver-And-Interface-ICs/Bipolar-Stepper-Motor-Drivers/A4aspx

[11] Bai J, La Rosa A H. 2017 Essentials of Building Virtual Instruments with LabVIEW and Arduino for Lab Automation Applications, International Journal of Science and Research (IJSR), https://www.ijsr.net/archive/v6i5/v6i5.php, **6** 640

[12] Bai J, Chen J, Freeouf J, La Rosa A H. A 4-layer method of developing integrated sensor systems with LabVIEW, Journal of Measurement Science and Instrumentation. 10.3969/j.issn.1674-8042.2013.04.001

[13] Bai J, La Rosa A H 2017 An SPM Stage Driven by 3 Stepper Motors. Journal of Measurement Science and Instrumentation. 10.3969/j.issn.1674-8042.2017.03.010

[14] Simon Haykin, Chapter I, Kalman Filters, Kalman Filtering and Neural Networks, Wiley, 2001

[15] Wang X, Fernandez R, Li N, Hung H, Venkataraman A, Nordstrom R, and La Rosa A H 2016 "Near-field acousto monitoring shear interactions inside a drop of fluid: The role of the zero-slip condition." *Physics of Fluids* **28** 052001 doi: 10.1063/1.4947597.

[16] La Rosa A H, Li N, Fernandez R, Wang X, Nordstrom R, and Padigi S K 2011 "Whispering-gallery acoustic sensing: Characterization of mesoscopic films and scanning probe microscopy applications, *Rev. Sci. Instrum.* **82**, 093704

IOP Conf. Series: Journal of Physics: Conf. Series **1143** (2018) 012014 doi:10.1088/1742-6596/1143/1/012014

Finite element method analysis of whispering gallery acoustic sensing

T Le[1], H Tran[1], R Fernandez[1], C J Solano Salinas[2], N Laal[1], R Bringas[2], J Quispe[2], F Segundo[2] and A H La Rosa[1]

[1]Portland State University, Portland, Oregon 97201, USA
[2]National University of Engineering, Facultad de Ciencias, Lima, Perú

E-mail: thanh4@pdx.edu, andres@pdx.edu

Abstract. Whispering Gallery Acoustic Sensing (WGAS) has recently been introduced as a sensing feedback mechanism to control the probe-sample separation distance in scanning probe microscopy that uses a quartz tuning fork as a sensor (QTF-SPM) [1]. WGAS exploits the SPM supporting frame as a resonant acoustic cavity to monitor the nanometer-sized amplitude of the QTF oscillations. Optimal WGAS sensitivity depends on attaining an exact match between the cavity's frequency peak response and the TF resonance frequency. However, two aspects play against this objective: *i*) the unpredictable variability of the TF resonance frequency (upon attaching a SPM-probe to one of its tines), and *ii*) cavities of arbitrary geometry tend to display complicated (multiple peaks) frequency response, making difficult to identify which cavity dimension control which peak. Practical matching frequency procedures are needed then to operate the Shear-force Acoustic Near-field Microscopy (SANM) more efficiently. As a first step, here we undertake finite-element method (FEM) analysis to find out cavities of simple frequency response and, ideally, easy frequency tuning ability. Based on previous results we focus our studies in analyzing the frequency response of conical cavities within a range around the 32 kHz operating frequency. To first validate our numerical simulation studies, we reproduce the experimental results obtained from a specific conical cavity. Then we proceed to simulate the response of cavities of slightly different geometries, and investigate the dependence on the young modulus, poison ratio, and slight changes in dimensions. This initial success encourages to undertake studies of cavities having more sophisticated geometries.

1. Introduction

Whispering Gallery Acoustic Sensing (WGAS) [1] constitutes a simple, low-cost, and still effective, way to implement probe-sample distance feedback control in scanning probe microscopy (SPM). A critical step in SPM is to count with a sensing mechanism (the control signal) that varies monotonically with the probe-sample distance. In Scanning Tunneling Microscopy the control signal is the tunneling current across the probe-sample gap [2]; in Atomic Force Microscopy it is the contact force exerted by the surface on the probe [3]; in QTF-SPM it is the electrical piezoelectric current caused by the electrically driven, lateral oscillation of the quartz tuning fork (QTF) sensor [4]. All these approaches have their advantages and disadvantages. The simplicity of the WGAS, and hence its low cost implementation, lies in exploiting the SPM support-frame as a resonant acoustic cavity to monitor the oscillatory motion of a distant QTF. In the interpretation suggested in Fig. 1a and 1b, the nanometer-sized lateral oscillations of the QTF-tines

IOP Conf. Series: Journal of Physics: Conf. Series **1143** (2018) 012014 doi:10.1088/1742-6596/1143/1/012014

generates acoustic waves on the holding central shaft. Upon reaching the upper region, the sound waves then spread across the walls of the cavity, and standing waves are established. By judiciously placing an acoustic transducer on the perimeter of the cavity for maximum signal detection, WGAS is able to monitor the nanometer-size oscillations of the probe. Fig. 1c shows the cavity acoustic response (magnitude and phase) when the QTF is electrically excited (external source not shown) at different frequencies near the QTF resonance frequency. At resonance, typical values of the QTF amplitude is 10 nm.

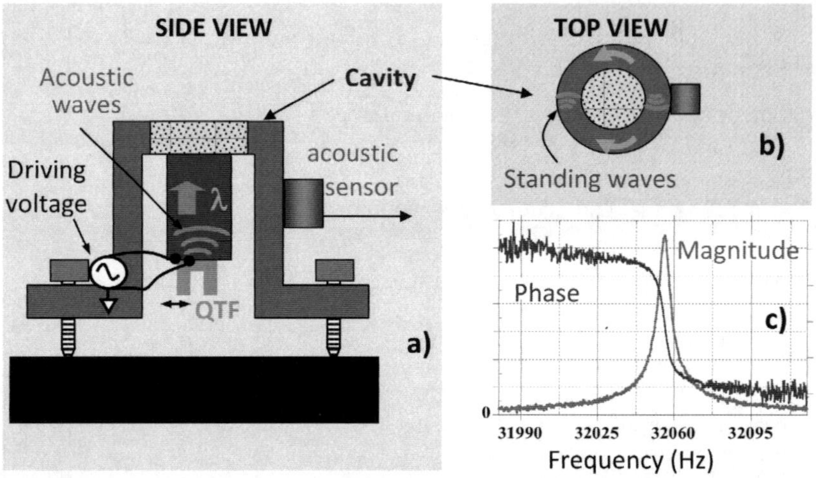

Figure 1. Suggested WGAS working principle mechanism: ***a)*** The nanometer-sized lateral oscillations amplitude of an electrically driven QTF establishes acoustic waves that propagate first upwards and ten towards the cavity; ***b)*** Standing waves are established on the cavity, and an acoustic sensor is placed at a position of maximum acoustic response; ***c)*** Cavity's acoustic response as the QTF is excited across its mechanical resonance frequency.

WGAS acoustic signal can be exploited in SPM by attaching a sharp stylus to one of the tines of a QTF. While driving the QTF (with the probe attached) at its resonance frequency, the probe is brought in close proximity to a sample (Fig. 2a). As a result of the probe-sample interaction, the QTF amplitude of oscillations decreases, which causes also the WGAS signal to decrease monotonically (Fig. 2b) as the probe-sample distance decreases. The latter feature makes WGAS suitable to be used as a feedback control signal. This capability—complemented with lateral scanning of the probe across the sample—allows characterization of the sample topography, as demonstrated in Fig. 2c.

Maximum sensitivity in WGAS would be obtained under ideal perfect match between the cavity's resonance frequency and the probe's mechanical resonance frequency. But the latter seldom occurs (see Fig. 3) because of *i*) the variability of the QTF resonance frequency upon attaching a SPM probe to one of its tines, and *ii*) the typically complicated spectral response of the cavity (displaying multiple peaks, and no clarity on which dimension affects which peak). A potential solution consists in adding a small mass to the QTF [5]; but it carries a consequent loss in sensitivity any time glue is added to the probe. Another potential solution is to model analytically the SPM frame and identify stable resonance peaks (*i.e.* those that do not change dramatically with the cavity dimensions), so a cavity could be built accordingly. But, lack of symmetry in typical SPM-heads makes the analytical equations very difficult to solve. In addition, a variety of bulk and surface acoustic waves can be excited in a given non-homogeneous and non-symmetrical cavity

IOP Conf. Series: Journal of Physics: Conf. Series **1143** (2018) 012014 doi:10.1088/1742-6596/1143/1/012014

[6]. As an alternative, we propose to use finite-element method (FEM) to analyze cavities of arbitrary geometries and dimensions.

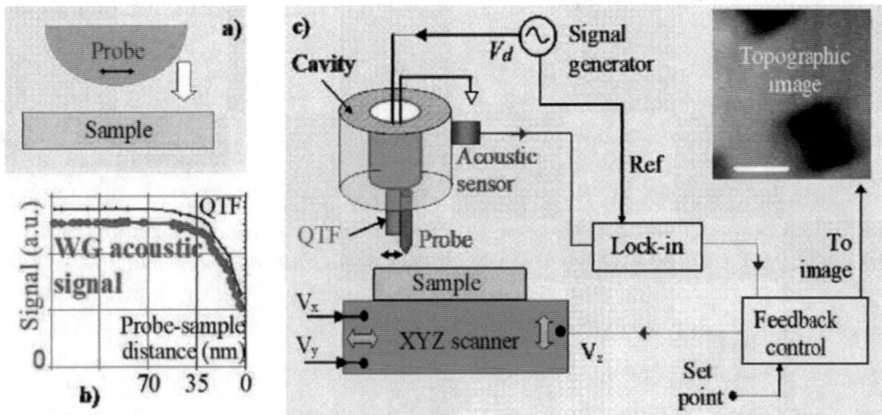

Figure 2. WGAS as a feedback signal-control for acquiring topographic images. When the probe approaches the sample **(a)**, the WGAS signal decreases monotonically when is close to the sample surface **(b).** The latter is exploited for feedback control and generate topographic image of the sample surface **(c)** [1].

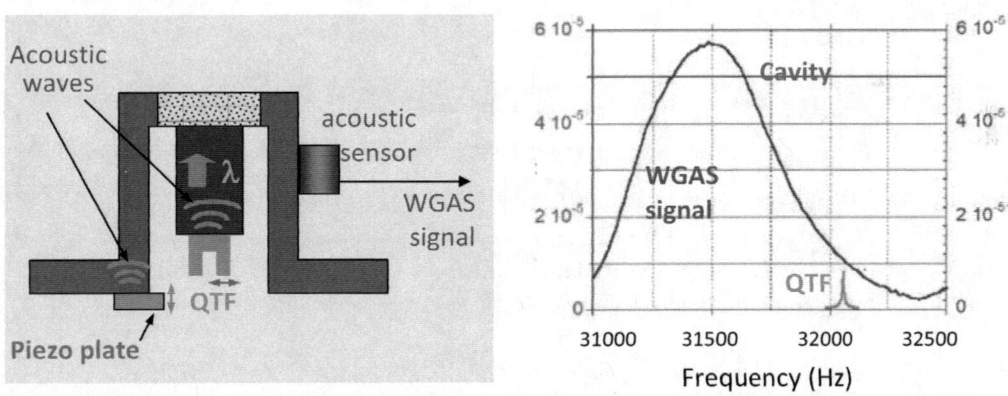

Figure 3. Left: Cavity excited by either a piezo plate or by the QTF within a narrow bandwidth (electrical driving sources not shown). **Right:** WGAS signal responses to the corresponding excitations. Notice the mismatch between the cavity's frequency response and the QTF resonance. In this paper we explore ways to slightly modify a cavity to match its frequency response with the QTF's resonance frequency.

Herein we report our preliminary results using FEM to analyze the frequency response of a conic-cylindrical cavity. The use of this geometry responds to the fact that in previous measurements [5] we found that such a cavity exhibits a well isolated frequency response near 32 kHz; which should simplify the FEM study approach. Still, it was unknown whether or not it was very sensitive to changes of the cavity's properties. This paper embraces such a pending task, and testing in addition the dependence of the frequency-peaks on the material's Young's modulus and Poisson's ratio. Further, to explore the possibility of enabling tunable-frequency capability into the cavity, the FEM analysis will include the effect of adding small rings of different thicknesses to the conical cavity.

2. Effects of Material Properties on the Spectral Response

2.1 Cavity geometry and the simulation procedure

Previous experimental measurements on cavities of cylindrical-type shape [5] revealed that cavities with conical shape display a relatively simpler spectral response (isolated peaks, lower number of spectral peaks). The conical taper appears to 'discard' some peaks that otherwise would be present if the cavity were vertically straight. A conical cavity results then suitable to investigate how a particular frequency peak depends on the physical parameters, size, and slight changes in geometry. Accordingly, we have selected a cylindrical conical cavity for the current studies. Fig. 4 shows a 3D view, vertical cross section and dimensions of the actual cavity used here for both experimental tests and simulation calculations.

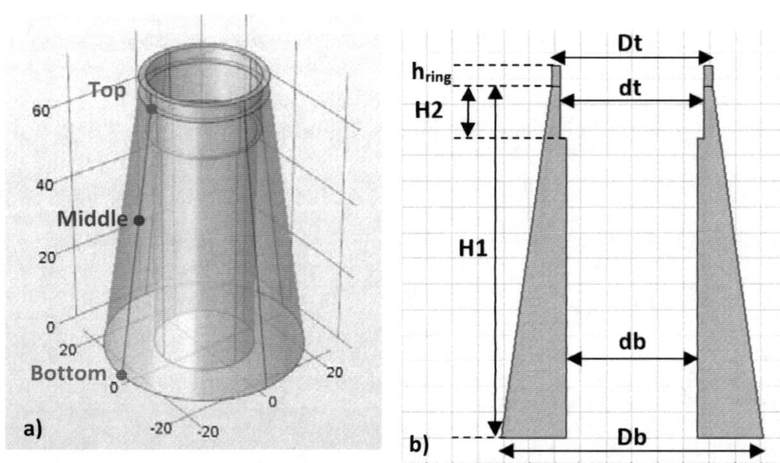

Dimension	Value (mm)
Db	53.6
Dt	31.2
db	27.8
dt	25.4
H1	68.5
H2	10.2
h_{ring}	vary

c)

Figure 4. Conical cavity used in FEM simulations: **a)** 3D view schematic of the actual cavity (top ring added) with the three positions where the acoustic sensor is located. **b)** Cross section view; **c)** Cavity dimensions table.

The structural mechanics module from COMSOL Multiphysics is used to solve the partial differential equations that govern the mechanical behavior of the cavity. The detailed geometry is designed in the COMSOL in order to minimize the meshing error in structural mechanics simulation. An appropriate discrete mesh for the simulation is selected, a task greatly simplified by the default physics-controlled meshing sequences available within the software package. Selection of the "fine" meshing option offers a good compromise between simulation accuracy and computing memory resources. The final free tetrahedral mesh has maximum element size of 5.64 mm and minimum element size of 0.70 mm.

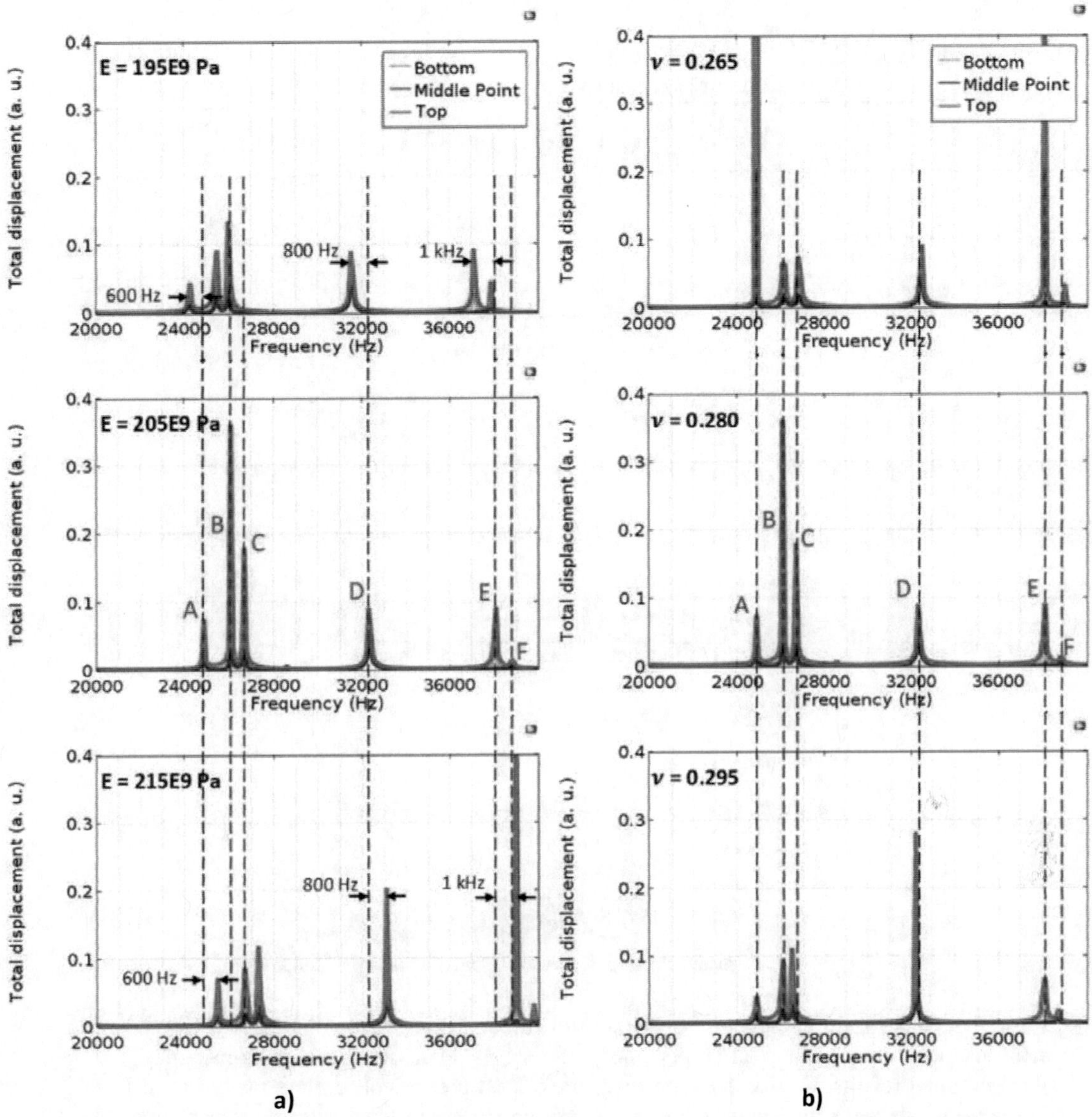

Figure 5. The frequency spectrum variation for 5% change in: **a)** Young's modulus and **b)** *Poisson*'s ratio.

The partial differential equations for the elastic wave inside solid structures were projected onto finite spaces using Solid Mechanic Module. A point load is added at the base of cavity to serve as the excitation source (mimicking the actual piezo plate use to excite the actual cavity.) The point excitation force is set to have equal amplitude along the Ox, Oy and Oz axes. The selected frequency domain of study is 20 kHz to 40 kHz. In all the results presented below (Figs. 5 to 7), three spectra are plotted together in the same graph, corresponding to three different positions of the sensor (top, middle and bottom). The general trends are that the intensity of the peaks changes with the sensor position, but there is no change in the peaks spectral position.

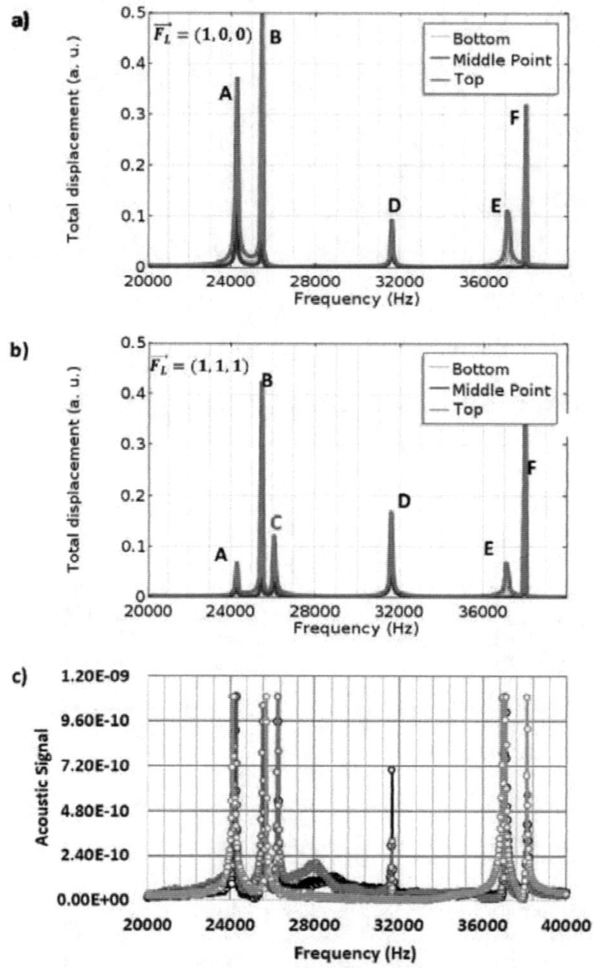

Figure 6. Superimposed spectra obtained with the acoustic sensor placed at different arbitrary positions on the external periphery of the conical cavity. **Bottom spectrum c)** Experimental results. Cavity is excited by an external piezoelectric plate placed at the base. Notice that in some traces peak "C" is missing. **Top spectra:** Simulations with a point source excitation (F_L) located at the cavity base **a)** $\mathbf{F_L}$ along the horizontal axis; **b)** $\mathbf{F_L}$ along (x,y,z) axes.

2.2 Sensitivity of the cavity's frequency response to changes in the material properties

Due to the different fabrication processes of the cavity material, there will be an uncertainty on the exact values of the cavity's physical properties, namely the Young's modulus and the Poisson's ratio. To evaluate the sensitivity of the cavity's spectral response to these properties, we perform simulations with nominal values of 205 $\times 10^9$ Pa for the Young's modulus and 0.28 for the Poisson's ratio and allowing $\pm$ 5% uncertainty [7]. The results are shown in Fig. 5. Notice all the peaks move consistently to the right with increasing values in the Young's modulus (Fig. 5a), with the peaks of higher frequency shifting by a slightly larger amount (peak A shifts by 600 Hz, peak D shifts by 800 Hz, and peak E shifts by 1000 Hz.) On the other hand, the spectra do not change substantially with increasing values of the Poisson's ratio (Fig. 5b).

2.3 Matching simulation calculations with experimental results

A good agreement between experimental and simulation results is obtained when using a Young's modulus of 195×10^9 Pa and a Poisson's ratio of 0.27 (Fig. 6). Again, the intensity of the peaks changes with different sensor positions, but the peaks spectral position stays the same. Notice in Fig. 6a that peak "C" is absent, which can be attributed to the fact that the excitation point source (placed at the base of the cavity) was selected to oscillates purely in the horizontal direction. In contrast, the results in Fig. 6b correspond to the point source oscillating along the (1, 1, 1) direction; the additional yz-component excitation results in the appearance of the extra peak "C" in the spectrum.

3. Analysis of the geometrical cavity dimension effect on the Spectral Response

3.1 Sensitivity of the frequency response to 100-micron changes in cavity dimensions

Here we report on the sensitivity of peaks' position in the spectra to changes in the cavity dimensions (Table 1). The values are considered to increase by 100 microns, which is a precision fairly achieved in a CNC lathe machine.

Table 1: Frequency changes (in Hz) when each dimensions is increased by 100 microns.

Dimension (mm)	Peak A 24.3 kHz	Peak B 25.5 kHz	Peak C 26.2 kHz	Peak D 31.6 kHz	Peak E 37.1 kHz	Peak F 38.0 kHz
Cone height (H1)	20	10	-30	-20	60	0
Top hollow height (H2)	20	-30	-60	-30	50	-50
Cone base diameter (Db)	-30	-40	-40	-20	-40	-40
Cone top diameter (Dt)	-10	0	0	0	-20	10
Hollow bottom diameter (db)	0	50	30	10	-40	40
Hollow top diameter (dt)	-100	-20	0	-40	-80	-20

3.2 Sensitivity of the frequency response to slight modification in cavity geometry

In order to obtain the effect of changing the geometry of the cavity on the spectral response, we performed two modifications. The first modification is simulations with different taper angle (1° change) geometries by keeping the same height (H1) then changing the conical base diameter (Db). Fig. 7a shows the results when Db is increased, which results in a decrease in the cone angle. The simulations with gradually smaller cone-angle (1° precision) allowed discriminating the peaks in two distinct groups: (A, B, C, E) with a tendency towards a "blue" shift, and (D, F) with a tendency towards a "red" shift. In the first group, peak E is the one that is sensitive the most (1.5 kHz per degree first, then 0.5 kHz per degree), while the group of peaks A, B, and C undergo a smaller shift (peak-A shift 1 kHz first and then 0.5 kHz per degree). On the other hand, the second group comprises peaks (D and F) that do not shift significantly. The position of peak F, for example, is practically immune to the cone-angle variations, while peak D displays at most a slight "red" shift of less than 200 Hz per degree. These results suggest *a*) peak E to be greatly influenced by the length along the cone (the steeper the cone, the smaller taper length, the greater the frequency), and *b*) peak frequencies D and F to be associated to the inner diameter of the cavity (a parameter that was not changed during these simulations).

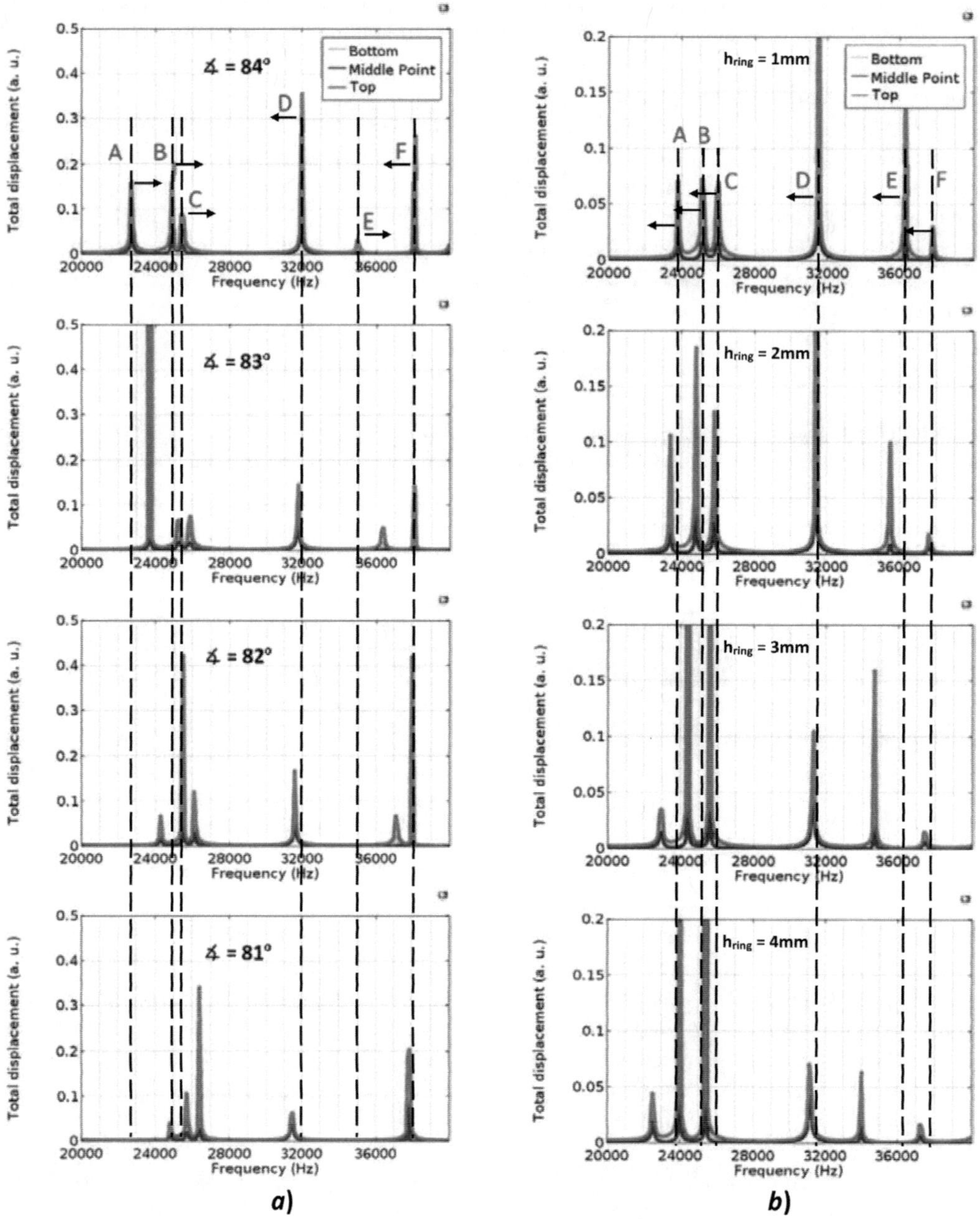

Figure 7. Spectral response variations as ***a)*** the cone angle of the cavity increases, and ***b)*** when adding rings of different thicknesses (h_{ring} in Fig. 4) to the top of the cavity.

Second, we modified the structure by adding hollow rings of different thickness to the top of the original cavity. Fig. 7b corresponds to the case in which a ring of variable thickness (h_{ring}) were added. Notice all the peaks undergo a "red" shift as the thickness of the ring increases. Frequency shifts of the order of 100

Hz were obtained for rings of 1 mm thickness. The peaks are discriminated in a similar fashion as when the cone angle changes: the (A, B, C, E) group has peaks E and A highly sensitive to the added perturbations, while the peaks D and F are minimally perturbed. Since peak D appears immune to the changes in the base and the height of the rings, it is plausible to associate this peak to the internal diameter.

4. Conclusions

Finite-element method (FEM) was used to study the structural frequency response of a stainless steel conical-cylinder cavity of dimensions typically used in actual Shear-force Acoustic Near-field Microscopy (SANM) settings. The cavity was purposely separated from the full SANM-frame design to simplify our preliminary FEM studies. As a preliminary step we verified changes in the cavity's spectral response for *i*) 5% uncertainty in the materials physical properties (Young modulus and Poisson's ratio), and *ii*) 100 m uncertainty in the different cavity dimensions.

First, we validated our simulation calculations by being able to reproduce experimental measurements performed on a hollow truncated-cone cavity. In preliminary simulations, the cavity was excited by a point-source oscillating along one-component shear excitation axis. In those cases, an initial discrepancy existed between the simulated and experimental results. Indeed, there was a particular peak consistently missed in the simulated spectra, which was also missing in some other spectra measurements. The discrepancy was resolved by incorporating both a shear and longitudinal excitation in the simulation, after which the simulations reproduced closely all the experimental peaks. This finding illustrate one of the advantages offered by FEM, helping to discriminate peaks based on the excitation modality.

Next, tests with different cone angles provide insights to take into consideration in the design and construction of SANM cavities. However, it is also necessary to explore solutions that can permit SANM users to in situ match the cavity's frequency response to the resonance frequency of the QTF (the uncertainty in the latter is caused by the stylus attached to one of its tines). A suggested practical solution considers adding small rings with different thickness on the top of the cavity and the simulation result support this approach. This way the SANM user would be able to modify the cavity in the act. Complementary tests for further verification of these hypotheses are currently underway.

In short, compared to the task of exploring analytical solutions of frequency response to arbitrary modifications of a cavity, the FEM analysis offers an alternative simpler approach. The ability of FEM to reproduce experimental results, as demonstrated here, constitutes an encouraging first step.

Bibliography

[1] La Rosa A H, Li N, Fernandez R, Wang X, Nordstrom R, and Padigi S K 2011 *Rev. Sci. Instrum.* **82**, 093704
[2] Chen C J, *Introduction to Scanning Tunneling Microscopy*, 2nd edition. Oxford: Oxford University Press, 2016
[3] Lanza M *Conductive Atomic Force Microscopy: Applications in Nanomaterials*. John Wiley & Sons, 2017
[4] Polesel-Maris J, Lubin C, Thoyer F, and Cousty J 2011 *J. Appl. Phys.* **109** 74320
[5] Hung H-C, La Rosa A H, Fernandez R, Comnes B, and Nordstrom R 2014 *SOP Trans. Appl. Phys.* **1** 22
[6] Ventsel E and Krauthammer T 2001 *Thin Plates and Shells: Theory: Analysis, and Applications*. CRC Press
[7] Overview of materials for Stainless Steel. Available online: http://www.matweb.com/search/datasheet.aspx?MatGUID=71396e57ff5940b791ece120e4d563e0&ckck=1.

IOP Conf. Series: Journal of Physics: Conf. Series **1143** (2018) 012015 doi:10.1088/1742-6596/1143/1/012015

Combining atomic force microscopy and shear-force acoustic near-field microscopy to characterize confined mesoscopic fluids

Monte Kozell[1,2], Theodore Brockman[1] and Andres H La Rosa[1]

[1] Physics Department, Portland State University, Physics Department, Portland, OR, USA
[2] Lam Research, Tualatin, OR, USA

E-mail: montekozell@gmail.com, andres@pdx.edu

Abstract. An atomic force microscopy (AFM) cantilever is integrated into to a quartz tuning fork (QTF) to probe the viscoelastic properties of mesoscopic fluid layers confined between two solid surfaces under shear. Two procedures to fabricate the AFM/QTF probe are described herein. In the first, a nano-manipulator is used to transport a commercially available afm cantilever from its chip holder to the edge of a QTF tine. In the second, an afm cantilever is fabricated at the edge of the QTF tine itself. In both cases we exploit the capabilities of a dual-beam system (focused ion beam/scanning electron microscope), equipped with Omni-Probe nano-manipulator and a Gas Injection System (GIS). The new device improves the ability of shear-force acoustic near-field microscopy (SANM) to monitor the constraining normal and damping shear forces exerted by the solid boundaries, concurrently with the acoustic emission from the trapped fluid.

1. Introduction

Shear-force Acoustic Near-field Microscopy (SANM) was recently introduced as a metrology tool to characterize the dynamic behavior of mesoscopic fluids trapped between two solid boundaries under relative shear motion. SANM measurements emphasize, in particular, the near-field detection of acoustic waves emitted from a trapped mesoscopic fluid[1,2], which offers complementary information to the simultaneously acquired shear-force also intervening in the probe-fluids interactions. This scenario is akin to interfacial friction phenomena (wear-less friction[3]) where the dissipation of energy is attributed to inelastic interactions between the solid boundaries and the confined fluid layer (with no solid-probe and solid-substrate contact.) The added acoustic detection capability brought by the new SANM technique allows contrasting the acoustic signal emitted by the fluid with the information obtained from the simultaneously acquired damping shear-force acting on the probe, which illustrates the improved metrology capabilities offered by SANM. Still, at its current state of development, this new technique lacks the ability to measure the normal force component of the solid-fluid interactions. This shortcoming is addressed herein.

In SANM, a tapered stylus (size of the apex radius of curvature typically tens of nanometer) is adhered to a quartz tuning fork (QTF), which is electrically driven at resonance frequency condition (~ 10 nm amplitude.) The probe is then allowed to interact with a substrate. The lateral oscillation amplitude of the tuning fork is routinely used as a feedback signal to control the probe-sample distance[4,5]. A benefit of the shear motion operation (probe oscillating parallel to the surface boundaries) is that the tip is kept at a fixed height above the sample surface, thus removing a direct tapping (*i.e.* intermittent contact) of the probe against the sample. Removing such a direct mechanical interaction extends the probe lifespan and reduces sample damage. However, one issue in the SANM field is the inconclusive knowledge of the absolute value of the probe-substrate separation distance, *i.e.* there is an uncertainty about the exact location of the substrate [6]. This causes ambiguity when interpreting the probe-fluid interaction mechanisms as a function of probe-sample separation distance. For the particular cases where the probe and substrate are electrically conductive, the exact location of the substrate can be established by monitoring the onset of the tunneling current across the probe-substrate gap. An exponentially increasing current is initiated when a biased probe comes into mechanical contact with the surface. But the method is applicable only to conductive samples.

Atomic Force Microscopy, by contrast, allows a direct vertical spatial mapping of the normal force components of the interaction as a function of probe-sample separation distance. The AFM consists of a beam cantilever that has a pyramid shape apex. AFMs typically operates by reflecting a laser off the cantilever and onto a position-sensitive photodiode (pspd). As the cantilever gets closer to the substrate, the intervening normal forces bend the cantilever, which results in a deflection of the laser spot centered on the pspd. The recorded normal force traces typically display two well-defined interactions regions. In the first region one can interpret the probe interacting with a mesoscopic water layer; in the second, the probe certainly undergoes a direct mechanical contact given linear response tracked by the pspd[7].

In the present work, we incorporate the functionality of an afm beam cantilever into the SANM. We focus in the description of an AFM/QTF probe fabrication method. In subsequent work we plan to use this system to obtain approach curves that monitor normal force, lateral shear force, lateral displacement and the acoustic amplitude to spatially map probe-sample interactions as a function of separation distance.

2. Probe Device Fabrication

A commercial non-contact (NanoSensor, NCH-W) afm cantilever was directly adhered onto the tine of a commercial 32,768 Hz QTF (Abracon, AB26TRQ-32.768KHz-T). A XE-120 PSIA (Park) AFM system was used for this experiment. The QTF was first adhered (using cyanoacrylate) to a commercial stainless steel mount in such a way that the tines oscillate parallel to the sample surface.

2.1 Method 1: Attachment of an AFM Cantilever to a QTF

A dual-beam FIB-SEM FEI Helios 400s, equipped with Omni-Probe nanomanipulator and gas injection system (GIS), is used to adhere an afm cantilever to the QTF tine. GIS gases comprising platinum and carbon are used. As starting material, a commercial afm cantilever comes already attached to a metal chip holder (for easy handling) as shown in Fig. 1.

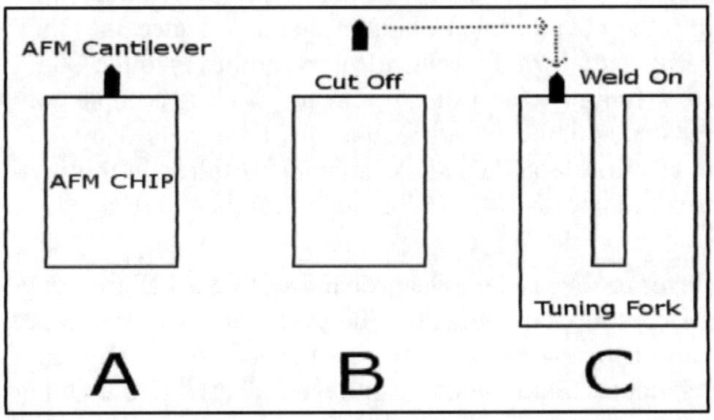

Figure 1: Schematic procedure to prepare an AFM/QTF probe**.** **(A)** Unmodified afm cantilever attached to a metal chip, as received from the manufacturer. **(B)** Afm cantilever cut free from the chip. **(C)** Cantilever afm welded onto the tuning fork tine.

As a first step, the afm chip and the QTF are mounted flat inside the FIB-SEM system. The Omni-Probe nanomanipulator is lowered to a height of approximately 1 μm above the cantilever surface (Fig. 2A and

2B), followed by the insertion of the platinum GIS needle into the chamber (Fig. 2C.) In the presence of a gas source inserted through the GIS needle, a gallium ion beam (30 kV and 300 pA) allows welding the cantilever to the Omni-Probe (Fig. 2D). In this process, while the gas is injected into the chamber, the ion beam rastered over a selected area generates secondary electrons, which subsequently break down the precursor gas (platinum or carbon) that ends up deposited onto the surface[8].

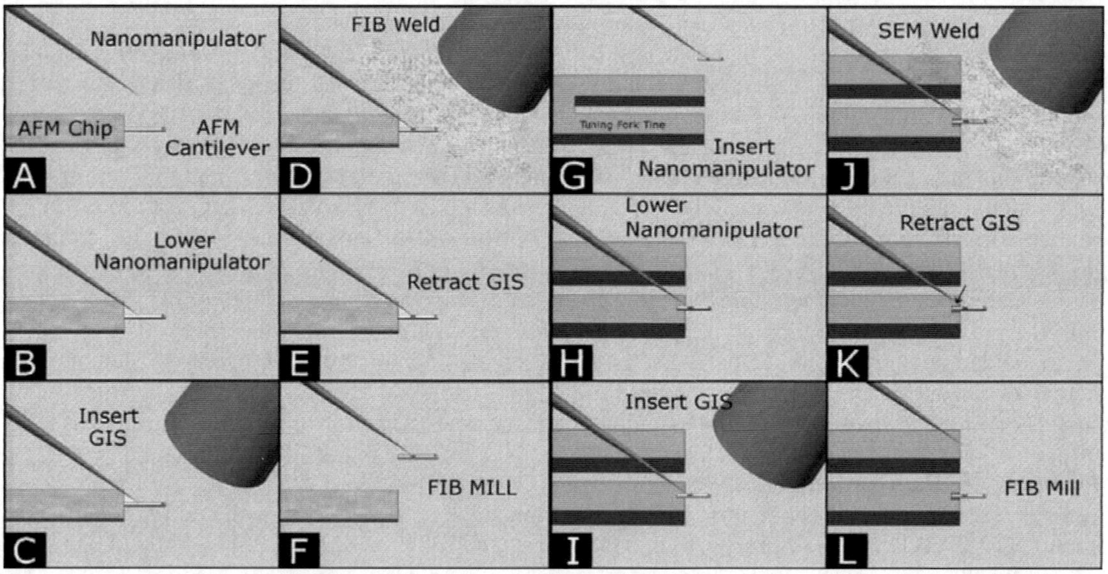

Figure 2: Sequential procedure of extraction of an afm cantilever and its attachment to the nanomanipulator (A to F). **(A)** Omniprobe nanomanipulator is inserted. **(B)** Nanomanipulator placed approximately 1µm above the afm cantilever. **(C)** Insertion of the platinum gas injection needle. **(D)** Platinum gas injected into the chamber while the ion beam weld is in pattern mode; this procedure welds the nanomanipulator to the afm cantilever. **(E)** Cantilever welded to the nanomanipulator. **(F)** FIB milling cuts the afm cantilever free from the chip. The nanomanipulator is raised above the sample and retracted. **Sequential procedure of attachment of the afm probe to a QTF-tine (H to L)**: **(G)** Nanomanipulator inserted. **(H)** The nanomanipulator is placed at approximately 1 µm above the QTF tine. **(I)** Platinum gas injection needle inserted into the chamber. **(J)** Platinum gas is injected into the chamber while the electron beam weld is in patterning mode, which welds the afm cantilever to the QTF tine. Several platinum welds points insure a secure attachment. **(K)** Electron beam welding completed. Electron beam deposition was used instead of ion deposition due to irregular gas flow caused by the large afm cantilever geometry impeding platinum gas flow. Ion deposition attempts resulted in milling instead of deposition. **(L)** Once welded, the nanomanipulator is FIB cut free, then lifted and retracted.

The welding is performed at the center of the cantilever to thus reduce the probability that a 125 µm long-arm afm cantilever collides with other internal SEM chamber components. The cantilever is then cut free near the afm chip-cantilever interface using the Ga-ion beam operated at 30 kV and 2.5 nA (fig. 2E). Subsequently, the GIS needle is retracted from the chamber. With the afm cantilever adhered to the Omni-Probe, the probe is raised to a safe height of approximately 200 µm above the afm chip and then retracted for safely moving the stage positions (Fig. 2F).

IOP Conf. Series: Journal of Physics: Conf. Series **1143** (2018) 012015 doi:10.1088/1742-6596/1143/1/012015

The FIB-SEM sample stage is positioned to work over to QTF tine. The Omni-Probe needle is lowered to a height of approximately 1μm above the tuning fork tine (Fig. 2G and 2H). The platinum GIS needle is inserted into the chamber (Fig. 2I). An SEM beam (10kv, 3.2nA) is used to weld the afm cantilever onto the end of the tuning fork tine (Fig. 2J). The Omni-Probe needle is cut free using the Ga-ion beam, 30 kV, 2.5 nA (Fig. 2K). Finally, the platinum GIS is retracted and the Omni-Probe raised to a height of approximately 200 μm and retracted (Fig. 2L).

Fig. 3 shows SEM images of the resulting AFM/QTF probe. Fig. 3A shows the actual tuning fork tines with an afm cantilever attached with electron beam platinum welds. Fig. 3B is a higher magnification image of the afm cantilever. Note the platinum weld at the center of the cantilever, which was used to temporarily adhere it to the nanomanipulator. The afm tip is at the top side.

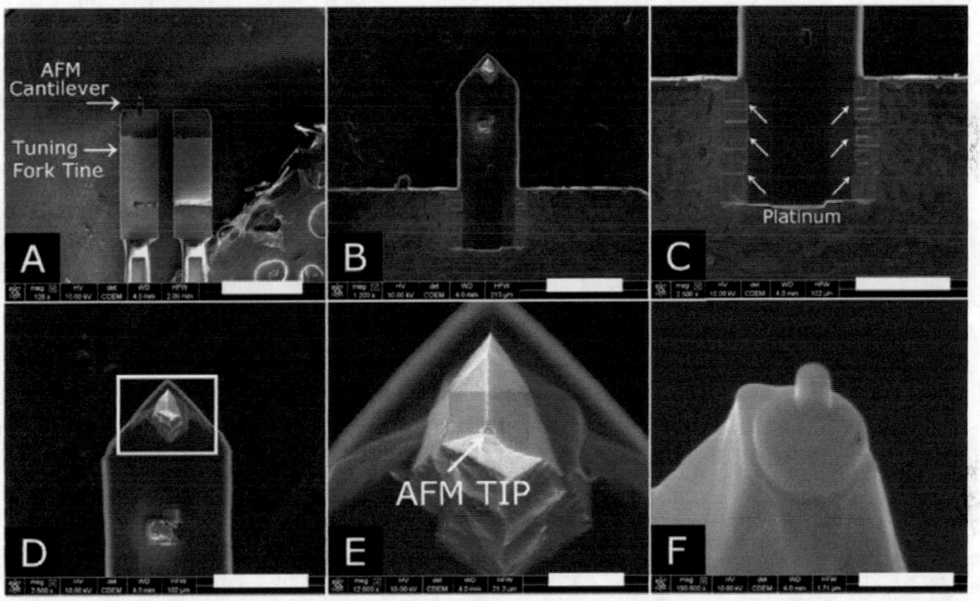

Figure 3: (A) QTF with an afm cantilever attached (via electron beam platinum welding) to one of its tines. Scale bar is 500 μm. **(B)** Image of the afm cantilever. Note at the center the platinum weld that was used to temporarily adhere the cantilever to the nanomanipulator. The afm tip is located near the edge. Scale bar 50 μm **(C)** Platinum welds adhering the cantilever to the tuning fork tine. Scale bar 30 erμm **(D)** High magnification image of the tip. Scale bar 30 μm **(E)** Carbon tip pyramid fabricated at the apex via Ga ion beam deposition. Scale bar 5 μm **(F)** Tilted image of the IBD carbon tip. Scale bar 500 nm.

Fig. 3C shows the platinum welds securing the attachment of the cantilever to the QTF tine. Fig. 3D is a higher magnification image of the cantilever apex. Fig. 3E is a top down (zero tilt) image of the pyramid structure of the afm, with a Ga ion beam deposited carbon tip. Fig. 3F shows a 52 degree tilted image of the ion beam deposited (IBD) carbon tip of base of radius 500 nm, tip height of 250 nm, and tip radius ~110 nm. By using a FIB mill/deposition method, old tips that have been damaged can simply be FIB milled away and a new carbon tip can be grown. Other materials can be used for tip growth, such as platinum, tungsten or insulator SiO_2. Electron beam deposited carbon tips[9] and ion beam deposited tips[10] have been shown to be effective for imaging purposes. IBD carbon probes have 70 Gpa shear modulus[11], 9-10 Gpa hardness, and 123-130 GPa Young's modulus[10].

2.2. Method 2: Focused Ion Beam Milled Cantilever Fabrication

We have also explored an alternative fabrication method, where a beam cantilever is built in at the corner of QTF tine. Fig. 4 shows the overall strategy. First, the QTF tine is properly FIB carved to create a thinner section; a subsequent ion mill of a groove results in the formation of a flat rectangular cantilever. A small area of one side of the cantilever is coated with a metallic pad (to increase its light reflectivity). On the other side, a pyramid shape tip is fabricated. The fabrication process is described in more detailed next.

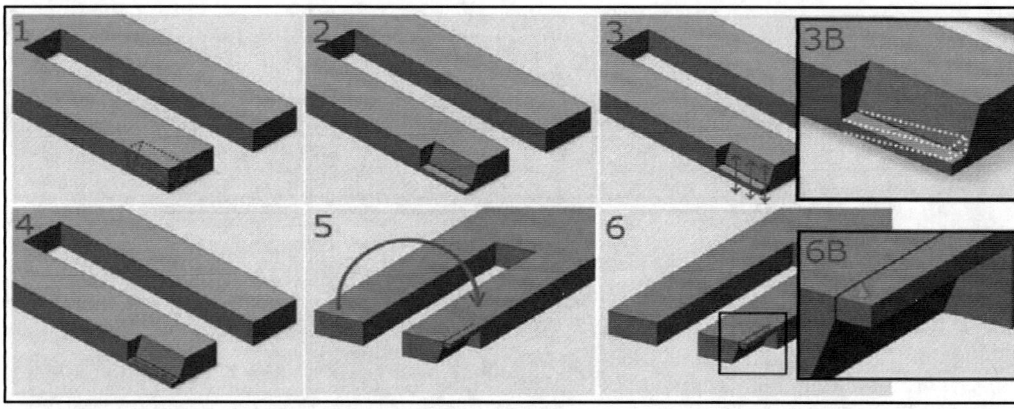

Figure 4 Sequence of the fabrication procedure of a cantilever built-in into a QTF-tine using a focused ion beam (FIB). **Image 1:** Volume section (red dots) of a QTF selected for ion beam milling. **Image 2:** QTF tine post FIB milling. Red dots illustrate another milled region that allows the cantilever the ability to flex in the vertical direction. **Image 3:** Cantilever flexing vertically. **Image 3B** Cantilever flexing range motion. **Image 4** Cantilever after ion-beam-deposited (IBD) platinum pad (purple color) deposited on the top of the cantilever, for improving laser reflection. **Image 5** Cantilever flipped over so a carbon probe can be fabricated. **Image 6** and **6B** illustrates the IBD probe (green color) onto the end of the cantilever.

We fabricate a pyramidal-shape platinum base followed by carbon tip at the apex. For constructing the base, a gallium ion beam and an in situ platinum gas injection system were used to deposit platinum onto the QTF tine. As a precursor gas we used (methylcyclopentadienyl) trimethyl platinum, $C_9H_{16}Pt$ (FEI, Helios 400s User Manual.) The platinum base (8x8 μm^2) was built-up with sequential deposition of platinum layers of reduced width reach time (current density of 2-6 pA/μm^2), as schematically illustrated in figures 5 A-C. After the deposition of 13 layers, the width of the apex was 1.5 μm. Then, carbon was deposited via gallium ion deposition (at current densities of 1-10 pA/μm^2); see figures 5 D-F. IBD carbon deposition occurs at a faster rate than platinum, hence the increased taper slope of the pyramid tip [12]. The final tip was deposited under spot mode (where the raster scanning of the FIB beam is disabled) and its growth occurs rapidly (often taking less than 30 sec.) The total height (base plus tip) is ~5 μm.

Attempts to build an afm pyramid tip directly out of carbon deposition on the QTF proved to be difficult. The reason is basically due to electron charge that builds up when performing electron beam deposition on insulating materials (Quartz resistivity between 10^5 to 10^{18} Ωm)[13]. Charge buildup results in a repulsive electric field developing on the tuning fork tine, causing electron beam induced tip growth to be skewed in a lateral direction. Proper grounding of the tuning fork electrodes allowed charge build-up to be reduced, but not always eliminated. As an alternative, it was decided to produce Gallium ion deposited carbon tips for experimental use. Since gallium ions are 127,000 times more massive than electrons (Gallium 69.723 u, Electron 5.485×10^{-4} u, unified atomic mass unit), charge build-up has a lesser deflection effect on gallium ion beam path. In addition, the positive gallium ion are attracted to surfaces with built-up electron charge, and has an effect to neutralize the substrate. Limited by the resolution of the ion beam, tip radius of ~100 nm were reliably produced. The table 5 below shows the deposition parameters used for the final carbon tip fabrication.

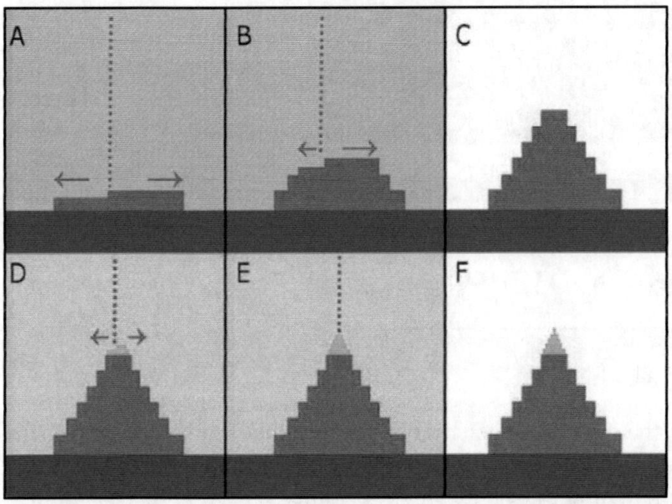

Figure 5: Schematic procedure to fabricate a pyramidal tip near the edge of a beam cantilever. The probe-base is fabricated by sequentially depositing rectangular patterns of IBD platinum as shown in **images A-C**. The ion beam is rastered while platinum gas is injected into the chamber. **Images D-F** illustrates the sequential pattern deposition of the IBD carbon probe tip. The ion beam is rastered while a carbon gas is injected into the chamber. **Image F** displays the final deposition of the probe apex; no FIB rastering was used to minimize probe radius.

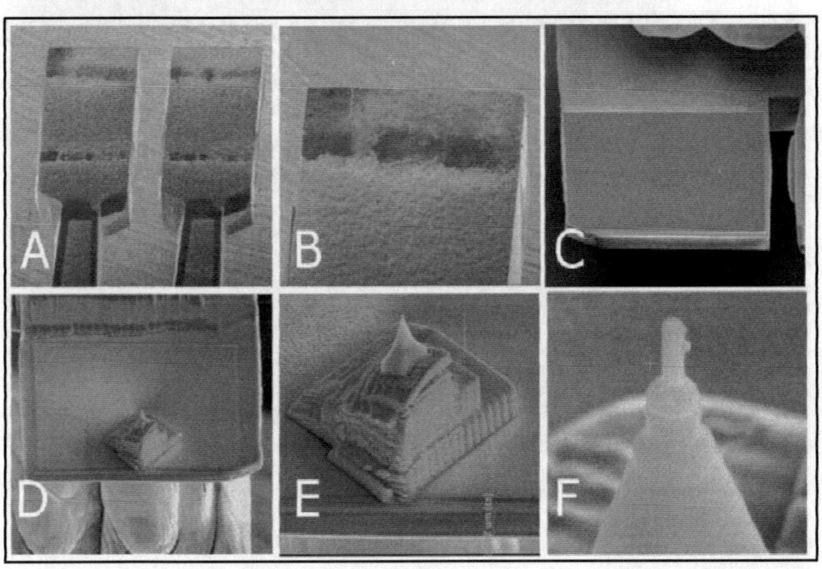

Figure 6. SEM images (with false colors to differentiate materials.) Horizontal field width (HFW) stands for image width. **(A)** QTF lower region; HFW=1.28 mm. **(B)** QTF bottom edge; HFW=320 µm. **(C)** Gallium-ion beam platinum pad (purple) deposited on top of cantilever; HFW=60 µm. **(D)-(E)** First iteration of a Ga-IBD deposited platinum base (purple) with carbon tip (green). HFW=60 µm and 19.5 µm respectively. **(F)** Ga-IBD carbon tip with radius of 100 nm; HFW=2.0 µm.

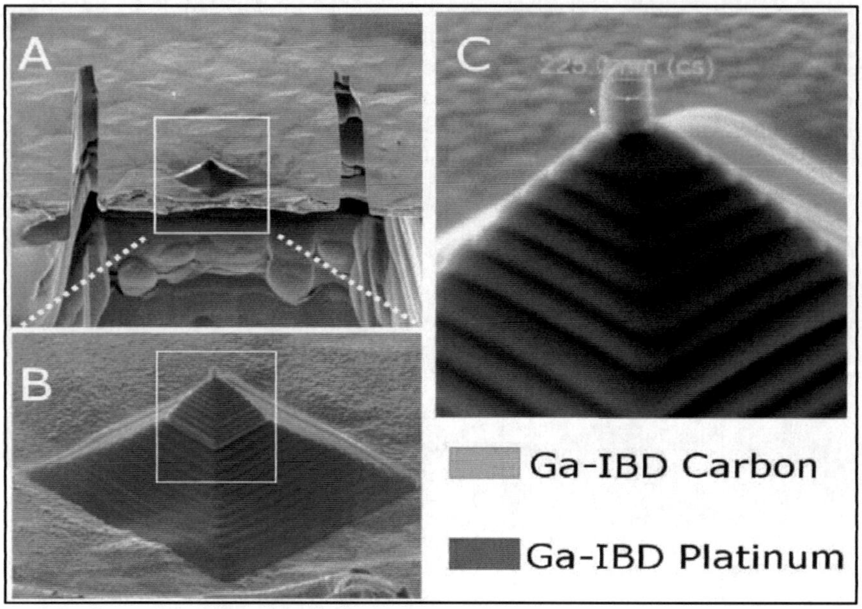

Figure 7. Refined fabrication iteration of a platinum-base and a carbon-tip. **(A)** FIB milled cantilever and platinum pyramid base fabricated with Ga IBD deposition. HFW= 73.1 µm. **(B)** Higher magnification with false colors of the platinum base (purple) and the carbon-tip (green); HFW=17.1µm. **(C)** Carbon tip of 112 nm radius; HFW = 2.0 µm.

2.3 Estimation of the probe's spring constants

The length, width, thickness and tip height of the cantilever determine its spring constants in the normal, lateral, and torsional directions. Table 1 shows pertinent equations showing the relation between spring constants and cantilever dimensions. Table 2 gives the dimensions of FIB milled cantilevers; 76.5 µm in length, 35 µm width, and 5 µm thickness. The estimated cantilever spring constant in the normal dimension was approximately 196 N/m, much stiffer than typical tapping mode cantilevers commercially available (20-70 N/m). This will allow the AFM/QTF cantilever to reliably probe closer to the sample surface before undergoing a jump-to-contact. The lateral spring constant is estimated to be 12,000 N/m, while the torsional spring constant is 23,800 N/m. Both torsional and lateral spring constants are sufficiently large that displacement in either dimension will be small compared to normal displacement. Table 3 shows the material properties used for determining the spring constants and contains material properties of the crystal quartz tuning fork (SiO_2), and the ion beam deposited carbon used to fabricate the tip. Table 4 summarizes the estimated spring constants of the fabricated cantilever.

Table 1: Equations for Spring Constants	
Normal Force Constant K_{ver}	$K_{ver}=w*E/4*(t/l)^3$
Lateral Force Constant K_{lat}	$K_{lat}=t*E/4*(w/l)^3$
Torsional Force Constant K_{tor}	$K_{tor}=w*(G/3)*(t^3/l)/(H+t/2)^2$
t=thickness, w=width, l=length, H = Tip Height; Source [14]	

IOP Conf. Series: Journal of Physics: Conf. Series **1143** (2018) 012015 doi:10.1088/1742-6596/1143/1/012015

Table 2: FIB Milled Cantilever Dimensions				
Length (l)	Width (w)	Thickness (t)	Tip Height (H)	Tip Radius (R)
76.5 μm	35 μm	5 μm	5.8 μm	110 nm

Table 3: Material properties		
Quartz Tuning Fork (SiO_2)	Young's Modulus E(Perp) & E(Para) Shear Modulus G	E(perp) = 78.7 GPa [15] E(para) = 98.2 GPa G = 31.14 GPa [16]
IBD Carbon (Tip material)	Shear Modulus G	G = 70 GPa [11]

Table 4: FIB Milled Cantilever Spring Constants		
K_{ver}	K_{lat}	K_{tor}
196 N/m	12,000 N/m	23,800 N/m

Table 5: Summary of Ion-beam Tip Deposition Parameters			
Deposition Parameters	Accelerating Voltage	Beam Current	Deposition time
IBD Carbon	30 kV	13 pA	15 sec

3. Conclusions

Two procedures for fabricating an AFM type pyramid at the edge of a quartz tuning fork (QTF) were successfully implemented. The project pursued obtaining a robust probe, less susceptible to "jump to contact" instability (characteristic of commercially available typical afm cantilevers), still sensitive enough to measure normal forces. The "jump to contact" on regular fragile cantilevers is caused by the coalescence of a water meniscus that occurs when the pyramid probe is placed near another substrate at ambient conditions. The latter is detrimental to a metrology system. The probes fabricated here are one order of magnitude stronger. In combination with a QTF, the pyramid tip can also be driven into lateral oscillations, which enable the trapped fluid layer/meniscus to emit acoustic waves. The ability to detect such near-field acoustic waves is a peculiar capability in Shear-force Acoustic Near-field Microscopy (SANM). The ability to measure normal forces, lateral forces, and acoustic emission from solid/fluid interfaces elevates the status of SANM as a nanometrology system.

Acknowledgements

This work was supported in part by Lam Research Corporation, CA USA.

Bibliography

[1] La Rosa A H, Cui X, McCollum J, Li N and Nordstrom R 2005 *Rev. Sci. Instrum.* **76** 093707
[2] Fernandez R, Wang X and La Rosa A H 2011 *Proceedings of the Nanotechnology* (*IEEE-NANO*) *11th IEEE International Conference* 903
[3] Karrai K, Tiemann I 2000 *Phys Rev B: Condens Matter Mater Phys.* **62** 13174
[4] Bharat Bhushan 2012 NANOSCIENCE and NANOTECHNOLOGY Scanning Probe Microscopy in Nanoscience and Nanotechnology, Volume 3 (Springer).
[5] Ruiter A G, van der Werf K O, Veerman J A, Garcia-Parajo M F, Rensen W H, van Hulst N F 1998 *Ultramicroscopy* **71** 149
[6] Cui X, La Rosa A H 2005 *Appl. Phys. Lett.* **87** 231907
[7] Cappella B, Dietler G 1999 *Surf Sci Rep.* 1**34**(1-3):1–104
[8] Lipp S, Frey L, Lehrer C, Demm E, Pauthner S and Ryssel H 1996 *Microelectron. Reliab.* **36** 1779
[9] Schiffmann K I 1993 *Nanotechnology* **4** 163
[10] Lemoine P, Roy S S, Quinn J P, Maguire P D and McLaughlin J A D 2007 *Appl. Phys. A: Mater. Sci. Process* **86** 451
[11] Nakamatsu K, Nagase M, Namatsu H and Matsui S 2005 *Jpn. J. Appl. Phys.* **44** L1228
[12] Yao N Focused Ion Beam Systems: Basics and Applications (Cambridge University Press).
[13] Jair H and Nowick A S 1982 *Journal of Applied Physics* **53** 477
[14] Eaton P, West P 2010 Atomic Force Microscopy (Oxford University Press)
[15] Castellanos-Gomez A, Agraït N, Rubio-Bollinger G. 2009 *Nanotechnology* **20** 215502
[16] Götze J 2009 *Mineralogical Magazine* **73** 645

Probe damage evaluation in frequency-modulation shear-force acoustic near-field microscopy

T Brockman[1], R Fernandez[1], J Bai[1], M Kozell[1,2] and A H La Rosa[1]

[1]Portland State University, Physics Department, Portland, OR, USA
[2]Lam Research, Tualatin, OR, USA

E-mail: tbrock2@pdx.edu

Abstract. Shear-force acoustic near-field microscopy (SANM) and Whispering Gallery Acoustic Sensing have recently been introduced as a tandem system to characterize the viscoelastic response of fluids confined between two solid-boundaries in relative oscillatory lateral motion. SANM uses *a*) a laterally oscillating tapered probe (attached to a quartz tuning fork QTF) as one of the trapping boundaries, and *b*) an acoustic sensor (attached to the other flat-substrate boundary) that independently monitors the fluid's acoustic emission. On the other and, WGAS is another technique that uses an acoustic transducer (attached to the frame holding the probe) to monitor the probe's lateral motion amplitude. Steps towards improving the standing of the SANM/WGAS system as a nanometrology tool include *i*) to integrate frequency modulation methods, to thus be able to discriminate the elastic from the inelastic component of the probe-fluid interactions, and *ii*) evaluate the integrity and robustness of the probe to endure such surface interactions, since eventual deformation of the probe affect the reproducibility of the measurements. Here we report using the WGAS signal (or alternatively the QTF electrical response) to control the probe's motion in frequency modulation modality. In addition, systematic evaluation of eventual probe damage (acquiring scanning electron micrograph before and after the probe exerts one approach/retraction trip to/from the substrate) are also presented. The non-monotonic behavior of the fluid's acoustic emission (first increasing and then decreasing) as the probe approaches the sample—which occurs before the probe touches the surface, as revealed by the lack of damaged observed on the probe—constitutes one of the main findings reported here. The acoustic signal provides clues on the fluid's dynamic response when subjected to nanoscale confinement.

1. Introduction

The dynamic behavior of confined mesoscopic fluid films displays striking properties [1-3] that often departs from classical Newtonian liquids. The relevance of confined mesoscopic fluid studies extend to technological areas like adhesion [4, 5], wetting processes [6], and interfacial friction phenomena [7]. Understanding the dynamic response of confined fluids can shed light on the dynamics of proteins' folding/unfolding processes [8]; given the fact that proteins are often characterized by surfaces comprised of extended nonpolar regions, the aggregation and subsequent removal of water between these hydrophobic surfaces are believed to be central to their self-organization [9,10].

Current methods widely used to characterize confined fluids include the surface force apparatus [11] and atomic force microscopy (AFM) [1], in which the fluid is basically sandwiched between two solid boundaries (referred here as the probe and the substrate) each attached to mechanical springs. In these traditional methods the fluid's properties are inferred from the springs reactions [1,11]. In contrast our group has recently introduced a new method, shear-force acoustic near-field microscopy (SANM) [12], where the fluid under shear is independently monitored by its acoustic emission. SANM uses *a*) a laterally oscillating tapered probe (attached to a quartz tuning fork) as one of the trapping boundaries, and *b*) an acoustic sensor in contact with the other flat-substrate boundary to—independent from the probe—monitor the fluid's dynamic response to the shear interactions (see Fig. 1a).

The ability to detect acoustic waves even when the probe is not in mechanical contact with the substrate (*i.e.* the sound is engendered at the mesoscopic fluid and not the result of the probe tapping onto the substrate) constitutes one of the main features of the new technique. SANM brings then a new sensing mechanism—near-field acoustic detection—to help characterizing confined fluids under shear. Herein we emphasize on SANM instrumentation improvements, namely evaluation of potential probe damage during test, and incorporating frequency modulation detection method. The latter would allow discerning elastic from inelastic components of the probe-fluid interactions.

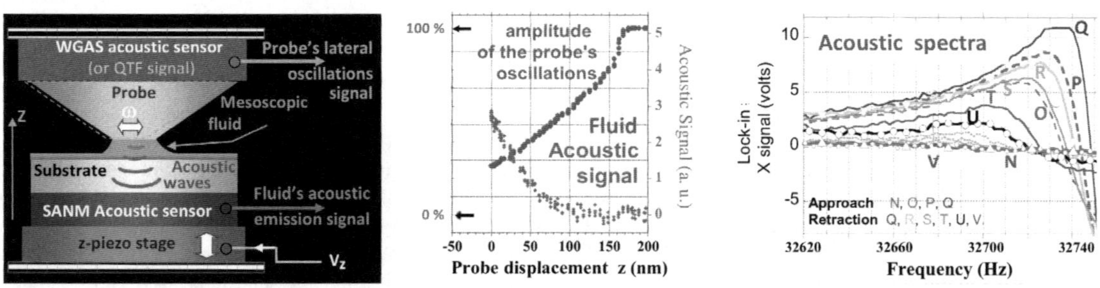

Figure 1. Left: SANM setting for monitoring the acoustic response from a confined mesoscopic fluid subjected to shear by a laterally oscillating (at 32 kHz) nanoprobe. The probe is held at a fixed vertical position; the substrate and the acoustic sensor rest on a z-piezo stage (MCL Nano-Z50HS). A V_z voltage controls the probe-substrate separation distance. **Center:** Acoustic emission from the trapped fluid (detected in the near-field by an acoustic transducer) at different probe-substrate distances (pink trace), while the probe is driven at constant frequency (32 kHz). The probe's lateral oscillations is also, concurrently and independently, monitored (green trace). The zero coordinate of the horizontal axis is arbitrarily chosen [12]. **Right:** Frequency sweep traces (x-signal of synchronous lock-in detection) acquired at different probe-sample distances, during approach (N, O, P, Q) and retraction (Q, R, S, T, U, V) trips of the probe towards (and from) the substrate. Notice the frequency-peaks undergo a "blue" shift as the probe gets closer to the sample.

2. Current SANM status: tests at single driving frequency

SANM uses a sturdy quartz tuning fork (QTF) [13] held by the frame of the microscope. A sharp stylus (a tapered metal wire, or the pyramid of an AFM cantilever) is attached at the bottom of one of the QTF tines (see Fig. 2), altogether constituting the probe.

In a typical test, the probe is electrically driven at its resonance frequency (10 nm amplitude, measured when the tip is far away from the substrate). While keeping this frequency constant, the probe apex is gradually brought into the proximity of the substrate. At a very close probe-sample distance (which varies stochastically, from few nm to tens of nm, depending on the surfaces wetting properties, the environmental humidity, and the temperature) a water meniscus is suddenly formed [14] (figure 1, on the left side); this gives rise to the emission of an acoustic signal, as well as to a decrease in the probe's amplitude. Both signals vary with the probe-sample distance, as well as whether the probe is advancing or retracting from the substrate. Fig 1(on the center) shows a typical retraction trace. Through a linear model of the probe's motion (the QTF tine is typically modelled as a mechanical simple harmonic oscillator) the elastic and damping properties of the fluid are inferred [7,15]. In the new SANM, the fact that the fluid's properties are inferred directly from its acoustic emission—and not indirectly inferred from the probe's motion—shall contribute to a better understanding of these mesoscopic systems.

A shortcoming of measurements performed at constant driving frequency is that a decrease in the probe's oscillation amplitude (as observed when the probe-substrate distance decreases) is actually caused by two factors: the damping forces and the "blue" shift in the probe's resonance frequency. Such a frequency shift is also reflected in the acoustic signal, as seen in Fig. 1(on the right side). Hence, a decrease in the probe's amplitude is not an exclusive signature of a damping force. It would be ideal to discriminate the damping and elastic components of the intervening solid-fluid interactions. One option is to use Phase Lock Loop (PLL) control feedback, which is explained below.

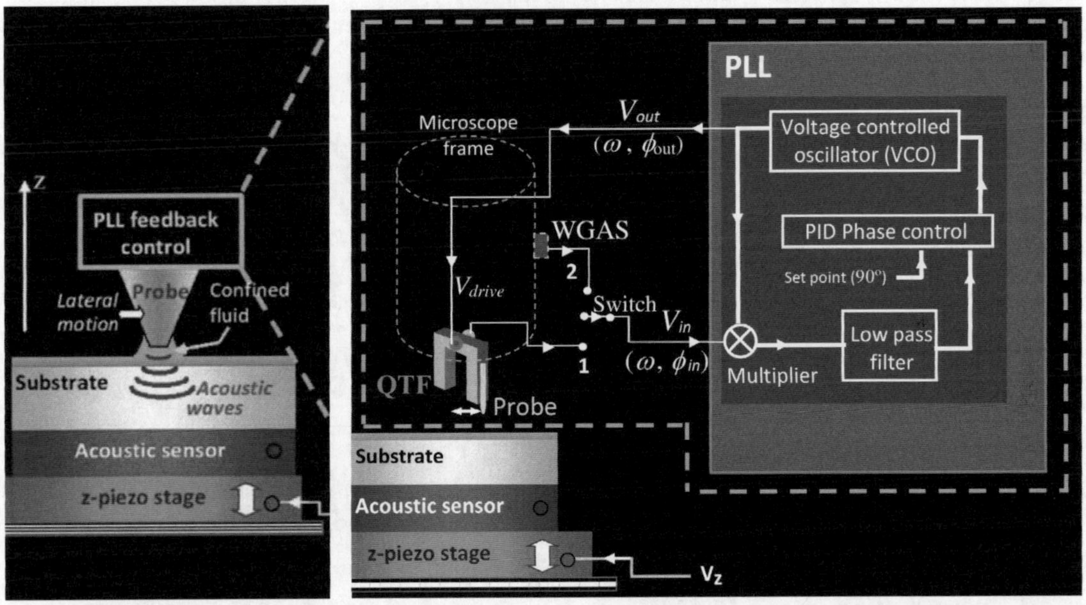

Figure 2. Phase-locked-loop (PLL) feedback-control integrated to the WGAS/SANM system to discriminate the damping from the elastic components of the probe-fluid-substrate interactions. The control signal can be set to be given by either the QTF electrical response (switch at position-1) or to the WGAS signal (switch at position -2 and placing point 1 to ground.) The QTF is attached to the frame of the microscope (dashed lines cylinder). The substrate (attached to an acoustic transducer) rest on a z-piezo scanner, which establishes its vertical position. Once the water bridge meniscus is formed, the probe's resonance frequency changes as the probe-substrate distance decreases; the PPL system responds by driving the QTF (V_{drive}) at its new resonance frequency. Accordingly, during an approach/retraction process, the probe is always oscillating at resonance condition. A decrease in the lateral oscillation amplitude can then be attributed solely to damping component and the frequency shift to elastic component of the probe-fluid-substrate interaction.

IOP Conf. Series: Journal of Physics: Conf. Series **1143** (2018) 012016 doi:10.1088/1742-6596/1143/1/012016

3. Operation with phase lock loop (PLL) control feedback using either WGAS or QTF signal

Figure 2 shows schematically an adaptation of the PLL feedback control [16] built into our WGAS/SANM system. The QTF (with a sharp stylus attached to one of its tines) is electrically driven at its resonance frequency (~ 32 kHz, 10 nm amplitude). At resonance conditions *i)* with the switch in position-1, the QTF electrical sinusoidal signal response is in phase with respect to the sinusoidal driving voltage (V_{drive}). Alternatively *ii)* with the switch in position-2, the whispering-gallery acoustic signal (WGAS) can also be used); WGAS, another acoustic base sensing mechanism recently developed in our lab [17], offers an alternative way to more accurately monitor the QTF mechanical oscillations (the QTF electrical response which not always correlates with the QTF mechanical response [16].)

During a probe approach, once the apex starts interacting with a mesoscopic fluid (assumed viscoelastic), the QTF mechanical response undergoes a phase change because it is now out of resonance. Such a change in phase (detected electrically by the QTF or the WGAS sensor) is utilized by the PLL section of the circuit (sitch in position 1 or position-2), where a voltage controlled oscillator (VCO) responds by adjusting the frequency (ω) of the driving voltage (V_{out}) until the laterally oscillating probe reaches again resonance condition (*i.e.* the phase of the signal at pin #1 or pin #2 is back in phase with the driving voltage) at the new probe-sample distance. This way, *i)* changes in the probe's amplitude can be ascribed to damping effects; and *ii)* the change in the resonance frequency gives a direct measurement of the elastic component of the probe-fluid interaction.

4. Implementation

Fig. 3 shows schematically the WGAS/SANM experimental setup. Subsections of its implementation are described next.

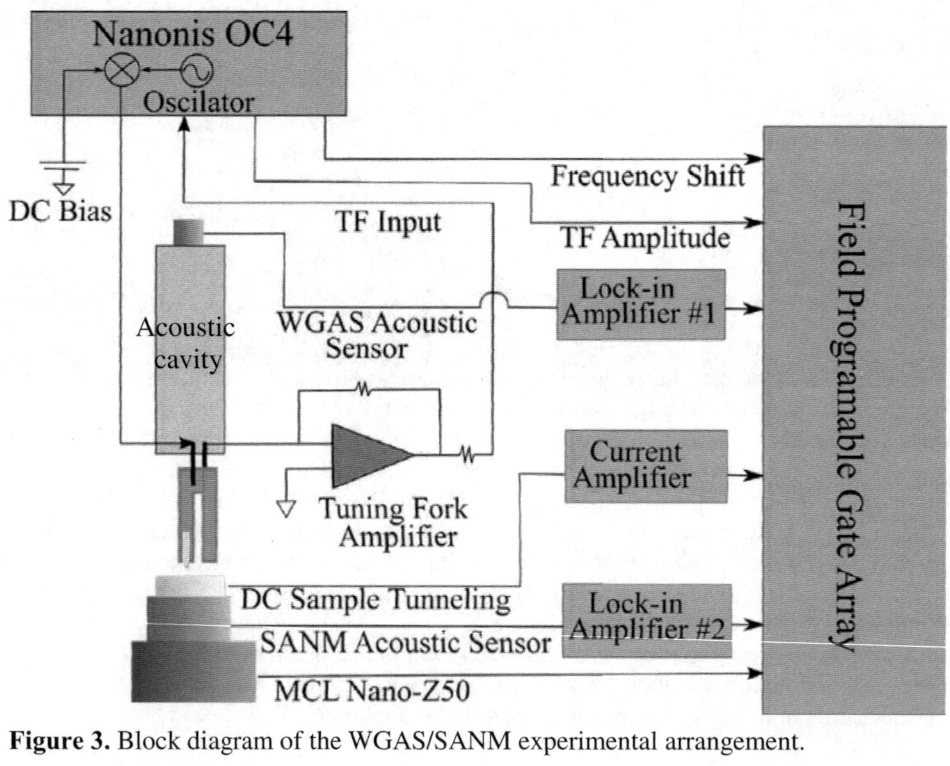

Figure 3. Block diagram of the WGAS/SANM experimental arrangement.

4.1 Piranha cleaned sample

A substrate with hydrophilic characteristics was prepared by cleaving a 1 cm square section of silicon wafer with its native oxide intact. The substrate was first ultrasonically cleaned in a solvent bath of acetone for 15 minutes to remove particulates and oils from the surface. It was then rinsed thoroughly in deionized water before soaking for 30 minutes in a Piranha etch solution. Cleaning procedure with piranha solution is expected to renders OH groups thus giving the surface a hydrophilic character [18]. Piranha is composed of three parts of concentrated sulfuric acid and one part of 9.8M hydrogen peroxide, which removes any remaining organic matter on the surface of the substrate. It is worth noting that great care must be taken when adding the hydrogen peroxide to the sulfuric acid; it must be done drop by drop, as the reaction is extremely exothermic. Finally, the sample was rinsed again with deionized water, and a ~20° water contact angle was verified.

4.2 Fabrication of Gold Tapered Probes

The SANM sensor comprises a tapered probe adhered to a quartz tuning fork (QTF). Alternative to attaching an AFM cantilever (although also used herein), a more economical option is in-lab fabrication of tapered metal wires. As starting material, we used gold wire of 125 μm diameter (99.95% purity). A 4 cm long segment of gold wire was connected to an anode contact, oriented vertically, and positioned in such a way that 5 mm of the wire protrudes below the edge of the plate holder station. An etching solution consisting of one part hydrochloric acid at 37% concentration, and one part ethanol at 95% purity grade was placed in a small beaker below the gold wire. A 3-cm long, 1/8" diameter, graphite rod was used as the cathode. Bubbles are created around the cathode during the etching process; as they rise and break near the surface of the solution, they have a detrimentally effects on the roughness of the probe and, eventually, on the probe's apex. To prevent this adverse effect, a plastic pipette (trimmed into a cylindrical shape) is placed concentrically over the cathode. This cover guides the bubbles up through the inside of the pipette walls, so that they can be vented. To generate the etching current, an ac voltage (2 Volts amplitude, 150 Hz) plus a 2 Volts dc bias was applied between the gold wire and the graphite cathode. The z axis positioner was lowered until approximately 2 mm of the gold wire is submerged, and a 25 mA peak-to-peak current is measured. After approximately 90 seconds, the etching current drops to zero, at which point the z positioner is abruptly raised. The gold wire is then removed and rinsed in acetone bath. Finally, the etched portion is cut from the remaining gold wire yielding a tapered probe approximately 700 μm in length, which is subsequently attached to a quartz tuning fork using minimal amount of epoxy glue.

4.3 Humidity chamber

Depending on the wetting properties of the probe and substrate, as well as the humidity and temperature of the surrounding environment, a water meniscus is suddenly formed when the probe apex is placed in the proximity of the substrate. To increase the probability of water meniscus formation in our measurements, an acrylic chamber was built to enclose the microscope head-stage, and establish relative humidity (RH) above the typical 35% existent in our lab ambient. The RH is established through a combination of *i*) injecting low pressure humidified air, and *ii*) placing inside the chamber a saturated salt solution. A Kestrel 4200 flow tracker monitors the humidity in real time. Low-pressure humidified air was created by first sending compressed air (regulated at 0.5 psi) through a tubing network, then through a humidifier, and finally injecting the vapor stream into the chamber through an open cell foam diffuser. The humidifier comprises a stoppered 1L Erlenmeyer flask containing deionized water and two tubes, one bubbling air through a ceramic stone and the other returning the humidified air to the chamber. On the other hand, a saturated sodium bromide solution was placed in a shallow watch glass (a round, concave glass dish used for evaporation in chemistry) inside the chamber to exploit its fixed-point humidity of 59.1% at 20 °C.

IOP Conf. Series: Journal of Physics: Conf. Series **1143** (2018) 012016 doi:10.1088/1742-6596/1143/1/012016

Employing concurrently both methods allows bringing the chamber to a RH dictated by the selected saturated salt solution.

4.4 Estimation of the probe's oscillation amplitude
Based on previous calculations [12], the QTF have an electro-mechanical coupling constant of 0.4 nA/nm. Upon acquiring a spectral response, the difference ΔI between the current measured at resonance and the one measured well off resonance allows estimating the probe's amplitude. In this experiment, for 70 mV driving voltage one measures ΔI= 23 nA, which gives an estimated value of 9 nm amplitude of oscillation.

5. Experimental results

5.1 SANM operation in frequency modulation mode and using an AFM/QTF probe
For these tests an AFM cantilever was glued to the bottom of one of the tines of a quartz tuning fork (520-TFC3X8-X, Mouser Electronics) of dimensions in length L=3.8 mm, thickness t=0.6 mm, and width w=0.35 mm. Modelling the tine of a QTF as a harmonic oscillator gives an effective spring constant $K = (E/4)w(t/L)^3$ = 26×10^3 N/m, and a piezoelectromechanical coupling constant α= 6.5×10^{-6} C/m. Here we just exploit the pyramid of the AFM cantilever, hence any type of commercially available cantilevers would work fine.

Figure 4 demonstrates the operation of SANM in frequency modulation mode, the latter outlined in Section 3 above. The experiments were carried out at 43% relative humidity. The probe performs an approach/retraction test at 2 nm/s. The sample is a silicon sample covered on top with a monolayer of hydrophilic Poly(2-ethyl-2-oxazoline) (PEOX); 35° contact angle. The probe's amplitude (monitored by the QTF and WGAS signals), the resonance frequency shift (tracked by the PLL feedback electronics), and the acoustic emission from the fluid (SANM signal) are all simultaneously measured. Here we highlight the robustness and integrity of the probe, as reveled by the SEM micrographs Figure 4d. The SEM images of the probe were taken before and after the approach/retraction process; notice the shape of the pyramid appear intact in the images. This shows evidence that, while acquiring the traces displayed in figures 4 a-c, the probe has not been in mechanically contact with the substrate.

Succinctly, figure 4a indicates the resonance frequency shift is of the order of a few Hz. Figures 4a and 4b show the similarity between the electrical-QTF and acoustic-WGAS signals; this encourages to keep improving the sensitivity of the acoustic-based WGAS signal detection (and eventually replace the electrical-QTF detection) given its more direct correlation with the actual mechanical motion of the QTF probe [17]. Figure 4c is quite revealing; it displays a close correlation between the fluid's near-field acoustic emission and the probe's resonance frequency shift (in the region where the acoustic signal monotonically increases during the approach.) Interpretation of acoustic emission as an elastic energy dissipation channel in interfacial interactions is reinforced by these measurements, which should have implications in nanotribology studies [19]. Admittedly, Figure 4c also indicates that such a correlation breaks down at shorter probe-substrate distances, where the SANM signal decreases while the frequency-shift keeps increasing as the probe-substrate distance decreases. A couple of plausible arguments can explain this result: On one hand, the decrease in the acoustic signal may be a consequence that the fluid, trapped between the probe and the substrate and initially resilient to remain stuck to the solid boundaries, gives up and gets quizzed laterally out by the vertically approaching probe. In that scenario, the lower the amount of trapped water, the lower the acoustic emission signal intensity. On the other hand, the fact that frequency shift keeps increasing suggests that the water layers closer to the substrate display a solid-like behavior. As complementary information that support these interpretations, our lab has observed (in other experiment not reported here) that in the region where the acoustic signal experience a peak during the approach the probe's apex has not made mechanical contact with the substrate yet.

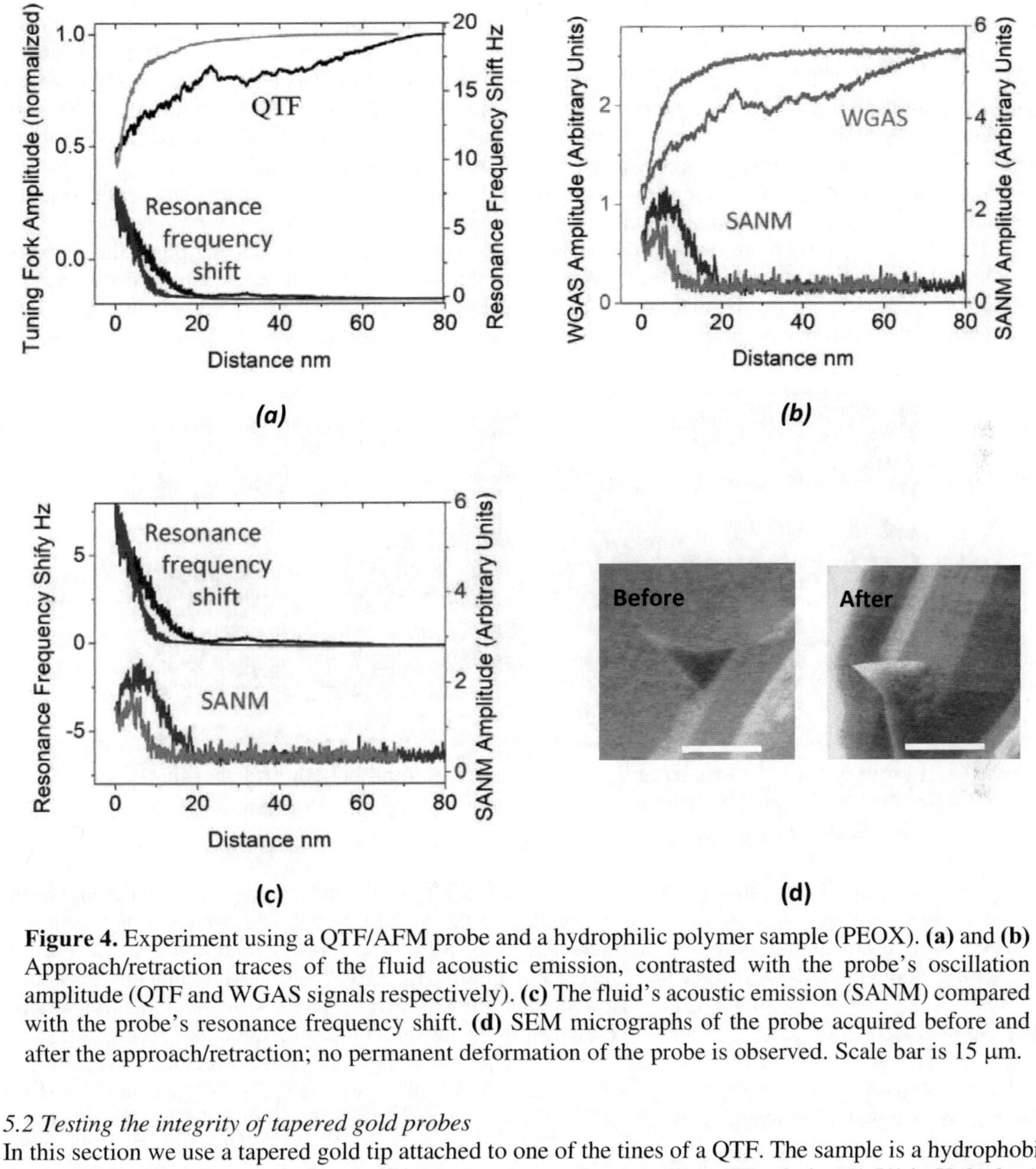

Figure 4. Experiment using a QTF/AFM probe and a hydrophilic polymer sample (PEOX). **(a)** and **(b)** Approach/retraction traces of the fluid acoustic emission, contrasted with the probe's oscillation amplitude (QTF and WGAS signals respectively). **(c)** The fluid's acoustic emission (SANM) compared with the probe's resonance frequency shift. **(d)** SEM micrographs of the probe acquired before and after the approach/retraction; no permanent deformation of the probe is observed. Scale bar is 15 μm.

5.2 Testing the integrity of tapered gold probes

In this section we use a tapered gold tip attached to one of the tines of a QTF. The sample is a hydrophobic silicon wafer, whose native oxide was removed by immersion in a dilute HF solution (1.5% in H_2O) for 15 min at room temperature following Piranha cleaning described above (this latter step is optional). This procedure gives a surface of hydrophobic character; approximately 90° contact angle.

The SANM is operated in PLL modality. When the tip is far away from the surface, the resonance oscillation amplitude is 9 nm. The tripod-like microscope head-stage (which houses the probe) is lowered

IOP Conf. Series: Journal of Physics: Conf. Series **1143** (2018) 012016 doi:10.1088/1742-6596/1143/1/012016

first through a coarse approach procedure, using three fine-thread screws activated with stepper motors; this way the tip is positioned down to approximately 20 μm distant from the sample. Subsequently, a fine approach (at 10 nm/s) is performed by up lifting the sample stage, which is implemented with a piezo stage (MCL Nano-Z50HS). The piezo controller operates in closed loop PLL feedback (to keep the probe in resonance at all times). The approach proceeds until the probe's amplitude (QTF signal) drops below two times the noise level of the detection system. This decrease in amplitude is indicative of interaction forces between the probe and the sample (more likely, between the tip and the incipient formation of a water meniscus); the probe is then immediately retracted until it just breaks away from any water bridge. Subsequently, a renewed fine approach starts but at lower speed (2 nm/s) until the amplitude decreases to a predetermined percent set point (90%, 80%, or 60%); upon reaching the particular set point, the probe is retracted at 2 nm/s. The approach/retraction process, with the same set point, is repeated 3 times, to verify reproducibility. Then the probe is removed for SEM imaging and evaluation of the probe's integrity. The full process is repeated for each selected amplitude decrease set point.

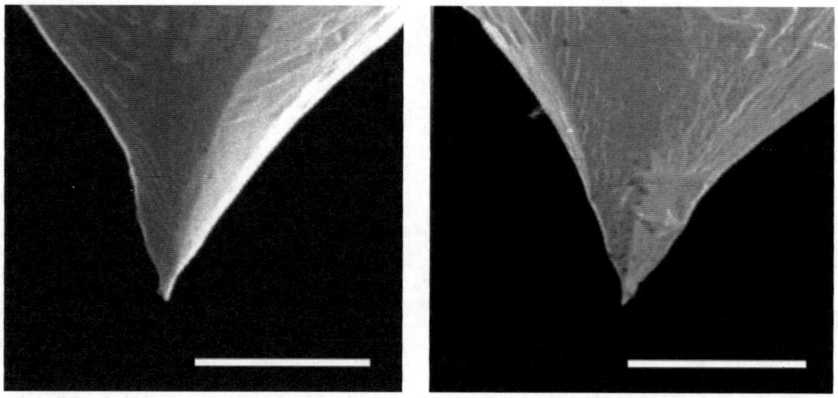

Figure 5. Left: SEM micrographs of a pristine gold probe. **Right:** The same probe after approaching the sample until its resonance amplitude decreases to 90%. Scale bar is 8.5 μm.

Figure 5 (on the left side) shows a SEM micrograph (15 kV accelerating voltage and 10 μA emission current) of a pristine gold probe. The probe is loaded into the SANM system and driven at its resonance frequency with 9 nm amplitude of oscillations. The integrity of the probe's morphology is tested by approaching the tip towards the sample until the QTF amplitude decreases to 90% (following the procedure described in the previous paragraph. *i.e.* fast approach/retraction first, then a slow approach/retraction, while the PLL feedback updates the probe's resonance frequency at all times.) An image of probe after this process is shown in figure 5 (on the right side), which verifies the probe's apex has experienced no damaged.

The procedure followed here (moving the probe back and forth between the SANM stage and the SEM system) is not exempt of accidental crashes of the tip. This eventuality may occur also during the initial fast approach rate (10 nm/s) included in the protocol (for the probe to find the substrate.) This is more problematic when testing hydrophilic samples, since, as seen in other tests in our laboratory, the meniscus formation occurs at shorter distance from hydrophilic substrates (compared to hydrophobic samples). Whatever the origin, we did verify deformation of the tip in one of the tests. But, when proceeding with the tests, we verify again that the tip morphology does not deform (see figure 6) when the probe approaches the sample until its resonant amplitude decreases to 60%.

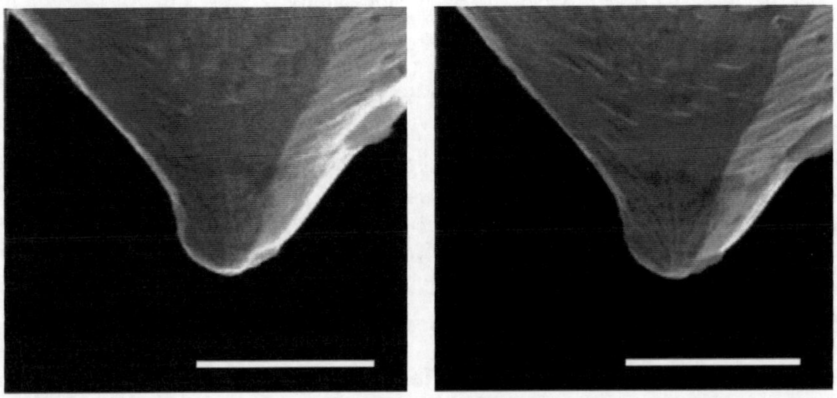

Figure 6. Left: SEM micrographs of a probe before (left) and after (right) performing one approach to the sample until its resonance amplitude decreases to 60%. Scale bar is 8.5 μm, 15 kV accelerating voltage and 10 μA emission current.

6. Conclusions

Frequency modulation (phase-lock-loop feedback) was successfully implemented into Shear-force Acoustic Near-field Microscopy. The ability to discern the elastic and inelastic components intervening in probe-fluid interaction enhances the versatility of the new acoustic technique to study the dynamic behavior of confined fluids under shear. Tests on the integrity of the probe (whether they undergo deformation or not after approaching the sample) suggest that AFM/QTF are very robust. Tests with gold probes needs further tests for a more definite conclusion, still two cases were presented where the probe remained intact after the probe's resonant amplitude decreased down to 90% and 60% of its initial resonance amplitude (the latter measured with the tip placed far away from the substrate). PLL/SANM using AFM/QTF probes revealed that the SANM acoustic signal provides more information than the monotonic-decreasing behavior of the probe's amplitude of lateral oscillations, as the probe approaches the sample. The acoustic signal, first increasing and then decreasing, suggest a following plausible dynamic behavior of the fluid at different stages of confinement: a fluid that is initially resilient to remain adsorbed to the substrate despite the shear motion of the upcoming probe (and thus being able to emit sound), but then being squeezed out of its confinement (at the final stages of the approach) due to the hard compression exerted by the approaching probe (hence, the less amount of water, the lower the intensity of the acoustic emission.) The solid like behavior of the fluid at small probe-sample distance (as revealed in the SANM by the monotonic increase in the probe's resonance frequency shift) offers an opportunity to test the behavior of fluids containing just few adsorbed monolayers.

Acknowledgements

This work was supported in part by Lam Research Corporation, CA USA.

IOP Conf. Series: Journal of Physics: Conf. Series **1143** (2018) 012016 doi:10.1088/1742-6596/1143/1/012016

Bibliography

[1] Khan S H and Hoffmann P M 2015 *Phys. Rev. E* **92** 042403
[2] Granick S 1991 *Science* **253** 1374
[3] Raviv U, Laurat P and Klein J 2001 *Nature* **413** 51
[4] Anjos P H A, Dias E O, Dias L and Miranda J A 2015 *Phys. Rev E* **91** 013003
[5] Squires T M and Quake S R 2005 *Rev. Mod. Phys.* **77** 977
[6] Schäffel D, Koynov K, Vollmer D, Butt H and Schönecke C 2016 *Phys. Rev. Lett.* **116** 134501
[7] Karrai K and Tiemann I 2000 *Phys. Rev. B* **62** 13-174
[8] Li I T S and Walker G C 2012 *Accounts of Chemical Research* **45** 2011
[9] Berne B J, J. D. Weeks J D and Zhou R 2009 *Annu. Rev. Phys. Chem.* **60** 85
[10] Zhou R, Huang X, Margulis C J and Berne B J 2004 *Science* **305** 1605
[11] Hammer M U, Anderson T H, Chaimovich A, Shell M S and Israelachvili J. 2010 *Faraday Discuss.* **146** 299
[12] La Rosa A H, Cui X, McCollum J, Li N and Nordstrom R 2005 *Rev. Sci. Instrum.* **76** 093707
[13] Karrai K and R. D. Grober 1995 *Appl. Phys. Lett.* **66** 1842
[14] Luna M, Colchero J and Baro A M 1998 *Appl. Phys. Lett.* **72** 3461
[15] Lee M, Sung B, Hashemi N and Jhe W 2009 *Faraday Discuss.* **141** 415
[16] Edwards E, Taylor L and Duncan W 1997 *J. Appl. Phys.* ***82***, 980
[17] La Rosa A H, Li N, Fernandez R, Wang X, Nordstrom R and Padigi S K 2011 *Rev. Sci. Instrum.* **82** 093704
[18] Kaya S, Rajan P, Dasari H, Ingram D C, Jadwisienczak W and Rahman F 2015 *ACS Appl. Mater. Interfaces* **7**, 25024
[19] Persson B N J 2000 *Sliding Friction* (Springer)

IOP Conf. Series: Journal of Physics: Conf. Series **1143** (2018) 012017 doi:10.1088/1742-6596/1143/1/012017

Confined Fluid Analyzed with Near-field Acoustic Detection

R Fernandez, T Brockman, J Bai and A H La Rosa

Portland State University, Portland, Oregon 97201, USA

E-mail: andres@pdx.edu

Abstract. Measurement of the damping and elastic interactions between two solids interfaces (one being the apex of a tapered probe that is attached to one tine of a quartz tuning fork while the other is a flat substrate) under relative lateral oscillatory motion are reported. The solid boundaries are separated by a nanometer sized gap, and emphasis is placed on the role played by the mesoscopic fluid trapped in between. The measurements were implemented using two new acoustic techniques that have been integrated into a tuning fork based scanning probe microscope; the whole metrology system offers sub-nanometer precision for controlling the probe's vertical position, as well as thermal drift correction. One of the acoustic techniques, Shear-force Acoustic Near-field Microscopy (SANM), monitors the acoustic emission from the trapped fluid subjected to shear stress, while Whispering Gallery Acoustic Sensing (WGAS) monitors the oscillation amplitude of the QTF. Additionally, the elastic and damping components of the probe-fluid interaction forces are decoupled by operating the SANM in frequency modulation mode, while the eventual contact between the probe and substrate is tracked by monitoring the electrical current signal. Results involving interaction in the 0 to 5 nN regime are reported. The measurements reveal a strong correlation between the monotonic shift of the probe's resonance frequency and the near-field acoustic emission from the trapped fluid. The SANM signal, then, accounts as one of the elastic energy dissipation channels involved in surface interactions at ambient conditions.

1. Introduction

The physical properties of mesoscopic fluids found adsorbed at substrates or confined between solid surfaces differ greatly from those measured in their bulk form[1-3]. For example, the effective shear viscosity is enhanced, viscoelastic relaxation times are prolonged [4], they reveal different phase diagram[5] (boiling at lower temperatures, ice formation at room temperature[6, 7]), the so called hydrophobic interaction manifests at distances well beyond commonly accepted molecular interaction ranges[8], interfacial adhesion[9], molecular transport in nanostructures[10, 11], molecular assembly of particles in liquid water[12, 13] and filtration[14]. But the physics of such a unique dynamic behavior displayed by mesoscopic fluid films still lacks a comprehensive description.

Herein we report the use of a new acoustic metrology tools, Shear-force Acoustic Near-field Microscopy (SANM)[15,16] and Whispering Gallery Acoustic Sensing (WGAS)[17], to characterize the dynamics of fluid films trapped between solid surfaces in relative motion. Figure 1a illustrates the particular physical situation being addressed. When the apex of a laterally-oscillating tapered probe is placed close enough to a flat substrate, a water bridge suddenly forms between the probe and the substrate; the latter occurs at a distance that varies stochastically (from few nanometers to tens of nanometers) depending on the wetting properties of the probe and the substrate, environmental humidity, and temperature. It is of interest to find out the elastic (k_{fluid}) and damping (γ_{fluid}) properties of the trapped fluid for different probe-

IOP Conf. Series: Journal of Physics: Conf. Series **1143** (2018) 012017 doi:10.1088/1742-6596/1143/1/012017

substrate separation distances. We report here a correlation between the fluid's acoustic emission and the frequency shift of the probe's resonance frequency.

2. About the new SANM and WGAS acoustic techniques

The development of SANM and WGAS was initially triggered by an interest for understanding the principles underlying *wear-free friction phenomena,* also referred here as *Interfacial friction.* The use of a tapered probe, as one of the interacting surfaces, is to study the problem at nanoscale dimensions.

i) Could it be that part of the energy is dissipated via acoustic emission (*i.e.* energy is dissipated in the form of sound waves emitted by the confined fluid, which then couples to the substrate underneath)? To address this question we place an acoustic transducer (the SANM acoustic sensor) in intimate contact with the bottom side of the substrate (as shown in figure 1b), to thus listen to the eventual fluid's acoustic emission[15-18]. To set the probe's apex into lateral motion, the probe is glued to one of the tines of a quartz tuning fork (QTF); an ac-voltage sets the tines of the QTF into lateral oscillatory motion. This way, adhesion interactions at the probe-fluid interface causes to set the fluid into shear motion.

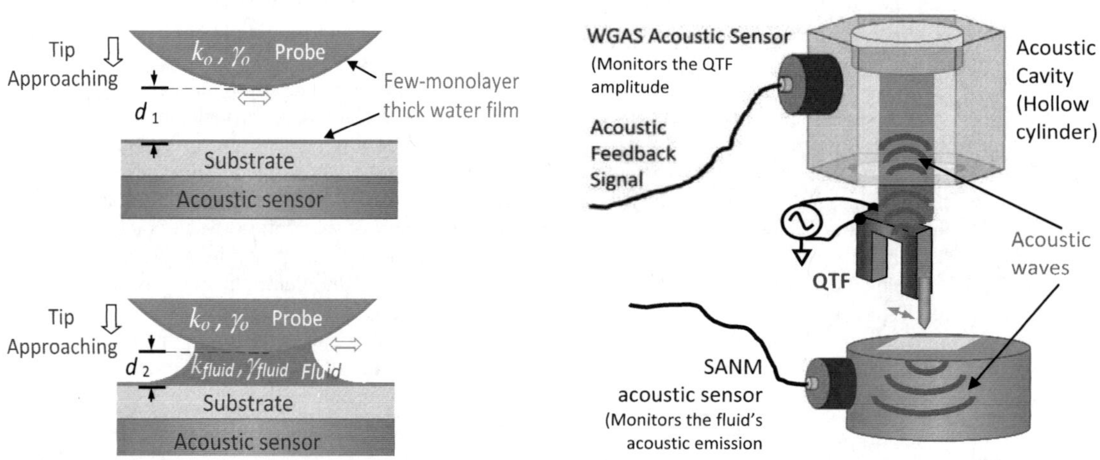

Figure 1. Left: Spontaneous formation of fluid bridge when a laterally oscillating probe approaches the substrate. The confined fluid undergoes shear interaction with the probe. Finding the fluid's viscoelastic properties (k_{fluid}, γ_{fluid}) of this fluid constitutes the subject of our studies. **Right:** The central part shows a probe attached to a QTF, which is electrically driven at its resonance frequency. In the upper part, the WGAS sensor monitors the probe's mechanical response. In the lower part, the SANM sensor monitors the confined fluid acoustic emission.

ii) Shouldn't such a fluid's elastic response (the acoustic emission) have a counterpart reaction on the probe, like a change in its resonance frequency, as well as a decrease in the probe's amplitude of oscillations? Such changes in the probe's amplitude and frequency shift can be addressed with any typical scanning probe microscopy. Here, however, we highlight a new contactless acoustic method (WGAS) to monitor the probe's motion. In effect, without disturbing the probe motion, we simply place an acoustic transducer at a strategic location around the microscope's frame, away from the probe (figure 1b)[17]. The suggested underlying mechanism is that the lateral motion of the tines sets an internal acoustic wave that travels up the central shaft holder (see figure 1b), and subsequently establishes stationary waves in the

IOP Conf. Series: Journal of Physics: Conf. Series **1143** (2018) 012017 doi:10.1088/1742-6596/1143/1/012017

cylindrical frame cavity. An acoustic sensor, strategically placed at one acoustic node on the periphery of the cavity, picks a signal that, it turns out, correlates linearly with the probe's amplitude. The WGAS measurements offer nanometer amplitude sensitivity.

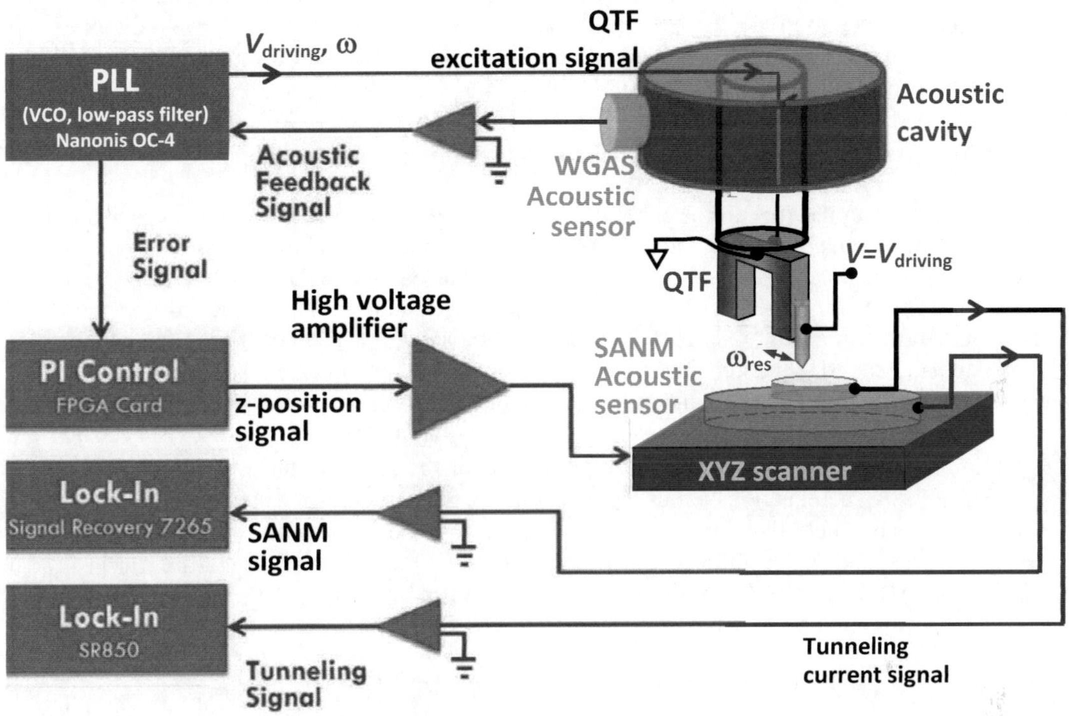

Figure 2. Probe-substrate distance regulation implemented with phase-lock-loop (PLL) frequency modulation. The PLL uses the WGAS as a feedback signal control. This setting allows driving the QTF at resonance at all times, while the probe's apex interacts with the fluid bridge. Several signals are monitored concurrently as the probe approaches the substrate: the probe's amplitude (WGAS), the acoustic emission from the trapped fluid (SANM), the shift in the probe's resonance frequency, and the tunneling current to establish the contact between the probe and the substrate. In the latter, the voltage driving the QTF is also used as bias voltage on the tip to measure the tunneling current.

3. Position control in frequency modulation modality using WGAS as a feedback control signal

Further, frequency modulation detection is also integrated into the SANM/WGAS system (figures 2 and 3) in order to discriminate the damping from the elastic forces components. This is an outgrowth of the single frequency excitation scheme depicted in Fig. 1, where a decrease in the probe's amplitude can be attributed to not only damping effects but also due the fact the resonance frequency of the probe changes as the probe-substrate distance decreases. A frequency modulation strategy overcomes this shortcoming. Succinctly, during a probe's trip approach, once the apex starts interacting with a mesoscopic fluid (assumed viscoelastic), the WGAS sinusoidal output signal (acoustic feedback signal in figure 2) undergoes a phase change because the probe is now out of resonance. Such a change in phase is utilized by the PLL section, where a voltage controlled oscillator (VCO) responds by adjusting the frequency (ω) of the output driving voltage (V_{driving}) until the laterally oscillating probe reaches again the resonance condition (*i.e.* the phase of the WGAS signal is back to its initial phase resonance condition) at the new probe-substrate distance. This way, *i*) changes in the probe's amplitude can be ascribed solely to damping effects; and *ii*) the change in the resonance frequency gives a direct measurement of the elastic component of the probe-fluid interaction. We emphasize the PLL-SANM settings shown in figures 2 and 3 use the WGAS acoustic signal as the control signal. Its selection over the electrical QTF response is that WGAS offers an alternative way to more accurately monitor the QTF mechanical oscillations (the latter not always correlates with the QTF electrical response.)

4. Retrieving viscoelastic properties from changes in resonance frequency shifts and variations in the probe's amplitude.

The use of a quartz tuning fork in the SANM is to capitalize on its very-low energy-dissipation property (one obtains a mechanical Q factor above 2,500 even after mounting a tapered probe), which also allows describing its mechanical response through a simple harmonic oscillator (SHO) model[19,20]. The predicted linear dependence between Q and the QTF oscillation amplitude (while a tip, attached to one of its tines, interact with a mesoscopic fluid at different probe-substrate distances) has been verified[21]. This has impelled using the SHO model also in frequency modulation methods to extract effective values of the fluid's viscoelastic parameters k_{fluid} and γ_{fluid} affecting the TF oscillatory motion[22]. This is obtained by comparing the equation of motion that describes the probe's apex interacting with a viscoelastic fluid layer $M\ddot{x}+ M(\gamma_o+\gamma_{fluid})\dot{x} +(k_o+k_{fluid})x = Fe^{i\omega t}$, with the one when the probe is far away from the substrate $M\ddot{x} +M\gamma_o\dot{x} +k_o x = Fe^{i\omega t}$. The equations are deceptively simple but k_{fluid} and γ_{fluid} can be a complicated function of several parameters, including temperature, pressure, and speed of the probe's motion[23].

In figure 1a, k_o and γ_o constitute, respectively, the elastic spring constant and the (intrinsic) damping constant of the QTF (with the tapered probe attached to one of its tines). Let's denote $x_\infty = x_\infty(t)$ as the lateral probe's position variable when the distance between the probe and sample is infinite,

$$M\,\ddot{x}_\infty + M\gamma_o\,\dot{x}_\infty + k_o\,x_\infty = Fe^{i\omega t} \tag{1}$$

$$x_\infty = A_\infty(\omega)\,e^{i\omega t} \tag{2}$$

$$[\,-M\,\omega^2 + i\omega M\,\gamma_o + k_o\,]\,A_\infty(\omega) = F \tag{3}$$

Using $\omega_{res,\infty}$, the resonance frequency of the probe when the probe-substrate distance is ∞,

$$\omega_{res,\infty}{}^2 \equiv k_o/M \tag{4}$$

one obtains,

$$M\,[\,-\omega^2 + i\omega\,\gamma_o + \omega_{res,\infty}{}^2\,]\,A_\infty(\omega) = F$$

At resonance condition, when $\omega = \omega_{res,\infty}$,

$$M\,[\,i\,\omega_{res,\infty}\;\gamma_o\,]\,A_\infty(\omega_{res,\infty}) = F \tag{5}$$

When the probe is placed near the substrate a meniscus is formed, whose effect on the probe's motion is quantified by an additional damping $\gamma_{fluid,d}$ and elastic $k_{fluid,d}$ parameters. When the probe-substrate separation is *d*, let's denote by x_d the variable monitoring the probe's lateral position,

$$M\,\ddot{x}_d + M\,(\gamma_o + \gamma_{fluid,d})\,\dot{x}_d + (k_o + k_{fluid,d})\,x_d = Fe^{i\omega t} \quad (6)$$

$$x_d = A_d(\omega)\,e^{i\omega t} \quad (7)$$

$$[\,-M\,\omega^2 + i\omega M\,(\gamma_o + \gamma_{fluid,d}) + (k_o + k_{fluid,d})\,]\,A_d(\omega) = F \quad (8)$$

Using $\omega_{res,d}$, the resonance frequency of the probe when the probe-substrate distance is *d*,

$$\omega_{res,d}{}^2 \equiv (k_o + k_{fluid})/M \quad (9)$$

one obtains,

$$M\,[\,-\omega^2 + i\omega\,(\gamma_o + \gamma_{fluid,d}) + \omega_{res,d}{}^2\,]\,A_d(\omega) = F \quad (10)$$

At resonance condition, when $\omega = \omega_{res,d}$

$$M\,i\,\omega_{res,d}\,[\,\gamma_o + \gamma_{fluid,d}]\,A_d(\omega_{res,d}) = F \quad (11)$$

From (4) and (9),

$$\omega_{res,d}{}^2 \equiv \omega_{res,\infty}{}^2 + k_{fluid,d}/M$$

$$k_{fluid,d} = M\,[\,\omega_{res,d}{}^2 - \omega_{res,\infty}{}^2]$$

$$= M\,\omega_{res,\infty}{}^2\,[\frac{\omega_{res,d}{}^2}{\omega_{res,\infty}{}^2} - 1]$$

$$k_{fluid,d} = k_o\,[\frac{\omega_{res,d}{}^2}{\omega_{res,\infty}{}^2} - 1] \quad (12)$$

From (5) and (11),

$$M\,[\,i\,\omega_{res,\infty}\,\gamma_o\,]\,A_\infty(\omega_{res,\infty}) = M\,i\,\omega_{res,d}\,[\,\gamma_o + \gamma_{fluid,d}]\,A_d(\omega_{res,d})$$

$$[\,\frac{\omega_{res,\infty}}{\omega_{res,d}}\,\gamma_o\,]\,\frac{A_\infty(\omega_{res,\infty})}{A_d(\omega_{res,d})} = [\,\gamma_o + \gamma_{fluid,d}]$$

$$\gamma_{fluid,d} = \gamma_o[\,\frac{\omega_{res,\infty}}{\omega_{res,d}}\,\frac{A_\infty(\omega_{res,\infty})}{A_d(\omega_{res,d})} - 1] \quad (13)$$

Alternatively, in terms of the *Q* factor measured when the probe is far away from the substrate,

$$\gamma_{fluid,d} = \frac{\omega_{res,\infty}}{Q}[\,\frac{\omega_{res,\infty}}{\omega_{res,d}}\,\frac{A_\infty(\omega_{res,\infty})}{A_d(\omega_{res,d})} - 1]$$

$$= \frac{k_o}{MQ\,\omega_{res,\infty}}[\,\frac{\omega_{res,\infty}}{\omega_{res,d}}\,\frac{A_\infty(\omega_{res,\infty})}{A_d(\omega_{res,d})} - 1] \quad (14)$$

Expressions (12) and (13) gives the elastic and damping constants of the fluid in terms of,

$$A_d(\omega_{res,d}) \qquad \textit{the probe's resonant amplitude,} \tag{15}$$

and

$$\Delta\omega = (\omega_{res,d} - \omega_{res,\infty}) \qquad \textit{the probe's frequency shift,} \tag{16}$$

measured at different probe-substrate distance *d*.

From expression (8), the contribution from the fluid to the elastic force acting on the probe, measured at resonance condition, is given by,

$$F_{elast} = k_{fluid,d}\, A_d(\omega_{res,d})$$

$$F_{elast} = k_o[\frac{\omega_{res,d}^{\ 2}}{\omega_{res,\infty}^{\ 2}} - 1]\, A_d(\omega_{res,d}) \tag{17}$$

From expression (8), the contribution from the fluid to the damping force acting on the probe, measured at resonance condition, is given by,

$$F_{damp} = i\omega_{res,d}\, M\,\gamma_{fluid,d}\, A_d(\omega_{res,d}) \tag{18}$$

Alternatively, using (14),

$$F_{damp} = i\omega_{res,d}\,\frac{k_o}{Q\,\omega_{res,\infty}}[\frac{\omega_{res,\infty}}{\omega_{res,d}}\,\frac{A_\infty(\omega_{res,\infty})}{A_d(\omega_{res,d})} - 1]\, A_d(\omega_{res,d})$$

$$F_{damp} = i\,\frac{k_o}{Q}\,\frac{\omega_{res,d}}{\omega_{res,\infty}}\,[\frac{\omega_{res,\infty}}{\omega_{res,d}}\,\frac{A_\infty(\omega_{res,\infty})}{A_d(\omega_{res,d})} - 1]\, A_d(\omega_{res,d}) \tag{19}$$

When measuring the full width at half maximum $\Delta\omega$ from a $A_\infty^2(\omega)$ vs ω spectrum, the mechanical quality factor is estimated to be $Q = \omega_{res,\infty}/\Delta\omega$. If the full width at half maximum $(\Delta\omega)'$ is measured from a $A_\infty(\omega)$ vs ω spectrum, then, it turns out,

$$Q = \sqrt{3}\,\frac{\omega_{res,\infty}}{(\Delta\omega)'} \tag{20}$$

Since in the definition of the quality factor, $Q=\omega_{res,\infty}/\Delta\omega$, the value of $\Delta\omega$ turns out to be equal to γ_o, then expression (2) can be written as $F= M\,[\,i\,\omega_{res,\infty}^{\ 2}/Q\,]\,A_\infty(\omega_{res,\infty})$, or $F=ik_o/Q\,A_\infty(\omega_{res,\infty})$. This expression allows estimating then the magnitude of the driving force,

$$F= \frac{k_o}{Q}\,|A_\infty(\omega_{res,\infty})| \tag{21}$$

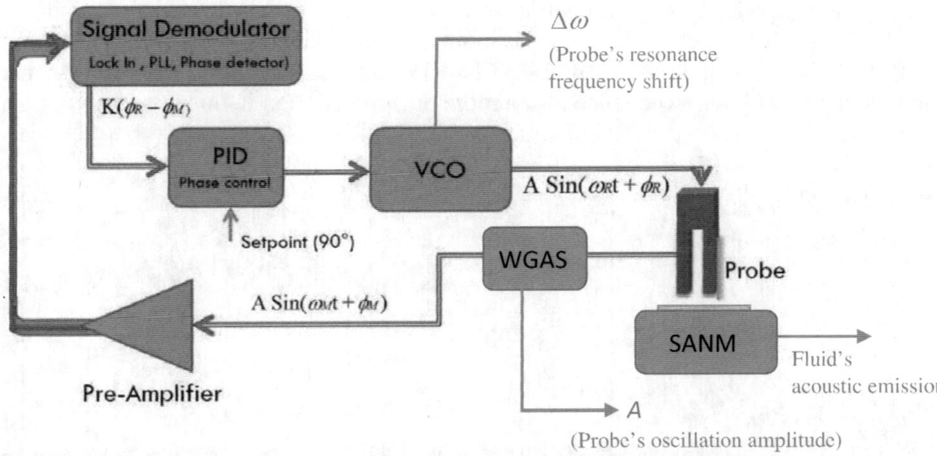

Figure 3. Implementation of frequency modulation to discriminate the elastic (via Δf detection) and damping component (via amplitude A detection) of the probe-fluid interactions. The amplitude and phase of the probe's motion are monitored by the WGAS system. The PLL demodulates the phase information, which is compared with set point of a PID phase control. Any deviation of the probe's phase from 90° triggers the VCO response to drive the QTF at different frequency until resonance condition is re-established.

5. Testing probe-fluid interactions in the low interaction forces regime (up to 5 nN)

A highly ordered pyrolitic graphite (HOPG) sample (7 mm X 7 mm x 1 mm SPI-2 grade, of resistivity values $\rho_{\perp}$ = 0.5 ohm-cm and $\rho_{//}$ = 0.5 x10^{-3} ohm-cm, from SPI Supplies, West Chester, PA) was secured with carbon conductive tape to the sample stage of the SANM. The sample was then cleaved with scotch tape. The QTF (with the probe attached) displayed a mechanical quality factor of Q=5,700; estimated oscillating amplitude of 26 nm (using a driving voltage V_{driving}=10 mV rms, see Fig. 2); resonance frequency $f_{res,\infty}$ = 31804 Hz; 117 nN driving force, as calculated from equation (21). The experiments were performed at 35 % relative humidity and 26 °C temperature.

The results reported here correspond to one of the first approaches of the probe towards the sample. On purpose, the approach stopped once the elastic interaction reached 5 nN, just to ensure an interaction of the probe with just the mesoscopic fluid (and aiming to avoid potential crash with the substrate). This implied, as displayed by the results, a decrease of the probe amplitude by 10% of its initial maximum amplitude. From previous results in our lab, the expected SANM signal at this shallow interaction condition should be very small (sometime undetected); hence the reason for driving the signal with a large amplitude (26 nm) larger than the usual 4 nm to 10 nm range used in previous experiments.

Figure 4 compares the WGAS signal (probe's amplitude of oscillations, expressed in percentage values), the normalized acoustic SANM magnitude (the normalization factor is the probe's amplitude of oscillations,) the acoustic SANM phase, and the damping force exerted by the fluid on the probe (calculated from expression (19),) all simultaneously obtained as the probe approaches the sample. The sudden small decrease in the probe's amplitude at 100 Å in Fig. 4 suggests the incipient (precursor) formation of finger-like bridges between the asperities on probe's apex and the sample [24]; the latter constituting also a vehicle for establishing a current (as corroborated in the right side trace in Fig. 5.) No acoustic signal is detected at this stage of the approach, probably because a fully coalesced meniscus is not formed yet. This situation

IOP Conf. Series: Journal of Physics: Conf. Series **1143** (2018) 012017 doi:10.1088/1742-6596/1143/1/012017

remains until the apex reaches z = 35 Å. Upon further approach, the acoustic SANM starts to change more significantly. This is interpreted as the transition from an incipient to a full coalescence of a meniscus between the probe and the sample [25]; hence a more coherent acoustic signal is engendered. This is consistent with the more sustained value of the SANM phase signal recorded for z<35 Å. The probe's amplitude and the damping force experience also a more drastic variation in this last stage of approach.

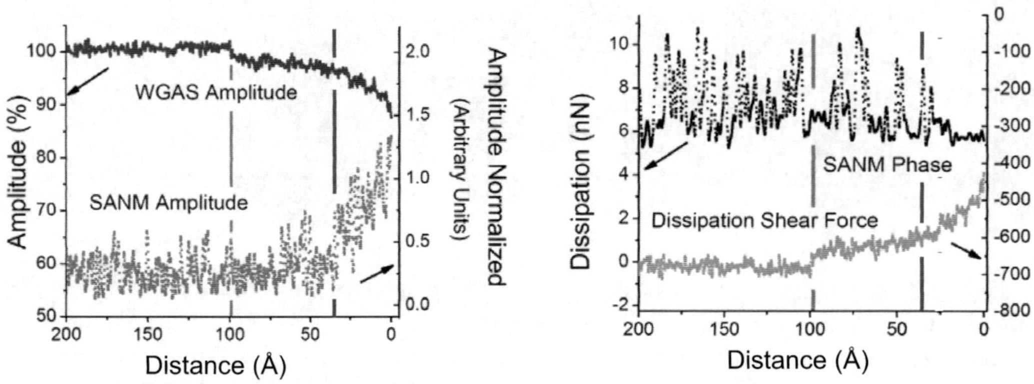

Figure 4 Approach traces. **Left**: WGAS signal (probe's amplitude of oscillations) and SANM amplitude (near-field acoustic emission from the meniscus fluid.) **Right**: Phase of the acoustic SANM signal (a consistent value is obtained only for z < 35 Å) and the damping force acting on the probe. The vertical lines separate regions of no change, small change, and drastic changes in the signals.

Figure 5 shows variations in the current flowing through the water bridge formed between the probe's apex and the substrate. The dashed lines along the contact current traces are to suggest that a different slope characterizes the trace. The greater slope for z < 35 Å is consistent with a denser (coalesced) meniscus bridge, compared to its incipient formation for z > 35 Å. The elastic force exerted by the fluid on the probe (as calculated from expression (17) also increases significantly for z < 35 Å. The traces on the right side show that the probe experiences an increase in its resonance frequency as it approaches the sample, which correlated very well with the elastic force exerted by the fluid on the probe.

Since we associate the generation of an acoustic wave to a non-damping energy dissipation channel of energy involved in the probe-fluid interaction, it is natural then to compare the behavior between the SANM signal and the frequency shift signal, since they correlate with the elastic force measured in these experiments. This is shown in figure 6. The near-field acoustic emission from the fluid, the SANM signal, correlates well with frequency shift experienced by the probe, and the elastic force exerted by the fluid on the probe. Our results support then near-field acoustic emission as one of the elastic energy dissipation channels involved in surface interactions.

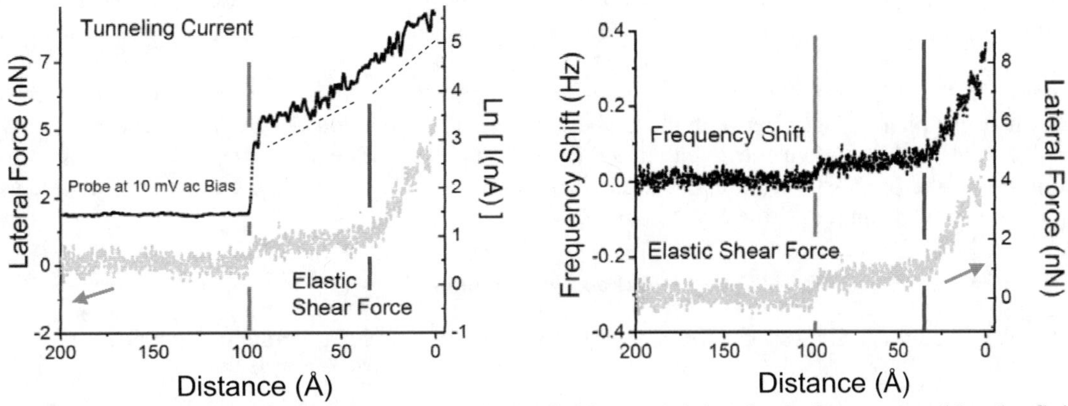

Fig. 5 Approach traces. **Left**: Current through the fluid gap and the elastic force exerted by the fluid on the probe. **Right**: Increase of the probe's resonance frequency compared with the elastic force.

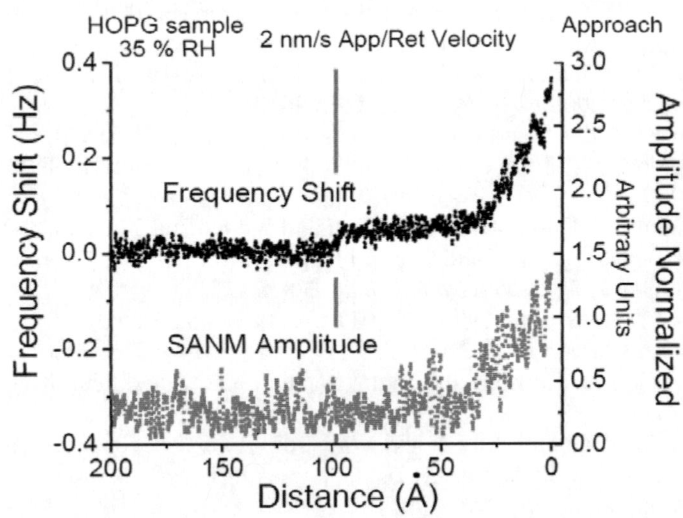

Fig. 6 Traces display a direct correlation between the near-field acoustic emission from the meniscus fluid (SANM signal) and the shift experienced by the probe's resonance frequency, as the probe approaches the sample.

6. Conclusions

Two near-field acoustic techniques, SANM and WGAS were used to characterize the fluid trapped between a sharp probe and a flat solid surfaces at ambient conditions. The lateral motion of the probe, at about 32 kHz and 26 nm amplitude, sets the fluid under shear motion and, as a consequence, causes acoustic emission. SANM is sensitive enough to detect such acoustic waves in the near field region, while WGAS monitors the probe's lateral oscillations amplitude with nanometer-size sensitivity. Implementation of the SANM in frequency modulation modality allowed to discriminate the damping and the elastic component of the probe-fluid interaction. One of the main findings constitutes the correlation between the SANM acoustic emission from the fluid and the shift in the probe's resonance frequency. The variations of the multiple signals acquired simultaneously (contact current, damping lateral force, elastic lateral force, frequency shift) as the probe approaches the sample were interpreted in term of *i*) first an stochastic formation of incipient bridges between the probe and the substrate, followed by *ii*) the coalescence of a water meniscus, whose effects (on the probe's amplitude and resonance frequency shift are more pronounced as the probe-substrate distance decreases. Overall, the new SANM accounts for an elastic energy dissipation channel in surface interactions.

Acknowledgements

AHLR acknowledges partial support for this work from Lam Research Corporation, CA USA.

Bibliography

[1] Granick S 1991 *Science* **253** 1374
[2] Klein J and Kumacheva E 1995 *Science* **269** 816
[3] Ortiz-Young D, Chiu H -C, Kim S, Voïtchovsky K and Riedo E 2013 *Nat. Commun.* **4** 2482
[4] Jinesh K and Frenken J 2006 *Phys. Rev. Lett.* **96** 166103
[5] Nosonovsky M and Bhushan B 2008 *Phys. Chem. Chem. Phys.* **10** 2137
[6] Jinesh K and Frenken J 2008 *Phys. Rev. Lett.* **101** 036101
[7] Choi E -M, Yoon Y -H, Lee S and Kang H 2005 *Phys. Rev. Lett.* **95** 085701
[8] Hammer M U, Anderson T H, Chaimovich A, Shell M S and J. Israelachvili 2010 *Faraday Discuss.* **146** 299
[9] Voïtchovsky K, Kuna J J, Contera S A, E. Tosatti and F. Stellacci 2010 *Nat. Nanotechnol.* **5** 401
[10] Chiavazzo E, Fasano M, Asinari P and Decuzzi P 2014 *Nat. Commun.* **5** 4565
[11] Duan C and Majumdar A 2010 *Nat. Nanotechnol.* **5** 848
[12] Ahmad M, WGu W, Geyer T and Helms V 2011 *Nat. Commun.* **2** 261
[13] Lv W and Wu R 2013 *Nanoscale* **5** 2765
[14] Huang H, Song Z, Wei N, Shi L, Mao Y, Ying Y, Sun L, Xu Z and Peng X 2013 *Nat. Commun.* **4** 2979
[15] La Rosa A H, Cui X, McCollum J, Li N and Nordstrom R 2005 *Rev. Sci. Instrum.* **76** 093707
[16] Fernandez R, Wang X, and La Rosa A H 2011 *Proceedings of the Nanotechnology (IEEE-NANO) 11th IEEE International Conference* 903.
[17] La Rosa A H, Li N, Fernandez R, Wang X, Nordstrom R and Padigi S K 2011 *Rev. Sci. Instrum.* **82** 093704
[18] Wang X, Fernandez R, Li N, Hung H -C, Venkataraman A, Nordstrom R and La Rosa A H 2016 *Physics of Fluids* **28** 052001.
[19] Karrai K, Grober R D Near Field Optics 1995 in: Paesler M A and Moyer P J (Ed.) 1995 *SPIE Proceedings Series* **2535** Bellingham, WA, pp. 69–81.
[20] Grober R D, Acimovic J, Schuck J, Hessman D, Kindlemann P J, Hespanha J, Morse A S, Karrai K, Tiemann I and Manus S 2000 *Rev. Sci. Instrum.* **71** 2776
[21] Karrai K and Grober R D 1995 *Appl. Phys. Lett.* **66** 1842
[22] An S, Kim J, Lee K, Kim B and Lee M 2012 *Appl. Phys. Lett.* **101** 053114
[23] Karrai K and Tiemann I 2000 *Phys. Rev. B* **62** 13 174-181
[24] Barel I, Filippov A E and MUrbakh M 2012 *J. Chem. Phys.* **137** 164706
[25] Luna M, Colchero J and Baro A M 1998 *Appl. Phys. Lett.* **72** 3461

IOP Conf. Series: Journal of Physics: Conf. Series **1143** (2018) 012018 doi:10.1088/1742-6596/1143/1/012018

Synthesis of SnO_2 quantum dots by chemical precipitation assisted by ultrasound

P Alvarado[1], D Acosta[2], M M Gomez[1] and J L Solis[1]

[1]Faculty of Sciences, Universidad Nacional de Ingeniería, Lima 25, Peru
[2]Thin Film Laboratory, Instituto de Física, UNAM, Circuito de la Investigación Científica, Ciudad Universitaria, C.P. 04510, Mexico

E-mail: jsolis@uni.edu.pe

Abstract. SnO_2 quantum dots were successfully prepared via chemical precipitation assisted by ultrasound. X-ray diffraction analysis confirmed the rutile tetragonal structure of SnO_2 QDs with a crystallite size of 1.4 nm. The SEM micrographs of the obtained powder exhibited that the SnO_2 QDs are agglomerated particles of around 10 nm, while TEM images confirmed that the crystallite size are similar to the measured from XRD. The band-gap energy of the SnO_2 QDs measured from the UV–Vis reflectance spectrum of the SnO_2 QDs was 4.3 eV, displaying a significantly blue-shifting attributed to quantum confinement.

1. Introduction

Tin dioxide is an n-type semiconductor with band gap energy, E_g, of 3.6 eV and its calculated exciton Bohr radius a_B is approximately 2.7 nm [1], thus the quantum dots (QD) of SnO_2 have smaller sizes than 2.7 nm. Recently, there has been growing interest in the properties of wide-band gap oxide QD SnO_2 semiconductors, because it has outstanding applications in various fields, as electrochemical Cd sensor [2], gas sensor [3], anode for sodium ion battery [4], and photocatalyst [5].

There are different physical and chemical methods that have been used to synthetize SnO_2 QDs as chemical precipitation [6], laser ablation [7], sonochemistry [8], hydrothermal [9], and microwave heating method [10]. Hence, synthesis of crystalline SnO_2 QDs by a fast and facile method is a challenging job. Herein, we develop a synthesis of SnO_2 QDs using chemical precipitation assisted by ultrasound in a heated ultrasonic cleaning bath commonly used in a chemical lab.

2. Experimental

All chemicals were of analytical grade and used without any further purification. A 25 ml of aqueous solution of 0.13 M of tin tetrachloride pentahydrate, $SnCl_4 \cdot 5H_2O$ (Aldrich), was prepared and then placed in an ultrasonic bath (Branson Model MT 1510) at 42 kHz and 75W and heated to 65 °C. After the solution was warm, then 4.5 ml of ammonia solution (Merck, 25%) was then added drop by drop to the above solution while sonicating, so that the final pH value of the solution was 9.5. The irradiation time was 150 min. The formation of a white tin oxide nanosol, presumably via the chemical reaction:

$$SnCl_4.5H_2O_{(ac)} + 4NH_4OH_{(ac)} \xrightarrow{\Delta T=65°/Ultrasound} SnO_2\downarrow + 4(NH_4Cl)_{(ac)} + 7H_2O_{(ac)} \quad (1)$$

The nanosols were precipitated by centrifugation (Eppendorf Centrifuge 5810R) and washed several times using distilled water. Finally, the resultant white solid was dried at 70 °C for 24 h. The structure and domain size of the nanoparticles were determined by X-ray diffraction using a Rigaku Miniflex II Desktop diffractometer operating with CuKα radiation (0.15045 nm wavelength) at 30 kV and 20 mA with a scan speed of 3° min^{-1}. The X-ray diffraction data were subjected to a general convolution process (Topas-Academic) allowing, in principle, any combination of appropriate functions to be employed for modeling the whole powder profile. These functions can represent the aberrations of the diffractometer as well as various contributions from the specimens; the technique is known as a 'fundamental parameters approach' [11]. It was used together with the Rietveld refinement method [12, 13] and gave the crystallite domain size of the SnO_2 QDs. The morphology of the nanoparticles was investigated by Field Scanning Electron Microscopy (FESEM) using a Zeiss ULTRA plus instrument. A JEOL FEG 2010-F Transmission Electron Microscope operated at 200 kV was used for recording the Bright Field (BF) and High Resolution Electron Microscopy (HREM) micrographs of the samples. UV–vis Diffuse Reflectance Spectroscopy (DRS) was measured by UV visible spectrometer (Perkin Elmer Lambda 750 double beam) equipped with 60 mm integrating sphere and spectralon was used as a reference. The powders were filled in a hole (10 mm in diameter and 3 mm deep) of a sample holder, and the surface was smoothed. The layer can be regarded as infinitely thick, as required by the Kubelka-Munk theory.

3. Results and Discussion

Tin oxide has a tetragonal structure with the space group *P4₂/mnm* (136) and lattice parameter of a=4.737Å and c=3.185 Å [14]. Sn and O ions are located at (0, 0, 0) and (0.307, 0.307, 0), respectively. Figure 1 shows that all diffraction peaks can be indexed as belonging to the tin oxide phase. Domain sizes were determined from the broadening of the x-ray diffraction lines by assuming a Voigt function [15]. The average domain size for sample of SnO_2 QDs synthesized assisted by ultrasound radiation was found to be approximately 1.4 nm.

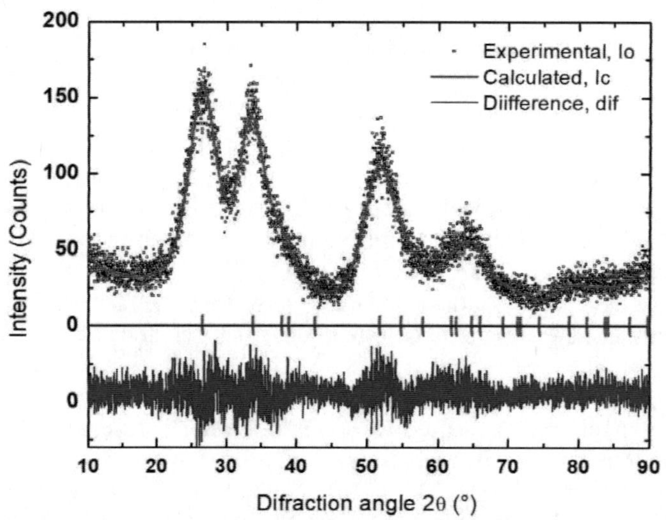

Figure 1. XRD pattern of SnO_2 QDs (squares). Fitted pattern is shown as solid curves and tick marks indicate expected position of diffraction peaks for a modeled tetragonal SnO_2 structure. The lower part of the panel shows the difference plot between the experimental and fitted patterns.

The sample dried was a translucent rock like appearance and the morphology of the SnO_2 QDs was studied by FESEM, and Figure 2(a) shows a typical micrograph at low magnification for the obtained sample, it looks a stone with a smooth surface. But at high magnification shown in Fig. 2(b) indicates the presence of agglomerated particles with a narrow size distribution and a particle size of around 10 nm.

IOP Conf. Series: Journal of Physics: Conf. Series **1143** (2018) 012018 doi:10.1088/1742-6596/1143/1/012018

a)

Figure 2. FESEM images at low (a) and high (b) magnification of SnO_2 QDs synthesized assisted by ultrasound irradiation for 150 min. Scale bars indicate magnifications.

Fig. 3 shows HRTEM images of the SnO_2 QDs, they are slightly agglomerated, and clearly show a single crystalline structure with a crystallite size of around 2 nm, which agrees with XRD measurement.

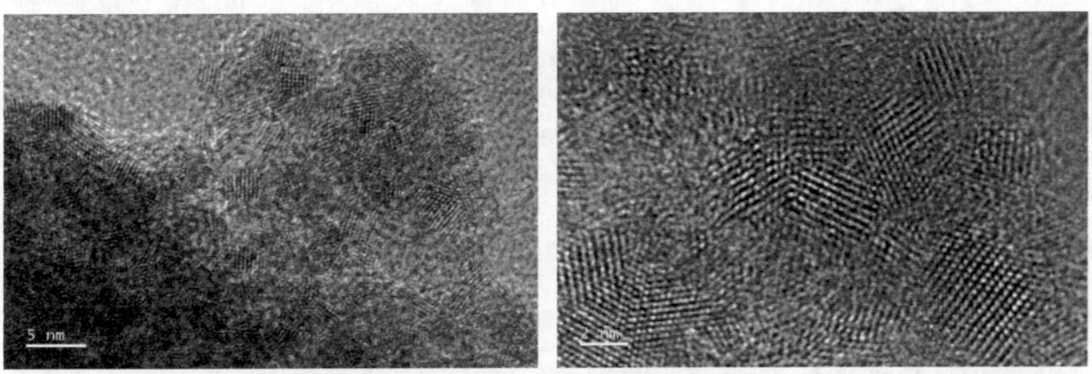

Figure 3. HRTEM images of SnO_2 QDs.

SnO_2 exhibited a direct band gap [16]. Accurate E_g estimation for studied SnO_2 QDs was obtained by using the equation for a direct transition [17]:

$$\alpha(h\nu) \approx B\left(h\nu - E_g\right)^{1/2} \quad (2)$$

A UV–Vis diffuse reflectance spectrum of the obtained SnO_2 QDs powder is shown in Fig. 4a. The band gap of SnO_2 QDs may be estimated from the Kubelka–Munk function *F(R)* which is proportional to the absorption coefficient (α), which is estimated by the following equation [18]:

$$F(R) = \frac{(1-R)^2}{2R} = \frac{K}{S} \quad (3)$$

where R denotes the reflectance. Figure 4b shows the modified Tauc plot of $(F(R)\,h\nu)^2$ vs photon energy ($h\nu$) for SnO_2 QDs. Extrapolation of this line to the photon energy axis yields the band gap as shown in Fig. 4b. The band gap of the SnO_2 QDs is about 4.3 eV. The band gap of semiconductor, with crystallite size smaller than the exciton Bohr radius, increases due to quantum confinement [1].

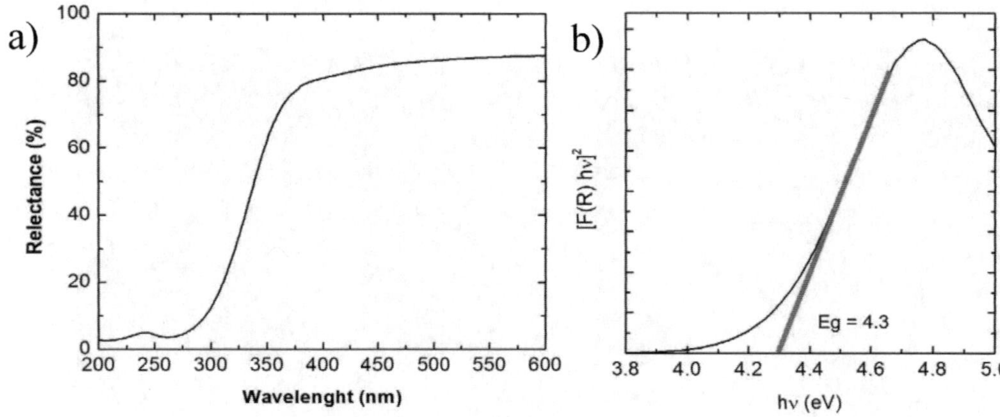

Figure 4. a) UV–Vis diffuse reflectance spectrum of SnO_2 QDs obtained by chemical precipitation assisted by ultrasound. b) Modified Tauc plot: $(F(R)\ h\nu)^2$ vs photon energy ($h\nu$) for SnO_2 QDs.

4. Conclusions

The method of the synthesis for the SnO_2 QDs is straightforward, inexpensive and consumes less material than other methods that eliminate chemicals to the environment. X-ray diffraction analysis confirmed the rutile tetragonal structure of SnO_2 QDs with a crystallite size of 1.4 nm. The SEM micrographs of the obtained powder exhibited that the SnO_2 QDs are agglomerated of particles of around 10 nm, while TEM images confirmed that the crystallite size are similar to the measured from XRD. The band-gap energy of the SnO_2 QDs measured from the UV–Vis reflectance spectrum of the SnO_2 QDs is 4.3 eV, so was significantly blue-shifted due to quantum confinement.

Acknowledgments

The authors wish to thank the General Research Institute of the National University of Engineering (IGI-UNI) for a financial support.

References

[1] Lee E, Ribeiro C, Giraldi T, Longo E, Leite E and Varela J 2004 *Appl. Phys. Lett.* **84** 1745.
[2] Bhanjana G, Dilbaghi N, Kumar R, Umar A. and Kumar S 2015 *Electrochimica Acta* **169** 97.
[3] Sedghi S, Mortazavi Y and Khodadadi A 2010 *Sensors and Actuators B* **145** 7.
[4] Wang Y, Jin Y, Zhao C, Duan Y, He X and Jia M 2017 *Materials Letters* **191** 169.
[5] Bhattacharjee A and Ahmaruzzaman V 2015 *Materials Letters* **139** 418.
[6] He Y, Tang P, Li J, Zhang J, Fan F and Li D 2016 *Materials Letters* **165** 50.
[7] Singh M, Mathpal M and Agarwal A 2012 *Chem. Phys. Lett.* **536** 87.
[8] Pang G, Chen S, Koltypin Y, Zaban A, Feng S and Gedanken A 2001 *Nano Lett.* **1** 723.
[9] Xu X, Zhuang J and Wang X 2008 *J. Am. Chem. Soc.* **130** 12527.
[10] Xiao L, Shen H, Hagen R, Pan J, Belkoura L and Mathur S 2010 *Chem. Commun.* **46** 6509.
[11] Cheary R and Coelho A 1992 *J. Appl. Crystallogr.* **25** 109.
[12] Rietveld H M 1967 *Acta Crystallogr.* **22** 151.
[13] Rietveld H M 1969 *J. Appl. Crystallogr.* **2** 65.
[14] Baur V 1956 *Acta Cryst.* **9** 515.
[15] Cheary R, Coelho A and Cline J 2004 *J. Res. Natl Inst. Stand. Technol.* **109** 1.
[16] Arlinghaus FJ (1974) *J. Phys. Chem. Solids* 35 931.
[17] Tauc J, Grigorovici R and Vancu A 1966 *Phys. Stat. sol. (b)* **15** 627.
[18] Kubelka P and Munk F (1931) *Z. Tech. Phys.* **12** 593.

IOP Conf. Series: Journal of Physics: Conf. Series **1143** (2018) 012019 doi:10.1088/1742-6596/1143/1/012019

Surface modification of TiO_2 nanostructured films by inserting Mg^{2+} ions applied to dye sensitized solar cells

A Huamán and M Quintana

Faculty of Sciences, Universidad Nacional de Ingeniería, P. O. Box 31-139, Avenida Tupac Amaru 210, Lima, Peru

E-mail: mquintana@uni.edu.pe

Abstract. Nanostructured electrodes were prepared based on titanium dioxide (TiO_2) modified by incorporating small amounts of magnesium oxide (MgO) to improve their efficiency in dye sensitized solar cells. J-V curve measurements show efficiencies increased from 2.32 to 2.55 % with the addition of $MgCl_2$. On the other hand, the incident photon-to-current efficiency (IPCE) measures obtained show an improvement with the addition of magnesium oxide. This is in agreement with absorbance results and short circuit current values. Values from cyclic voltammetry curves show that surface modification caused energy levels to increase below the TiO_2 conduction band by incorporating magnesium oxide. Electrochemical impedance spectroscopy confirmed the presence of these additional energy levels and gave evidence of increased recombination in the TiO_2 film and decreasing electron lifetime from 15.95 to 6.82 ms.

1. Introduction

In 1991, Gratzel et al. [1] set in motion for the first time a new device that uses electronic transfer mechanics similar to those that occur during photosynthesis in plants. Dye sensitised solar cells (DSSC) undergo optical absorption and charge separation processes due to the combination of a sensitizer, light absorbing material, and a wide band gap semiconductor of nanocrystalline morphology. These types of cell offer many different fabrication methods and are cheaper, flexible, non-toxic and also good for the environment. Because of the advantages they offer, DSSC's have been studied intensely since their inception and they have spawned a new multidisciplinary scientific research field. In this fieldTiO_2, a powder called P-25 (Degussa), containing anatase and rutile, two crystalline phases of titanium dioxide TiO_2, and a small amount of amorphous phases [2], is used. Degussa P-25 has been considered a standard material for electrode fabrication because it presents a relatively extensive surface (49 m^2g^{-1} [3]).

Many researchers have based their work on modifying TiO_2 electrodes through the assistance of other oxide materials, like alumina [4], ZnO [5], silica [6], Fe_2O_3 [7] and zirconia [8], by simply mixing them like a blocking layer or core/shell configuration. Due to their wide band gaps, incorporated materials slow electrons trying to follow a way back to the interface of cells and thus reduce charge recombination. On the other hand, these incorporated materials increase the dye charge in the TiO_2 electrode pores, thereby increasing the generated photocurrent.

This has led to an interest in studying optional materials that increase TiO_2 the photovoltaic potential. An interesting option is magnesium oxide (MgO), which typically has a wide band gap (6.0 – 7.8 eV); and has also already been reported as the best option among other materials for use in combination with TiO2 as electrodes to DSSCs with which various synthesis methods have been tried [9-12]. This article aims to continue the research of these insulating materials and to have a better understanding of their effects.

2. Experimental

Nanocrystalline titanium dioxide electrodes modified with magnesium oxide were prepared as follows. On a magnesium dioxide suspension in ethanol, 25% weight, small amounts of $MgCl_2$ were added to

IOP Conf. Series: Journal of Physics: Conf. Series **1143** (2018) 012019 doi:10.1088/1742-6596/1143/1/012019

obtain $0.25x10^{-3}$M, $1x10^{-3}$M and $4x10^{-3}$M concentrations in the resultant suspension. A reference suspension of only titanium dioxide was also prepared. In each case, the suspension was deposited on a substratum with fluorine doped tin oxide (FTO) conductive layer which was delimited using a special tape with a defined small area of 5 mm x 5 mm. Following the doctor blade method, the paste was homogeneously spread on this area using a glass rode, so that when removing the tape, we obtained a square thin film. Later, this substratum with the film was subjected to a thermic treatment of 450 ºC for about 45 minutes. Then the substratum was sensitized by submerging it in a solution of 0.5 mM of dye standard N719 ($C_{58}H_{86}O_8N_8S_2Ru$) for approximately 14 hours.

Figure 1. SEM images of mesoporous films made from suspension of (a) TiO_2 (b) TiO_2 + 4mM $MgCl_2$

In another glass conductor substratum, platinum was deposited using a solution of chloroplatinic acid in ethanol and a thermic treatment of 450 ºC for about 15 minutes. This substratum was used as a counter electrode. The cell was sealed with the help of a special thermoplastic sealant that had a space to add drops of electrolyte (I^-/I_3^-) before sealing the space.
Micrographs of samples were prepared using an EVO MA 10 scanning electron microscope (SEM) with an EDS system for analysis. Diffractograms of the photoelectrodes were obtained with a Phillips X'PERT MPD unit using Cu-Kα (λ=1.5418 Å) radiation and a 2θ range of 20 to 80°. The electrochemical system used to obtain different voltammograms for the fabricated electrodes was formed by a platinum electrode (counter electrode) and an electrode of Ag in a 0.1M KCl solution (reference electrode). A potentiostat, PG580 of Uniscan Instruments, was also used to obtain results.
A dichroic lamp of 50W was used as a light source to obtain I-V curves. The incident photon to electron conversion efficiency (IPCE) equipment used consisted of an Oriel Instruments xenon lamp of power 1000W coupled to a monochromator of the same manufacturer. We measured the power of the incident light with an Oceans Optics USB 4000 UV-Vis spectrophotometer. These measurements were made on wavelengths varying from 400nm to 700nm, in steps of 20 nm. Electrochemical impedance spectra of solar cells were obtained with an Autolab PGSTAT 10 potentiostat using a 150W Oriel xenon lamp.

Figure 2. EDS mapping of elements present in the film fabricated from a suspension of TiO_2 + 4mM $MgCl_2$. Each element was identified with its corresponding colour.

3. Results and discussions
SEM applied to fabricated electrodes (figure 1), confirm their nature as a nanoporous network. This network contains spheroid particles of approximately 50nm size. The scatter from magnesium oxide, resulting from the manufacturing method employed, and the homogeneous nature of the nanostructure network are shown in the EDS mapping of figure 2. However, the presence of clusters of amorphous phases of MgO cannot be ruled out as this might be suggested by the absence of the crystalline phase in X ray diffractograms, show in figure 3, although this could be due to the small amount of precursor used.

IOP Conf. Series: Journal of Physics: Conf. Series **1143** (2018) 012019 doi:10.1088/1742-6596/1143/1/012019

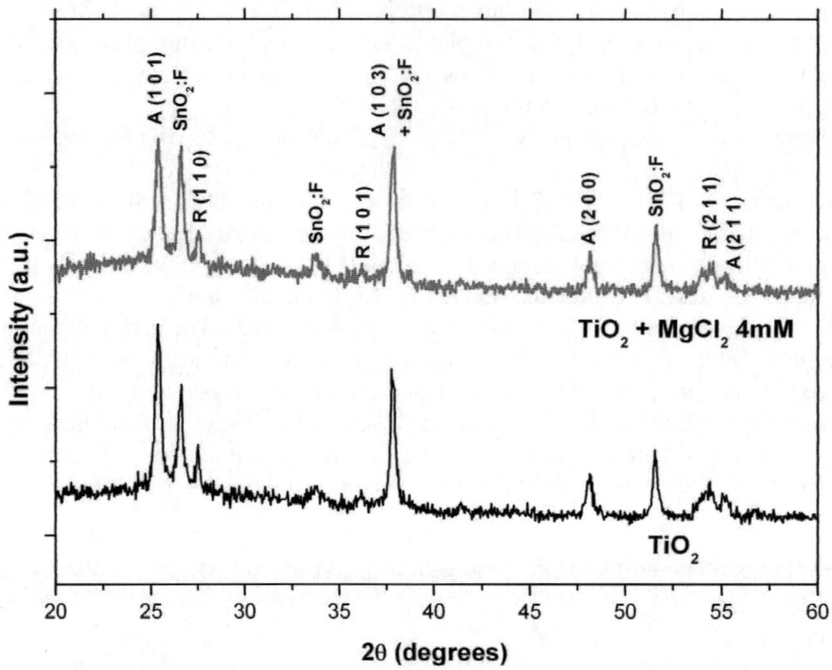

Figure 3. X ray diffraction pattern of electrodes prepared by the Dr. Blade' method with only TiO_2 and TiO_2 + 4mM $MgCl_2$. A: anatase, R: Rutile.

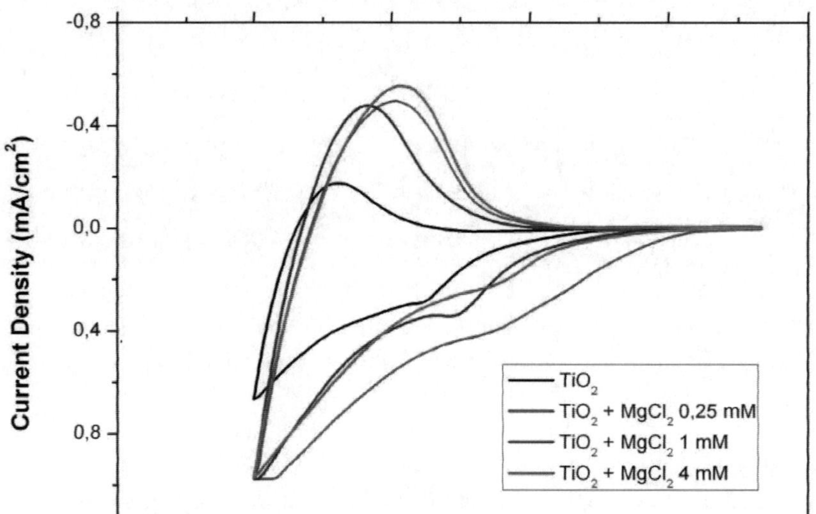

Figure 4. Cyclic Voltammetry DC obtained from at a scanning rate of 0.05V/s for TiO_2 electrodes made with different $MgCl_2$ amounts of $MgCl_2$.

IOP Conf. Series: Journal of Physics: Conf. Series **1143** (2018) 012019 doi:10.1088/1742-6596/1143/1/012019

Figure 4, represents typical curves of cyclic voltammetry for TiO_2electrodes with MgO obtained with a scanning rate of 50 mV/s in a 0.1M solution of KCl. As can be seen, potential electrodes in the negative direction produce relatively large cathodic current while in the reverse direction anodic currents are produced. These currents are higher for the case of electrodes containing MgO.

In the case of the electrode only prepared with TiO2, the anodic current begins to flow in -0.8V, while in the case of electrodes made with 0.25, 1 and 4 mM $MgCl_2$, the anodic current begins to flow at voltages of -0.59, -0.62 and -0.55 V, respectively. These values are higher than the reference TiO_2electrode. The presence of magnesium oxide has induced changes in the potential starting points of the cathodic current making them less negative. This indicates that the limit of the conduction band of the MgO modified TiO_2surfaces has suffered a variation that allows production of current from low potentials. This results in a decrease of the solar cell photovoltage. Small peaks that appear between -0.9 and -0.6 V are characteristic of nanostructured films [13]. These peaks have been interpreted as reversible filler products of the surface state located below the limit of the conduction band [14], which promotes conductivity of electrodes.

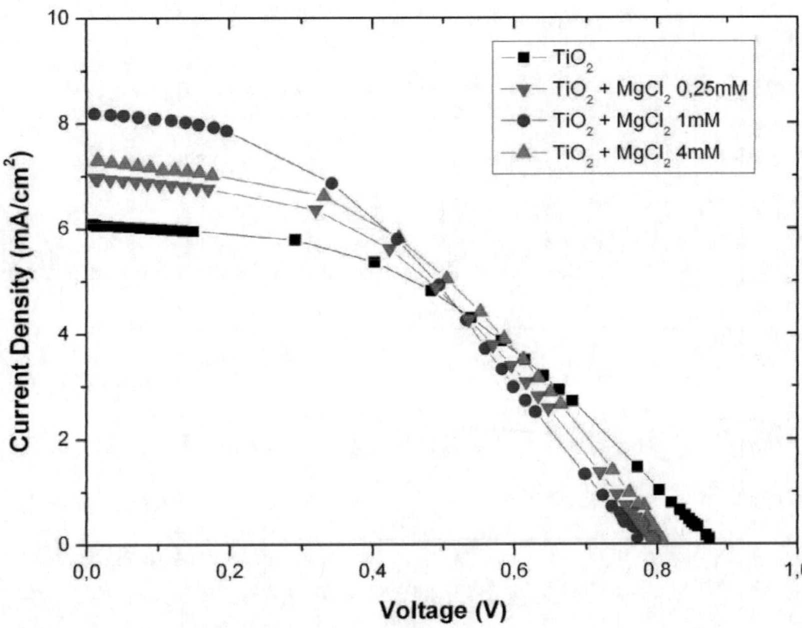

Figure 5. Current-Voltage curves of solar cells obtained with the addition of different amounts of $MgCl_2$ to the suspension. A light intensity of $1000W/m^2$ was used.

Table 1. Main parameters obtained from I-V curves of figure 5.

	V_{OC} (V)	J_{SC} (mA/cm^2)	FF	$\eta(\%)$
TiO_2 (P25)	0.87	6.06	0.44	2.32
TiO_2 + 0.25mM $MgCl_2$	0.79	6.99	0.43	2.39
TiO_2 + 1mM $MgCl_2$	0.78	8.43	0.40	2.55
TiO_2 + 4mM $MgCl_2$	0.81	7.30	0.43	2.44

IOP Conf. Series: Journal of Physics: Conf. Series **1143** (2018) 012019 doi:10.1088/1742-6596/1143/1/012019

J-V (current density vs. voltage) curves for solar cells fabricated with the incorporation of different$MgCl_2$ amounts of MgCl2 are shown in figure 5. Table 1 shows efficiencies that were achieved by each cell and also show short circuit current density (JSC), open circuit voltage (VOC) and form factor (FF). The same values are shown in figure 6 where their variation with respect to $MgCl_2$the amount of MgCl2 added to the suspension is highlighted.

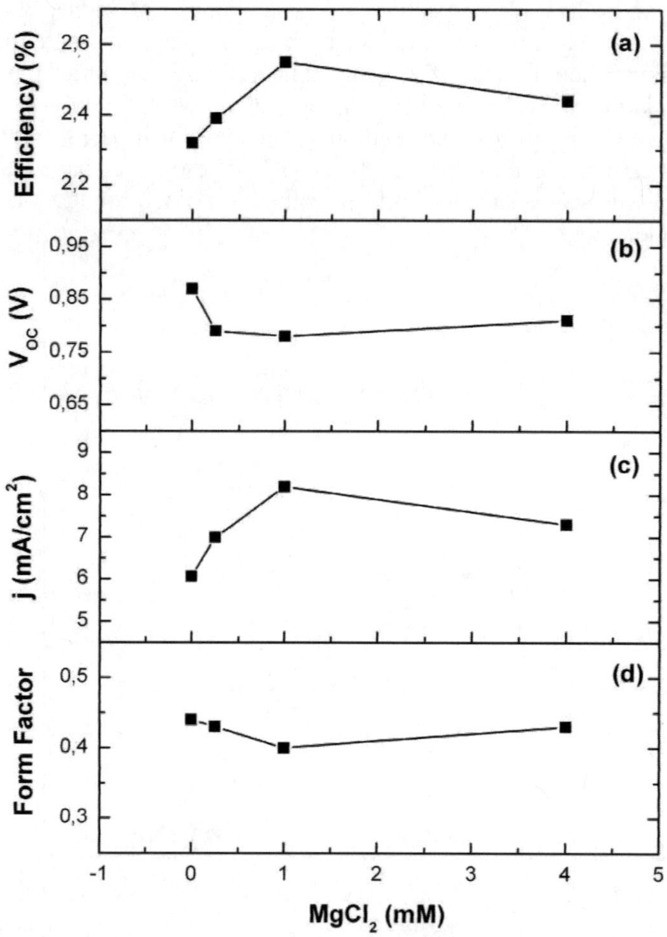

Figure 6. (a) Efficiency, (b) Open circuit voltage, (c) short circuit current density and (d) form factor of cells made with different amounts of $MgCl_2$.

As can be seen, the best efficiency, 2.55%, was obtained when 1mM of MgCl2 was added. The same sample achieved the best current density of 8.34 mA/cm2. Using more (4 mM) or less (0.5 mM) $MgCl_2$ did not result in higher current densities and efficiencies which have similar trends to each other. The open circuit voltage shows a reduction in cells prepared with $MgCl_2$, but beyond this, no major change occurred with increasing amounts of $MgCl_2$. The form factor doesn't show any major change except in the cell prepared with 1mM $MgCl_2$. where it was reduced to 0.40. Because of this, there was not a major increase in efficiency.

Figure 7 shows incident photons to current conversion efficiency (IPCE), or quantum efficiency, as a function of illumination wavelength for cells obtained using different amounts of $MgCl_2$. It is observed that cells prepared with $MgCl_2$ show better efficiency in comparison with the electrode prepared with only TiO_2. The IPCE curves are very similar reaching a peak at a wavelength of approximately 548 nm with an IPCE value of around 0.45. Beyond this point, the protrusion in the curve corresponds to the cell prepared with 1mM $MgCl_2$. In this way, IPCE curves have a similar relationship with current density as seen in figure 5.

IOP Conf. Series: Journal of Physics: Conf. Series **1143** (2018) 012019 doi:10.1088/1742-6596/1143/1/012019

Figure 8 shows spectra of electrochemical impedance obtained experimentally for cells prepared with TiO_2 and TiO_2+$MgCl_2$ 1mM $MgCl_2$. The curves were adjusted according to the equivalent circuit showed in figure 9 using ZView software version 2.9. The main parameters calculated by the software are shown in table 2.

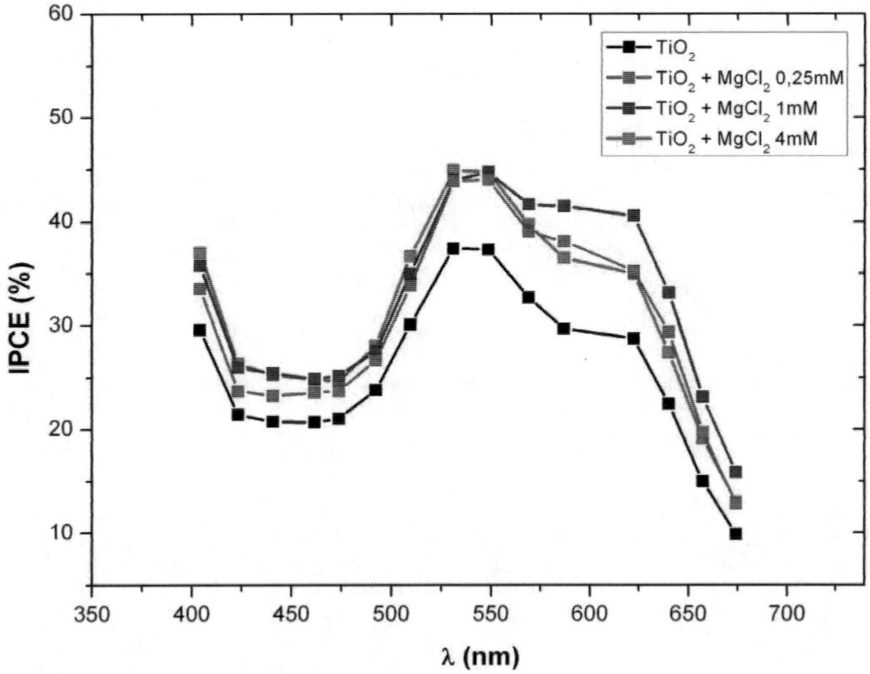

Figure 7. IPCE spectrum of cells prepared with TiO_2 with 0.25, 1, 4 mM of $MgCl_2$ added.

Evident differences between the spectra can be seen. For the TiO_2 case, three semicircles are observed in the shape of the graph (the third semicircle is barely noticeable), while for the case of the electrode with TiO_2+1mM $MgCl_2$ two semicircles are observed. The Warburg element of diffusion in the electrolyte is not appreciable; and so the potential taken from impedance measurements is not sufficient to see this semicircle. As for this latter case, the diffusion ionic resistance is not large and it is possible that there has been a major movement of electrolyte species during charge transport. The change in the impedance spectra correspond to values obtained in table 2.

The series resistance of FTO is the only parameter that remains almost invariant because we used the same conductor substrate in both cells. All other parameters feature variations between the two cells that are related to each other. The only difference between the cell components is theTiO_2 film. For this reason we should focus our attention on the resistance due to electron diffusion in the film and theTiO_2 chemical capacitance. It can be seen that the first of these parameters decreases in magnitude while at the same time the second shows an increase when $MgCl_2$ is added. These changes are directly related to the electrochemical potential of electrons [15], which was affected when magnesium oxide was incorporated in the nanoporous network. The electron lifetime, calculated from parameters already mentioned, suffers a decrease from 15.95 to 6.82 ms. This signifies recombination in the interface TiO_2or electrolyte. This higher recombination is responsible for the steeper fall in the J-V curve of the solar cell, which contributes to a reduced form factor [16].

IOP Conf. Series: Journal of Physics: Conf. Series **1143** (2018) 012019 doi:10.1088/1742-6596/1143/1/012019

a)

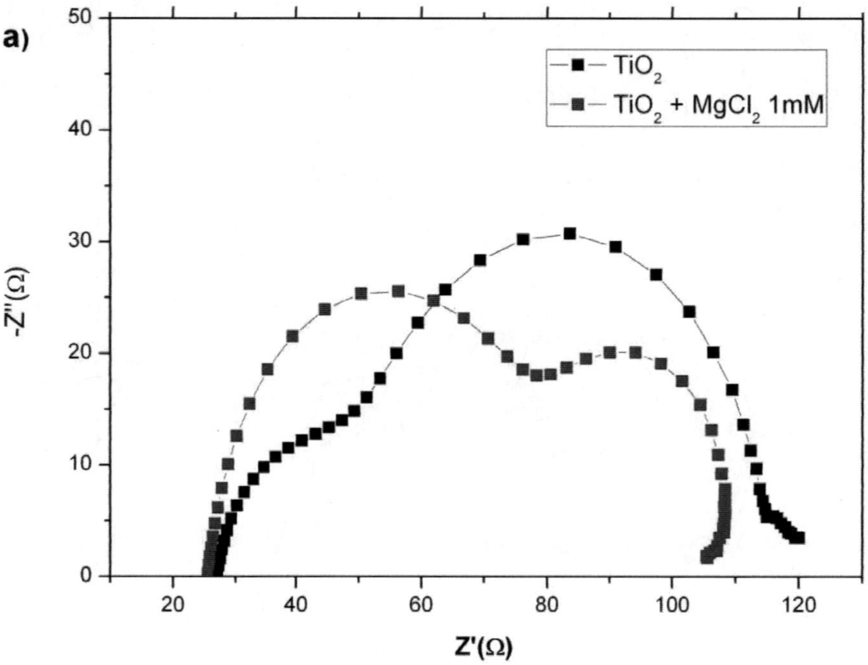

b)

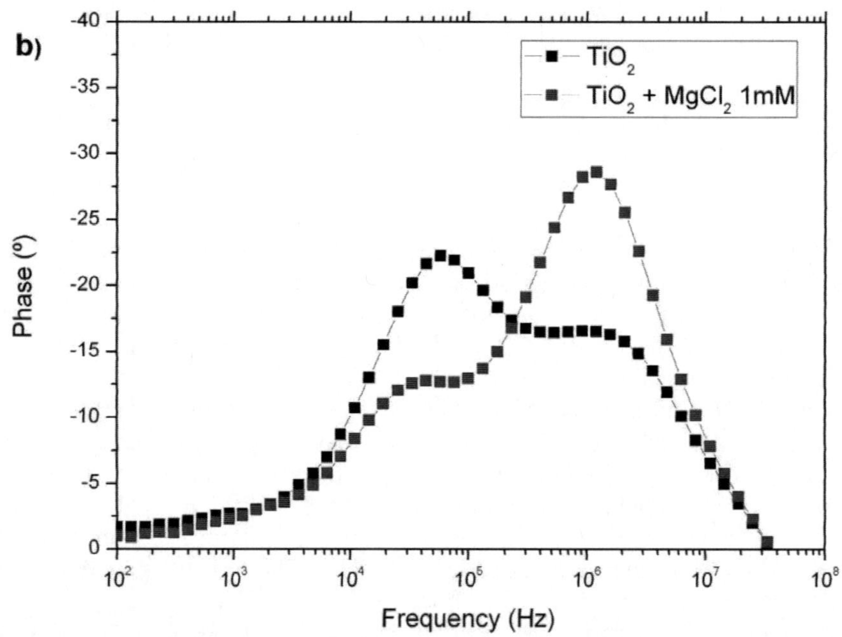

Figure 8. Spectra of electrochemical impedance represent in diagrams of Nyquist (a) and Bode (b) for TiO_2 and TiO_2 + 1mM $MgCl_2$ cells.

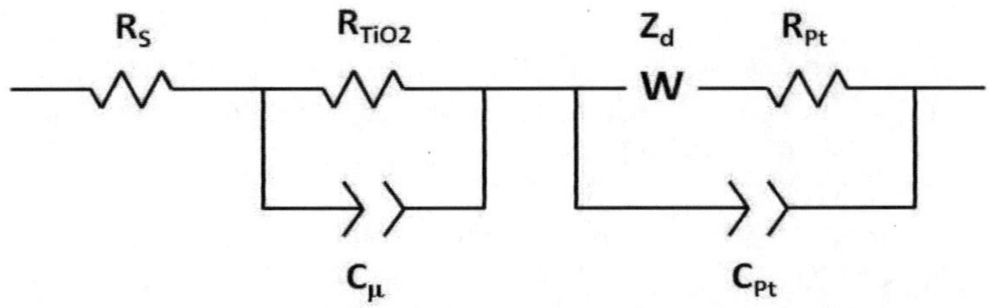

Figure 9. Equivalent circuit used in the ZView program to adjust the impedance spectra of fabricated DSSC .

Table 2. Main parameters obtained from adjusted impedance spectra.

Description	Parameters	TiO_2	TiO_2 + $MgCl_2$ 1mM
Series resistance of the substrate conductor	R_S	26.82 Ω	25.66 Ω
Charge transfer resistance in the counter electrode	R_{Pt}	23.85 Ω	48.49 Ω
Constant Phase Element (CPE) in the interface Pt/electrolyte	C_{Pt}	24.4 μF	7.3989 μF
Index CPE for the capacitance in the interface Pt/electrolyte	n_{Pt}	0.84	0.95
Resistance of electron diffusion through TiO_2	R_{TiO2}	63.68 Ω	42.83 Ω
CPE correspond to chemical capacitance of TiO_2	C_μ	107.05 μF	372.4 μF
Index CPE corresponding to chemical capacitance of TiO_2	n_μ	0.92773	0.82461
Ionic diffusion resistance in the electrolyte	Z_d	7.755 Ω	-
Electron lifetime	τ_n	15.95 ms	6.82 ms

4. Conclusion

The surface of the nanostructured TiO_2 film was modified by incorporating of MgO from small amounts of $MgCl_2$. The fabricated DSSC showed a slight increase in the power conversion efficiency with the incorporation of magnesium oxide that reaches a maximum of 2.55%, which represents an increase of 9.9% with respect to the cell made only with TiO_2. An increase in the short circuit current density was observed in the current density - voltage (J-V) characteristic curve. The IPCE spectra obtained for the modified cells reported improvements, the highest values being obtained by the cell made with TiO_2 + 1mM $MgCl_2$. The incorporation of magnesium oxide in the TiO_2 network caused surface defects which led to an increase in the Electronic State Density. This was the direct cause of the decrease in open circuit voltage and an increase in electron recombination. Electrochemical Impedance Spectroscopy performed on the cell made with 1mM TiO_2 + $MgCl_2$ provided information on the recombination processes, which were mostly with the ionic species in the electrolyte. These recombines caused a low form factor in the J-V curves and a shorter electron-time in the TiO_2 film.

References

[1] Aderhold J, Davydov V Yu, Fedler F, Klausing H, Mistele D, Rotter T, Semchinova O, Stemmer J and Graul J 2001 *J. Cryst. Growth* **222** 701
[2] O'regan, B, and Grätzel, M. 1991 *Nature* **353** 737
[3] Ohtani B, Prieto-Mahaney O, Li D and Abe R 2010 *Journal of Photochemistry and Photobiology A: Chemistry*, **216** 179-182
[4] Ohno T, Sarukawa K, Tokieda K and Matsumura M 2001 *Journal of Catalysis* **203** 82-86.
[5] Wang C, Yu Z, Bu C, Liu P, Bai S, Liu C and Guo S 2015 *Journal of Power Sources* **282** 596-601
[6] Kanmani S and Ramachandran K 2012 *Renewable Energy* **43** 149-156
[7] Al Dahoudi N, Xi J and Cao G 2012 *Electrochimica Acta* **59** 32-38
[8] Kiliç B, Gedik N, Mucur S P, Hergul A S and Gür E 2015 *Materials Science in Semiconductor Processing* **31** 363-371
[9] Menzies D, Dai Q, Cheng Y B, Simon G P and Spiccia L 2005 *Materials Letters* **59** 1893-1896
[10] Avellaneda C O, Gonçalves A D, Benedetti J E and Nogueira A F 2010 *Electrochimica Acta* **55** 1468-1474
[11] Li B, Lü G, Luo L and Tang Y 2010 *Wuhan University Journal of Natural Sciences* **15** 325-329
[12] Photiphitak C, Rakkwamsuk P, Muthitamongkol P and Thanachayanont C *2012 Proceedings of the International Conference on Electrical, Computer, Electronics and Communication Engineering* 485-489
[13] Wu S, Han H, Tai Q, Zhang J, Xu S, Zhou C and Zhao X Z 2008 *Nanotechnology*, **19**, 215704
[14] Wang H, He J, Boschloo G, Lindström H, Hagfeldt A and Lindquist S E 2001 *The Journal of Physical Chemistry B* **105** 2529-2533
[15] Kavan L, Kratochvilová K and Grätzel M 1995 *Journal of Electroanalytical Chemistry* **394** 93-102
[16] Bisquert J, Fabregat-Santiago F, Mora-Seró I, Garcia-Belmonte G, Barea E M and Palomares E 2008 *Inorganica Chimica Acta* **361** 684-698

Laser power influence on Raman spectra of multilayer graphene, multilayer graphene oxide and reduced multilayer graphene oxide

H Ferreira[1], **G Poma**[2], **D R Acosta**[3], **J Barzola-Quiquia**[4], **M Quintana**[2], **L Barreto**[1] and **A Champi**[*1]

[1]Centro de Ciências Naturais e Humanas, Universidade Federal do ABC, Santo André, 09210-170, Brazil
[2]Universidad Nacional de Ingeniería, Lima, Av. Túpac Amaru 210, P.O. Box 31-139, Peru
[3]Universidad Nacional Autónoma de México, Ciudad de México, A.P.20-364, Mexico
[4]Division of Superconductivity and Magnetism, University of Leipzig, Leipzig, D-04103, Germany

Abstract. In this work, the influence of laser power on Raman spectra on different carbon based materials was studied: multilayer graphene (MLG), multilayer graphene oxide (MLGO) and reduced multilayer graphene oxide (rMLGO). The presence of oxygen and hydrogen was verified by means of Fourier-transform Infrared Spectroscopy, indicating that our oxidation and reduction process was success in our multilayers graphene. Through the position and width of the band G, we discuss the effects of the incorporation of oxygen by the graphene matrix in terms of thermal expansion and doping effects. The ID/IG ratio was used to determine the presence of defects and showed a similar behavior for MLGO and rMLGO demonstrating a feature of partial reduction and reduction effects caused by the power supplied by laser.

*e-mail: ana.champi@ufabc.edu.br

Keywords: Multilayer Graphene, Multilayer Graphene Oxide, Reduced Multilayer Graphene Oxide, Raman Spectroscopy, Laser Power.

1. Introduction

Graphene has triggered new studies since its discovery by Novoselov and Geim [1], who used the process of mechanical exfoliation [2] to obtain the first two-dimensional stable material consisting of hexagonal networks of carbon with sp^2 hybridization. An example of these studies is the incidence of a laser beam on the surface of graphene, which makes this system to behave as an ideal sensor that exposes intrinsic properties of environmental perturbations, showing that the charge-carrier density of exfoliated graphene on a substrate of SiO2/Si can be transported reversibly between the hole and electronic doping with visible photons. This photo-induced doping occurs under moderate conditions and with a total control of the laser power (P_λ) that fall on the samples. These findings suggest that optically closed graphene devices that operate on a scale of less than one second time and expected from Raman spectroscopy is not always as noninvasive as is generally assumed [3].

Graphene has many superior interesting properties, such as high electronic and thermal conductivity, high mechanical strength [4], electronic quantum transport, gap band space tunable, high mobility, high elasticity, electromechanical modulation [5], as well as properties for applications in nano-electronic devices [6], water decontaminant [7], etc. Another material worth studying is graphene oxide (GO); for their applications in biosensors [8], water decontamination [9], thin film transistors [10], transparent conductive films [11], just to mention a few examples. Being the

Hummer-Offeman method [12] the most used, due to the fact that it allows us to work with different amounts of powdered graphite or flake type, as well as to modify quantities and solvents that are used during the synthesis process. In this way, we can obtain significant amount of graphite oxide and it is consequently reduced.

Raman spectroscopy is a technique used for the study of a wide range of materials, one of which is graphite and its derivatives such as graphene and graphene oxide. This technique allows the determination of the layers [13, 15] and crystallite size [16] among other studies widely reported in the literature [14]. In particular, the increasing in local temperature at the point where the Raman laser hits the material under study causing a shift of the D and G bands due to thermal expansion and phonon dispersion. Therefore, the local temperature at the measuring point can be represented as a function of the laser intensity absorbed by graphene [17]. In this direction, we have studied the effects produced by applying several laser power on three systems: multilayer graphene (MLG), multilayer graphene oxide (MLGO) and reduced multilayer graphene oxide (rMLGO) thermally at a temperature of 80 °C. In this way, clear relationships between the laser power than incident on MLGO and rMLGO were observed fixing an only laser energy. These results are shown and discuss.

2. Experimental procedure

For the synthesis of MLGO, the process started from a solid mixture of graphite and sodium nitrate. Under constant stirring sulfuric acid was added and then potassium permanganate was slowly poured, forming a reddish coloration corresponding to Mn_2O_7 [18].

The reaction was maintained for about 2 hours and then 50 ml of water was added, increasing the reaction temperature, which was set between 95-98 °C. After 40 minutes, the sample displays a greenish color.

After cooling the sample to room temperature, hydrogen peroxide was added to stop the oxidation reaction [19] with constant stirring, forming on the basis of a beige precipitate, and then it was washed 2 times with water, hydrochloric acid and ethanol. Finally, this solution was filtered and allowed to dry at room temperature.

Once graphite oxide was obtained by chemical technique; multilayer graphene oxide (MLGO) was obtained through the process of mechanical exfoliation [2]. Worked out on a substrate of SiO_2/Si, through the Zeiss optical microscope, the number of layers that the composite of graphene presents can be observed [20, 21]. An annealing process is then performed at low temperature, approximately 80 °C to improve its stability and obtain the reduced multilayer graphene oxide (rMLGO) [22], and then their properties were studied using the Raman spectroscopy manipulating the laser power. At each moment, the incidence of the laser on each sample was the same, and the intensity variation was from the lowest value to the highest without damaging the sample.

The equipment used was the Horiba T64000 Triplo Raman spectrometer on micro-Raman mode with a focal objective of 100x; studies were performed with an excitation green laser at 532 nm (photon energy of 2.33 eV) for a laser power controlled and checked by a potentiometer Laser check (ROHS), Coherent. Therefore, we can obtain the values of each of these intensities than incident on the sample. In this work, the values used were from 0.06mW until 3.00mW reaching the sample.

The morphological images of graphene derivatives were performed in a JEOL JCM-6000 Scanning Electron Microscope**.** Fourier-transform infrared spectroscopy (FTIR) performed in a Varian-Agilent 640 FTIR on ATR mode was used to determine the functional groups in multilayer graphene oxide.

3. Results and discussion

In figures 1 (a) and (b), SEM micrographs display different morphological structures, before and after the oxidation process performed by the method of Hummers [12]. The different morphology between samples is an indicative of the presence of oxides. Since multilayers graphene was not chemically

IOP Conf. Series: Journal of Physics: Conf. Series **1143** (2018) 012020 doi:10.1088/1742-6596/1143/1/012020

treatment appreciating a smooth surface; while for obtaining the multilayers of GO, it was worked with sulfuric acid to the intercalation of the graphite layers and other process as mentioned in the experimental procedure forming oxygenated functional groups showing an irregular surface as can be observed in figure 1 (b).

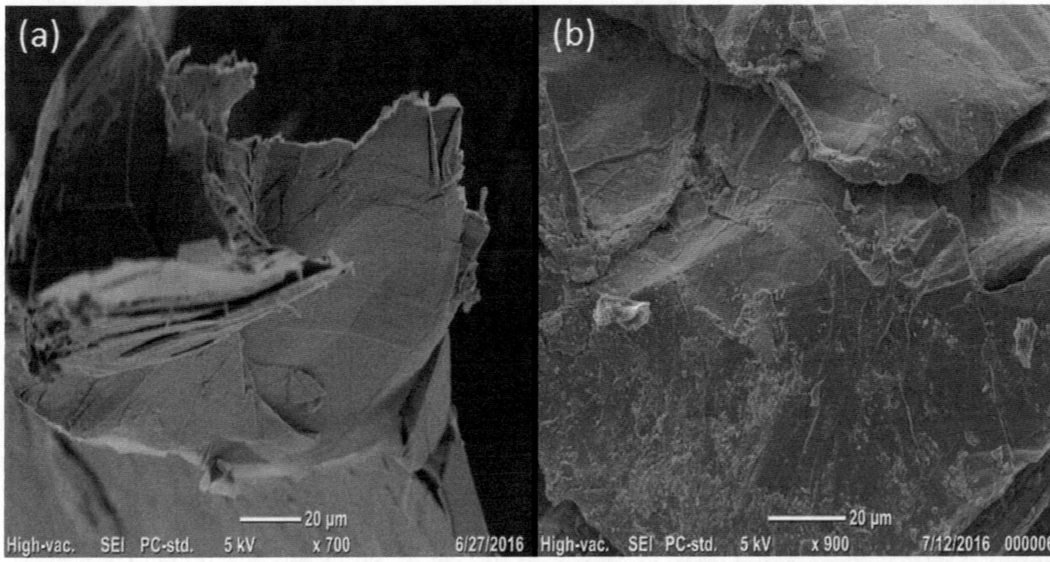

Figure 1. SEM image to pure graphite (a) and (b) after chemical process to obtain graphite oxide.

After exfoliation the graphene multilayer, MLGO and rMLGO were deposited on SiO_2substrate to be studied by Raman spectroscopy as observed in the figure 2 (a), (b), (c), as a function of laser power. Thus, as we can see in the figure 2 (a) the G band appears well defined around 1585cm^{-1} and approximately not changes with the laser power (figures 4 (a) and 4 (b)). On the other hand, when oxygen is incorporated in graphite by Hummers–Offmen method and is obtained the MLGO by exfoliation, the D and D' bands appear in 1354 cm^{-1} and 1620 cm^{-1} respectively shown in the figure 2 (b). Figure 2 (c) shows the Raman spectra of rMLGO. In all cases, these bands increase even more with the increase of the laser power.

The amount of defects generated by the oxidation of graphite, as well as the degree to which the system of sp^2 hybridization is regenerated after the annealing process at 80 °C were studied by Raman spectrums as shown in the figure 2 (c). In the G band, shows an increase in the bandwidth due to the influence of sp^3 hybridization, defined path amorphous material. The appearance of the band D' occurs due to relaxation which allows high frequency phonons to be scattered [13-15]. On the other hand, when the peak D is observed in graphene and graphite systems is directly related to disorder or defects being used as a quality measure [14].

Even doping of electrons or holes can be detected from Raman measurement and it is reflected on the stiffness and the sharpness of the band G [13].

Figure 3 shows FTIR spectroscopy for MLGO, obtaining the characteristic bands of its structure, around 1645 cm^{-1} corresponding to the vibration of the C = C, in addition to the bands associated with vibration stretch of flexing of the OH bond around 3400 cm^{-1} [23], for a wavenumber of 1060 cm^{-1} C-O bond, at 1220 cm^{-1} vibration tension to C-OH and a band at 1720 cm^{-1} which is attributed to C = O bonds of the carboxyl corroborate the formation of oxygenated functional groups.

IOP Conf. Series: Journal of Physics: Conf. Series **1143** (2018) 012020 doi:10.1088/1742-6596/1143/1/012020

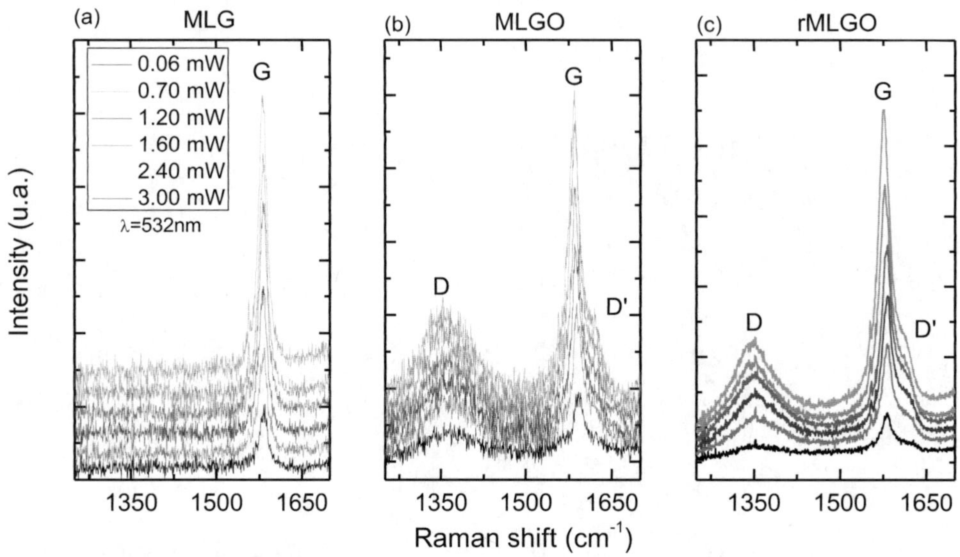

Figure 2. Raman spectrum of the (a) multilayer graphene (MLG) (b) multilayer graphene oxide (MLGO) and (c) reduced multilayer graphene oxide (rMLGO) to the different laser power P_λ.

Figure 4 shows the properties of the G band of the Raman spectrum as a function of laser power for the three studied systems MLG, MLGO and rMLGO. We decided to study the G-band because it is present in all systems and is one of the most important bands for the characterization of carbon materials. On the impact of the laser intensity, it is important to consider that for some materials, it is possible to damage or alter the sample with this parameter without modification the laser energy. The effect of relatively small changes in the laser intensity, the effect induces small changes of temperature in the samples. Most carbon nanomaterials are black and absorb significant amounts of visible light, most of this absorbed energy is converted into heat, and thus changing the sample temperature in the area where the laser beam is focusing.

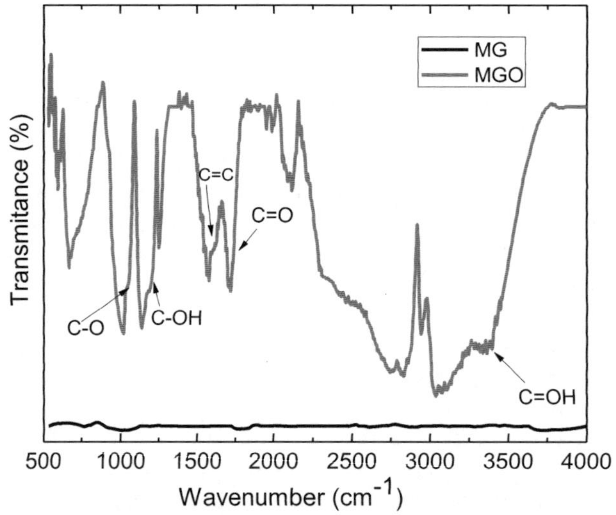

Figure 3. FTIR-ATR spectrum for multilayer graphene oxide.

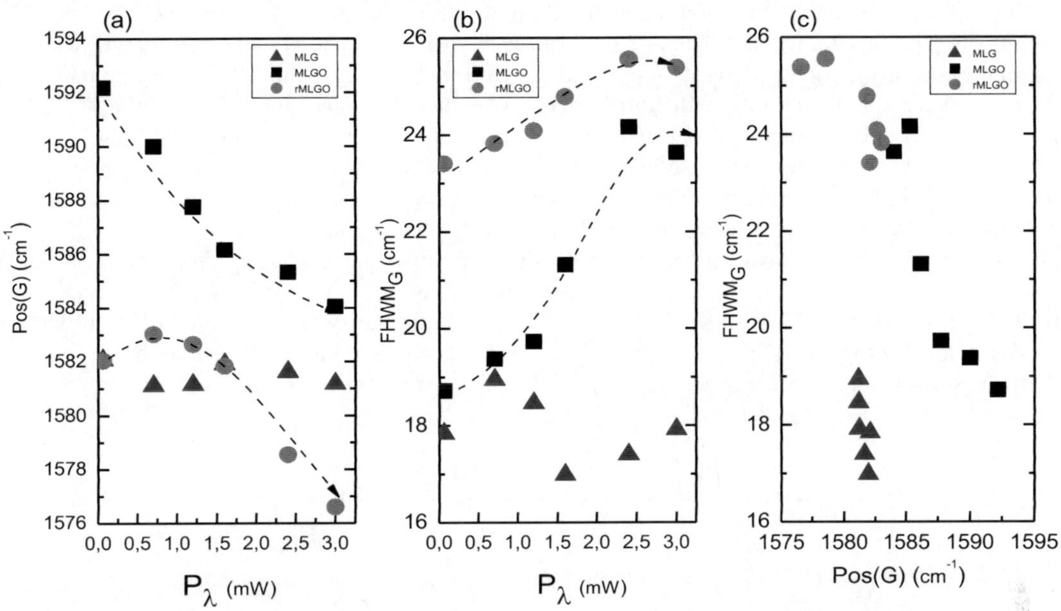

Figure 4. Properties of G band: (a) Width band (FHWM$_G$) and (b) band position (Pos(G)) worked for different laser power at a wavelength of 532 nm; (c) Width band (FHWM$_G$) as a function of the band position (Pos(G)). The dashed curves are only to better visualize the behavior in (a) and (b).

The different intensities of the laser should not necessarily cause changes in the energy of the scattered phonon due to the quantization effect of the network vibrations, but they work to improve the signal-to-noise ratio in obtaining the Raman spectrum.

In figure 4 (a) the position of the G band represents the average energy of phonon generated in the system by the scattered laser photon. For the MLG this energy does not vary considerably depending on the different laser power. However, for MLGO there is considerable dependence. In the latter case, a tendency to return to the same behavior of the MLG due to the elimination of the functional groups by the heat treatment process is observed for the rMLGO for low laser power. It is likely that physicochemical changes caused by oxygenation of graphene alter the modes of attachment of the crystalline lattice of the material, inserting defects and / or altering the length of the bonds and the crystalline lattice. This configuration of changes makes the system more susceptible to influence the intensity of the incident laser. By influencing the laser on MLG, MLGO and rMLGO to different powers, gives changes in temperature causing a change in the peak positions of the Raman spectra. As temperature increases, there is a linear shift of the G peak (Pos(G)) due to increased anharmonic phonon coupling increased thermal expansion in the lattice. The lattice expansion changes the energy center of dispersion of the phonons Pos(G) and is much more significant in the presence of defect elements (functional groups such as oxygen) in the crystalline lattice. In addition, the presence of these defects increases the dispersion of the phonons (figure 4 (b)) and is not fully reversed with the thermal reduction. Because this, there remains an oxygen doping in the rMLGO matrix, which causes nonlinear effects with the temperature in the regime of lower power of the laser (figure 4 (a)). In figure 4 (c) finally, we observed the difference in the response of each studied property of the system to the variation of laser power. It is observed that the MLG system maintains the characteristics of the vibrational mode only by increasing the phonons population (increase of the dispersion), the MLGO alter the vibrational characteristics and add dispersion, and the rMLGO system presents a tendency to return to the MLG vibrational modes but still maintaining higher dispersion due to non-complete elimination of the functional groups by the annealing process.

IOP Conf. Series: Journal of Physics: Conf. Series **1143** (2018) 012020 doi:10.1088/1742-6596/1143/1/012020

The ratio of the intensity (area) of the peaks D and G (I_D/I_G) in figure 5 is used to characterize the level of disorder in samples. It exists a regime of density of "low" defects where I_D/I_G will increase as a higher density of defects creates more elastic scattering, this happens until a regime of "high" density of defects, at which I_D/I_G will start to decrease as there is a growth of the density of defects, this results in a more amorphous structured carbon, and attenuating the Raman peaks [13, 24]. Besides to a peak D occurs, a charge carrier should be excited and be inelastically scattered by a phonon, and then a second elastic scattering by a defect or boundary region must occur to result in recombination. As both processes are of the same order, it was expected that the area relationships would remain constant for different P_λ. The observed differences indicate that each scattering mechanism produces a slightly different response with the increase in the number of photons arriving on the material, even presenting a similar ratio I_D/I_G for MLGO and rMLGO. This similarity may indicate that the reduction process is not able to eliminate the defects of the matrix entirely. In addition, the behavior of the graph (figure 5) shows that there is a process of apparent clutter reduction when more power is supplied to the sample.

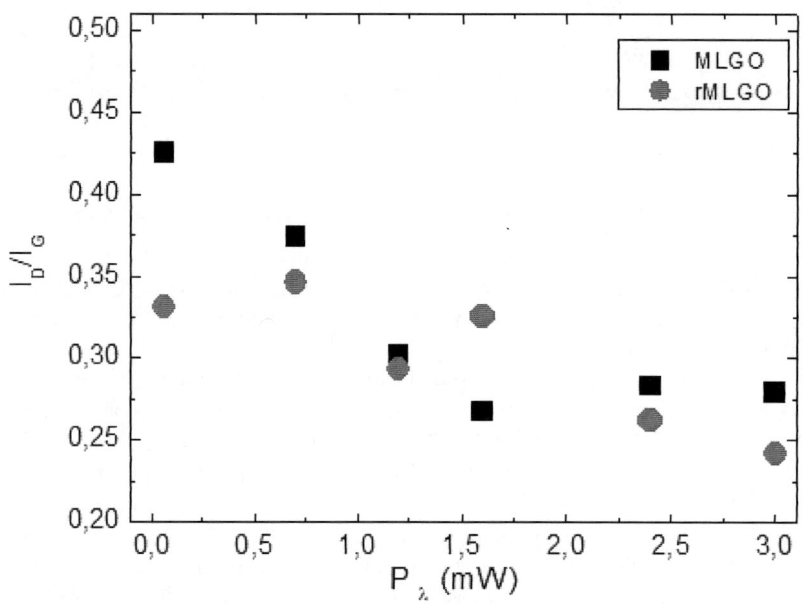

Figure 5. The relation between D and G bands (I_D/I_G) ratio for different laser power P_λ. In all cases, to an only laser energy of λ = 532nm.

4. Conclusion

In this present work we did prepared multilayer graphene and by chemical reaction we have produced graphene oxide and reduced. Using FITR spectroscopy we have verified that an oxidation process was obtained. Raman spectroscopy was also performed, but we have focused our study on the influence of the laser power on the spectra using moderate range of power. According to Raman results, the produced multilayer is free of defects, while the oxidized and reduced multilayers show clearly a Raman peak related to the presence of defects. The defects we attribute to the crystal structure modification produced during the oxidation process.

We show that the laser power does not affect the position of the G band peak to the MLG, as expected, but it affects the position in different ways for the MLGO and rMLGO. We believe that these effects are related to the effects of thermal dilation and oxygen doping of the matrix. In addition,

IOP Conf. Series: Journal of Physics: Conf. Series **1143** (2018) 012020 doi:10.1088/1742-6596/1143/1/012020

the oxidation process increases the dispersion of the phonons, and, conversely, the reduction process increases further. This increase in dispersion is possibly associated with the fact that the reduction does not totally eliminate the defects of the crystalline lattice that may persist even after the release of the associated functional groups, and the presence of more dispersed oxygen in a smaller quantity. These defects and the not completely eliminated oxygen alter the heating mechanisms of the matrix, demonstrating different vibrational properties with the power provided by the laser. In this way, we show how that the range in laser power used in this work affect these materials (MLG, MLGO and rMLGO), which can be used as reference in the future to investigate similarly produced samples.

5. References

[1] K. Novoselov, A. Geim, V. Morozov, D. Jiang, Y. Zhang, V. Dubonos, V. Grigorieva, A. Firsov, *Electric Field Effect in Atomically Thin Carbon Films*, Science, 306 (2004) 666-669

[2] L. Ciric, A. Sienkiewicz, B. Nafradi, M. Mionic, A. Magrez, and L. Forro. *Towards electron spin resonance of mechanically exfoliated graphene*, Phys. Status Solidi B 246 (2009) 2558–2561

[3] A. Tiberj, M. Rubio-Roy, M. Paillet, J. R. Huntzinger, P. Landois, M. Mikolasek, S. Contreras, J. Sauvajol, E. Dujardin and A. Zahab, *Reversible optical doping of graphene*, Scientific Reports, 3 (2013)

[4] Y. Shen, Q. Fang, B. Chen, *Environmental applications of three-dimensional graphenebased macrostructures: adsorption, transformation, and detection*, Environ. Sci. Technol. 49 (2015) 67–84

[5] T. Ohta, A. Bostwick, T. Seyller, K. Horn, E. Rotenberg, *Controlling the electronic structure of bilayer graphene*, Science 313(2006) 951–954

[6] Z. Ni, Y. Wang, T. Yu, and Z. Shen, *Raman Spectroscopy and Imaging of Graphene*, Nano Res 1 (2008) 273-291

[7] S. Chowdhury, R. Balasubramanian, *Recent advances in the use of graphene-family nanoadsorbents for removal of toxic pollutants from wastewater*, Adv. Colloid Interface Sci. 204 (2014) 35−56

[8] Y. Wang, Z. Li, J. Wang, J. Li and Y. Lin, *Graphene and graphene oxide: biofunctionalization and applications in biotechnology*, Trends in Biotechnology 29 (2011) 205-211

[9] W. Gao, M. Majumder, L. Alemany, T. Narayanan, M. Ibarra, B. Pradhan, P. Ajayan, *Engineered graphite oxide materials for application in waterpurification*, ACS. Appl. Mater. Interfaces 3 (2011) 1821–1826

[10] I. Karteri, S. Karatas and F. Yakuphanoglu, *Electrical characterization of graphene oxide and organic dielectric layers based on thin film transistor*, Applied Surface Science 318 (2014) 74-78

[11] X. Li, Y. Zhu, W. Cai, M. Borysiak, B. Han, D. Chen, R. Piner, L. Colombo, R. Ruoff, *Transfer of large-area graphene films for high-performance transparent conductive electrodes,* Nano Lett. 9 (2009) 4359-4363

[12] W. Hummers and R. Offeman, *Preparation of Graphitic Oxide*, Journal of the American Chemical Society80 (1958) 1339

[13] A. Ferrari, *Raman spectroscopy of graphene and graphite: disorder, eletron-phono coupling, doping and nonadiabatic effects*, Solid State Communications 143, (2007), 47-57

[14] L. Malard, M. Pimenta, G. Dresselhaus, M. Dresselhaus. *Raman spectroscopy in graphene*, Physics Reports 473 (2009) 58-71

[15] A. Ferrari, J. Meyer, V. Scardaci, C. Casiraghi, M. Lazzeri, F. Mauri, S. Piscanec, D. Jiang, K. Novoselov, S. Roth, and A. Geim. *Raman Spectrum of Graphene and Graphene Layers,* Phys. Rev. Lett.97 (2006) 187401-4

[16] L. Cançado, K. Takai, T. Enoki, M. Endo, Y. Kim, H. Mizusaki, A. Jorio, L. Coelho, R. Magalhães and M. Pimenta. *General equation for the determination of the crystallite size La of nanographite by Raman spectroscopy*, Applied Physics, (2006) 88-90
[17] W. Zhao, Y. Wang, Z. Wu, W. Wang, K. Bi1, Z. Liang, J. Yang, Y. Chen, Z. Xu and Z. Ni, *Defect-Engineered Heat Transport in Graphene: A Route to High Efficient Thermal Rectification*, Scientific Reports, 5 (2015) 11962
[18] M. Dimiev and M. Tour, *Mechanism of Graphene Oxide Formation*, ACS NANO 8 (2014) 3060–3068
[19] V. Kumar, M. Bardhan, S. Tongay, J. Wu, M. Belcher and C. Jeffrey, Grossman, *Scalable enhancement of graphene oxide properties by thermally driven phase transformation*, Nature (2013) 1820
[20] Z. Ni, H. Wang, J. Kasim, H. Fan, T. Yu, Y. Wu, Y. Feng, and Z. Shen, *Graphene thickness determination using reflection and contrast spectroscopy*. Nano Lett. 7 (2007) 2758
[21] F. Obelenis and A. Champi . *Determination of the Number of Graphene Layers on Different Substrates by Optical Microscopy Technique*. Brazilian Journal of Physics, 44, 682-684, 2014
[22] A. Nekahi, P. Marashi, D. Haghshenas, *Transparent conductive thin film of ultra large reduced graphene oxide monolayers*, Applied Surface Science, 295(2014), 59– 65
[23] E. Choi, T. Han, J. Hong, J.Kim, S. Lee, H. Kim and S. Kim, *Noncolvalent Functionalization of Graphene with End Functional Polymers*, 20 (2010), 305-353
[24] I. Calizo, A. Balandin, W. Bao, F. Miao, and C. Lau, *Temperature Dependence of the Raman Spectra of Graphene and Graphene Multilayers*, Nano Letters, 7 (2007) 2645-2649

Acknowledgments

The authors are thankful to the projects 138-FINCyT-IA-2013 (Peru) and PLAF/SBF (Brazil) for the financial support; also to the Nacional de Grafite Ltd from MG-Brazil for providing the graphite flakes to obtain the multilayer graphene; and to the UFABC Multiuser Central Facilities for letting us use the Raman spectroscope. We also thank and appreciate the financial support of UNAM-DGAPA IN 102816 project and would like to thank CAPES for the granted scholarship.

IOP Conf. Series: Journal of Physics: Conf. Series **1143** (2018) 012021 doi:10.1088/1742-6596/1143/1/012021

Molecular Energetic of Organic Species with Environmental Impact

Jhonatan Martínez[1] **and Juan Dávalos**[2]

[1]Universidad Nacional de Ingeniería-Perú
[2]QIFR-CSIC-Spain

E-mail: jhonatan.martinez.ore@gmail.com

Abstract. Many types of atmospherical compounds and radicals have a high enthalpy. Otherwise, they can react between them or with the O_3 of the ozone layer. Therefore, these reactions generate consequences, such as ozone layer depletion, acid rise of the atmosphere, acid rain, greenhouse effect, etc. In this studio, we present the computational calculation of the enthalpy of formation of some organic compounds which have been emitting since the Industrial Revolution by the human activity.

1. Introduction

The human activity has been emitting a several organic compounds and radicals, most of them can react in the atmosphere, because the atmosphere, specifically the troposphere, has the necessary environment for these reactions (temperature, pressure, etc.). Some reactions generate cycles, thus producing other radicals or compounds, which depends of their enthalpy of formation, the products can be more or less stable in this environment. Actually, most of them have reactions with the O_3 or causing other kind of negative effects in the atmosphere.
However, the emission of organic compounds and radicals not is only the consequence of the human activity, since in the environment exists plenty of sources of these compounds and radicals, but these sources have been being in equilibrium with the environment, so the emanation haven't been being a problem. The oxidative capacity of the atmosphere is a perfect environment for the reaction of the O_3 with HONO[1], NOx[2], NxO [3], CO_x, CH_x, N_x [4], N_xNO[5], etc. In case of CO_2, which is product of the emanation of many sources and contributes in the greenhouse effect, develops cycles which produces other radicals and compounds. Radicals produced in those cycles react with other compounds in that environment for example CH_2O, CH_3OH [6], CH_3CN, NH_4NO_3[3], NH_2NO_2[7], etc.
In general, compounds such as CO_2, CO, hydrocarbons, CH_2CN, etc. are called volatile organic compounds, because they have a low volatility, some of them can react in the atmosphere more easily than others. This characteristic depends of the conditions of the environment.
The present work has the following objectives:

(i) To calculate the enthalpy of formation using Gaussian-n theory, specifically G3 and G4 theory.
(ii) To compare the calculations with the best experimental reported data base.

IOP Conf. Series: Journal of Physics: Conf. Series **1143** (2018) 012021 doi:10.1088/1742-6596/1143/1/012021

2. Computational Methods

Gaussian-n theory is an alternative method for calculate specific thermochemistry like the enthalpies of formation. This theory combines different methods improving the calculations. Every computational method has its own margin of error, in the case of Gaussian-n theory is less of 4.18 KJ/mol, however this theory has had a lot of improvements, in consequence that margin is less. Plus, we optimized this method using ab initio calculations for the geometries for each compound and radicals.

3. Results and Discussions

For the comparison we use the experimental data base reported,

Table 1. Enthalpy of formation of organic compounds and radicals

	Experimental[8]	G3	G4
	$\Delta H_f 298K$		
trans HONO	-78.94±−0.6	-79.63	-77.17
NH_2O	66.9±−3.3	64.68	67.5
NH_2NO_2	4.9±−2	4.89	2.45
HO_2NO_2	-52.7±−4.1	-55.8	-51.01
HNOH	96.7±−3.3	116.52	118.46
trans HOCO	-184.5±−1.2	-180.27	-176.66
HC(O)O	-126.6±−1.7	-130.78	-116.28
CH_2O	-108.99±−1.76	-112.13	-111.25
CO	-110.53±−0.17	-114.55	-111.67
CH_3OH	-201±−0.6	-201.19	-201.38
CH_3O	21±−2.1	17.41	20.46
CO_2	-393.51±−0.13	-397.43	-397.43
HCN	132±−4	128.16	131.59
CN	440±−5	442.31	446.51
CH_3CN	74±−0.4	72.61	74.43
CH_2CN	252.6±−4	257.5	[r]258.57

It is easy to observe the closeness between the calculations and the experimental data base in most of the compounds, on the other hand some of them like NH_2NO_2(G4), HNOH(G3-G4), HC(O)O(G4) and CH_3O(G3)the calculations are not well, it can be consequence of the base used for the geometries, this is the case for HNOH whose the enthalpies of formation are approximately the same, however in comparison with the experimental data the result the error margin is considerable.

In the other cases the geometry no is the problem, because for NH_2NO_2(G3), HC(O)O(G3) and CH_3O(G4) the results are close to experimental data. Probably, this error is consequence of the methodology of its own procedure.

With this table is possible to know the correction factor for each method in the following figure:

IOP Conf. Series: Journal of Physics: Conf. Series **1143** (2018) 012021 doi:10.1088/1742-6596/1143/1/012021

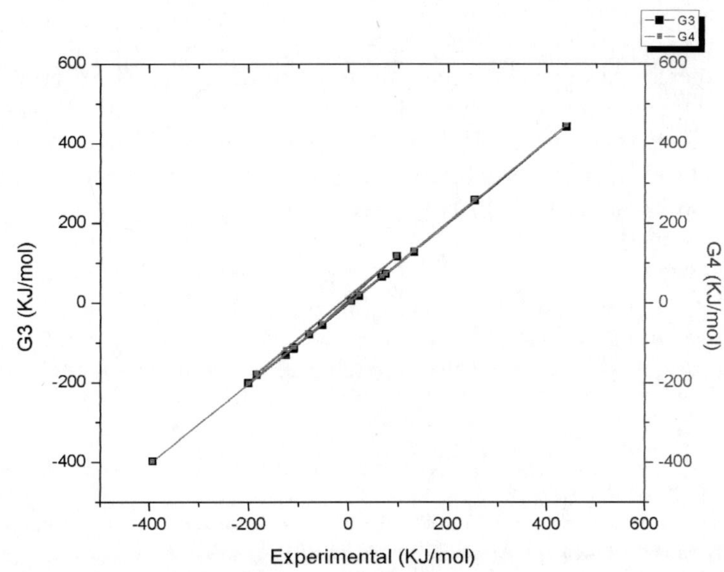

Figure **1**. Enthalpy of formation Experimental Vs G3 and G4($\Delta_f H$ 298K).

For G3 the correction factor is 1.0099 with R^2=0.9992 and for G4 is 1.0104 with R^2=0.999. However, if we take off the compounds and radicals whose results present a big margin of error and regraph, we obtain:

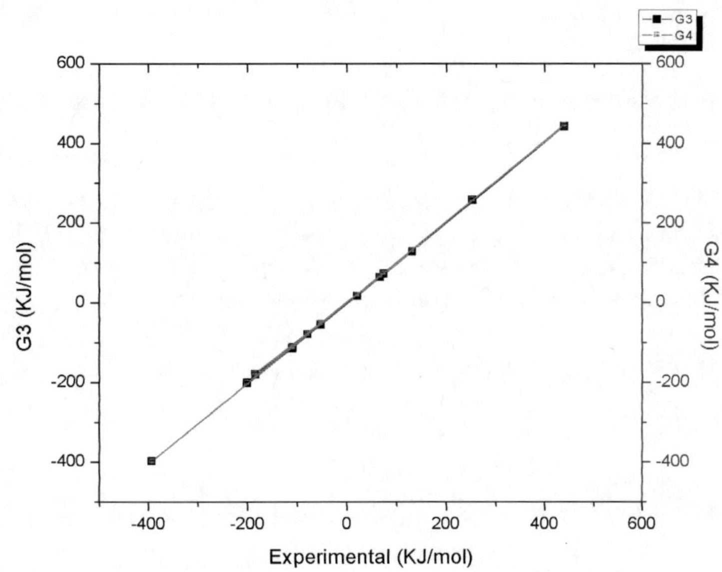

Figure **2**. Enthalpy of formation Experimental Vs G3 and G4($\Delta_f H$ 298K)(regraph)

As expected, the curves are better than the previous graphic. The reason is that, the deviation in average is less to 5%. Actually, most of the calculations for this graphics is 2% in average. Therefore, the correction factors are 1.0056 for G3 and 1.0089 for G4 with R^2=0.9998 for both curves.

IOP Conf. Series: Journal of Physics: Conf. Series **1143** (2018) 012021 doi:10.1088/1742-6596/1143/1/012021

4. Conclusions

With this methodology we could prove the efficiency of the Gaussian-n theory (n=3,4) for organic compounds and radicals. Also, we can recalculate the results using the correction factor for each method. Apparently, the calculations in G3 is better than G4, although the correction factor between them are close, thus the calculations, in general, are good with both methods.
Is necessary considerate the literature, in order to optimize the calculations such as the geometry structure (isomers and conformers), type of molecule, etc. Therefore, this data could be more efficient the calculations.
Consequently, this method is a good way for study and predict other radicals and compounds, which is impossible to know experimentally, since some of them have a high enthalpy of formation. With the obtained data is possible to classify the compounds and radicals according the enthalpy of formation or the reactivity of each one in the atmosphere.

References

[1] Xin Lu, J. Park, and M. C. Lin. Gas phase reactions of hono with no2, o3, and hcl: Ab initio and tst study. *Phys. Chem*, 104:8730–8738, 07 2000.

[2] Rubik Asatryan, Joseph W. Bozelli, and John M. Simmie. Thermochemistry for enthalpies and reaction paths of nitrous acid isomers. *Wiley InterScience*, pages 378–398, 12 2006.

[3] Michel J. Ross, Jeffrey C. Bottaro, and Donaldf. McMillen. The thermal decomposition of the new energetic material ammoniumdinitramide (nh4n(n02)2) in relation to nitramide (nh2n02) and nh4n03. *International Journal of Chemical Kinetics*, 25:549–570, 1993.

[4] F. Bach Maier, K. H. Ebernius, and Th. Just. The formation of nitric oxide and the detection of hen in premixed hydrocarbon-air flames at 1 atmosphere. *Combustion Science and Technology*, 7:77–84, 1973.

[5] Timothy d. Lee and Christopher E. Dateo. he heat of formation of hno. *Chem Phys*, 103(20), 11 1995.

[6] Robert C. Macdonald and Ray Fall. Detention of substantial emission of methanol from plants to the atmosphere. *Atmospheric Environment*, 27(11):1709–1713, 02 1993.

[7] Yu-Ichiro Izato, Mitsuo Koshi, and Atsumi Miyake. Identification of thermal decomposition products and reactions for liquid ammonium nitrate on the basis of ab initio calculation. *International Journal of Chemical Kinetic*, pages 83–99, 11 2016.

[8] J. B. Burkholder, J. P. D. Abbatt, R. E. Huie, M. J. Kurylo, D. M. Wilmouth, S. P. Sander, J. R. Barker, C. E. Kolb, V. L. Orkin, and P. H. Wine. Chemical kinetics and photochemical data for use in atmospheric studies evalution number 18. *JPL Publication 15-1*, 10 2015.

Acknowledgements

These calculations were calculated with the support of the clusters of IQFR-CSIC Spain. We optimized the calculations with the geometry data reported in the literature. Also, we use another computational program for the conformers and isomers.

IOP Conf. Series: Journal of Physics: Conf. Series **1143** (2018) 012022 doi:10.1088/1742-6596/1143/1/012022

Comparison between Monolayer and Bilayer Graphene energy bands using the Tight Binding model

Lucila Menacho, M. Carrasco and Z. Ayala

Universidad Nacional de Ingeniería (UNI), Faculty of Sciences, Lima, Peru

E-mail: alice.xto@gmail.com

Abstract. The Tight Binding model is used to describe the electronic energy band structures of the monolayer and bilayer graphene. The "creation" and "annihilation" quantum operators are employed to compute the respective Hamiltonians. From this, two and four energy bands were respectively obtained. These computations are an attempt to show a unifying approach representing both cases.

1. Introduction

Many special properties of Graphene (both monolayer and bilayer) have their origins in their regular hexagonal lattice structure [1]. Their band structure is also discussed in this article. Monolayer and bilayer graphene have the same generating base vectors and also the same reciprocal lattice. Monolayer graphene has no gap between the conduction and valence band. While, in the case of bilayer graphene, it has 2 equivalent planes in its lattice structure. Furthermore, electron band structure can be analysed following the plane symmetry of the bilayers, as it is made in the analysis of graphite (i.e. over 10 graphene layers) [1].

Both, the monolayer and bilayer electronic band structures have been worked out using the Tight Binding model by different authors [2,3,5,6 and 9]. This model assumes that particles are strongly bounded to the lattice. When an electron is captured by an ion during its motion through the lattice, the electron remains there for a long time, before leaking or tunnelling to the next ion [3,4]. It is well known, in the case of graphene, that relatively strong bindings are maintained for a few layers, which is consistent with the chosen model. Evidently for graphite, such model is not applicable [2].

2. Theoretical background: Tight Binding Model

The Tight Binding model is employed to compute the electronic band structure using the basis wave functions and the linear combination of these [4,5,6]. Considering that strong potential crystal, Φ_v as an atomic orbital, is characterized by subindices v.

$$\Psi_k(x) = \frac{1}{\sqrt{N}} \sum_{j=1}^{N} e^{ikX_j} \Phi_v(x - X_j) \qquad (1)$$

In this equation, the coordinate X_j especifies the i-th atom position. The function $\Phi_v\ (x - X_j)$ tells us the orbital is centred around the j-th atom. On the other hand, as it is a crystal structure $X_j = ja$, where a is the crystal lattice constant, the Bloch function can be expressed as:

$$\psi_k(x) = \frac{1}{\sqrt{N}} e^{ikx} \sum_{j=1}^{N} e^{-ik(x-X_j)} \Phi_v(x - X_j) \qquad (2)$$

In general, an electronic function can be obtained as a superposition of n different Bloch functions:

IOP Conf. Series: Journal of Physics: Conf. Series **1143** (2018) 012022 doi:10.1088/1742-6596/1143/1/012022

$$\Psi_j(x) = \sum_{l=1}^{n} c_{j,l}(k)\psi_l(k,x) \qquad (3)$$

The energy $E_j(k)$ of the j-th band is given as follows:

$$E_j(k) = \frac{\langle \Psi_j | H | \Psi_j \rangle}{\langle \Psi_j | \Psi_j \rangle} \qquad (4)$$

Where H is the Hamiltonian. Replacing (3) in (4):

$$E_j(k) = \frac{\sum_{i,l}^{n} H_{i,l} c_{j,i}^{*} c_{j,l}}{\sum_{i,l}^{n} S_{i,l} c_{j,i}^{*} c_{j,l}} \qquad (5)$$

Where:

$$H_{i,l} = \langle \Psi_i | H | \Psi_l \rangle \quad \text{and} \quad S_{i,l} = \langle \Psi_i \Psi_l \rangle$$

The Energy E_j is minimized with respect to $c_{*}^{j,m}$, deriving and setting to zero, where is obtained:

$$\sum_{l=1}^{n} H_{m,l} c_{j,l} = E_j \sum_{l=1}^{n} S_{m,l} c_{j,l} \qquad (6)$$

From this new equation we can find the energy as follows:

$$det(H - E_j S) = 0 \qquad (7)$$

3. Method

In this section, calculations are presented of the energy bands following the Tight Binding model for monolayer and bilayer graphene.

3.1 Tight Binding model applied to monolayer Graphene

In the monolayer graphene (Figure 1), which has two carbon atoms in its unit cell (A and B), the bond angle between these atoms is $2\pi/3$ or 120° [6], therefore, for each triangle sublattice, it can be written as vectors of the generating base:

$$a_1 = \frac{3}{2}a\begin{pmatrix} 1 \\ \frac{1}{\sqrt{3}} \end{pmatrix}, a_2 = \frac{3}{2}a\begin{pmatrix} 1 \\ -\frac{1}{\sqrt{3}} \end{pmatrix} \qquad (8)$$

Where $a \approx 1{,}42$ Å is the lattice constant. Around of each atom of the sublattice A, there are three atoms of the sub lattice B. They are connected by the following vectors (first neighbors):

$$\delta_1 = \frac{a}{2}\begin{pmatrix} 1 \\ \sqrt{3} \end{pmatrix}, \delta_2 = \frac{a}{2}\begin{pmatrix} 1 \\ -\sqrt{3} \end{pmatrix}, \delta_3 = a\begin{pmatrix} 1 \\ 0 \end{pmatrix} \qquad (9)$$

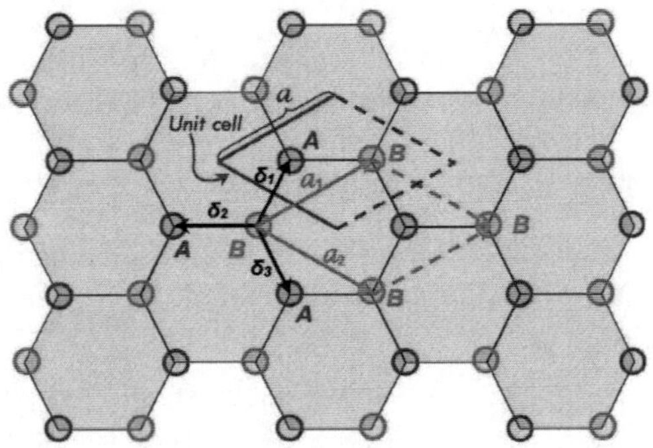

Figure 1. Structure of monolayer graphene.

Hence, the basis in the reciprocal lattice is:

$$b_1 = \frac{2\pi}{3a}\begin{pmatrix}1\\ \sqrt{3}\end{pmatrix}, b_2 = \frac{2\pi}{3a}\begin{pmatrix}1\\ -\sqrt{3}\end{pmatrix} \qquad (10)$$

We form a Hamiltonian based on the "creation" and "annihilation" operators of particles in each one site of sublattices, as follows:

$$H = -t \textstyle\sum_{\langle i,j\rangle} \sum_{s=\pm 1} (a^{\dagger}_{j,s} b_{j,s} + c.h.) - t' \displaystyle\sum_{\langle\langle i,j\rangle\rangle} \sum_{s=\pm 1} (a^{\dagger}_{i,s} a_{j,s} + b^{\dagger}_{i,s} b_{j,s} + c.h.) \qquad (11)$$

Where the term "c.h." the hermitic conjugate. Parameters t *and* $t^{\dagger}$ are related to the probability amplitude of the electrons to effect transitions between the first and second neighbors, respectively, $a^{\dagger}_{i,s}$ y $a_{i,s}$ are "creation" and "annihilation" electron operators, in each one site i with spin s belonging to triangle sublattice A of the hexagonal lattice, similarly to the sublattice B.

By writing down the "creation" and "annihilation" operators in terms of its Fourier transform and diagonalizing it, the bands structure of the system is obtained. Boundary conditions are imposed on the Halmiltonian's auto vectors, in such a way that $\psi(x + a_i) = \psi(x), i = 1{,}2$.

$$a^{\dagger}_i = \frac{1}{N}\sum_{\vec{k}} e^{-i\vec{k}.\vec{x_i}} A^{\dagger}_{\vec{k}} \qquad (12)$$

$$a_i = \frac{1}{N}\sum_{\vec{k}} e^{-i\vec{k}.\vec{x_i}} A_{\vec{k}} \qquad (13)$$

$$b^{\dagger}_i = \frac{1}{N}\sum_{\vec{k}} e^{i\vec{k}.(\vec{x_i}+\vec{c_3})} B_{\vec{k}} \qquad (14)$$

$$b^{\dagger}_i = \frac{1}{N}\sum_{\vec{k}} e^{-i\vec{k}.(\vec{x_i}+\vec{c_3})} B_{\vec{k}} \qquad (15)$$

Since the spin does not play an important role, the subindex will be omitted and the sum over s = ± 1, only the contribution of the first Hamiltonian neighbors are considered. Then, it is obtained:

$$\sum_{\langle i,j\rangle} a_i^\dagger b_j = \sum_{\langle i,j\rangle} \frac{1}{N} \sum_{\vec{k}} e^{-i\vec{k}.\vec{x_i}} A_{\vec{k}}^\dagger \frac{1}{N} \sum_{\vec{k'}} e^{i\vec{k'}.(\vec{x_j}+\vec{c_3})} B_{\vec{k'}} \quad (16)$$

From here, performing a suitable computation, we get:

$$\sum_{\langle i,j\rangle} a_i^\dagger b_j = \sum_{\vec{k}} (1 + e^{i\vec{k}.\vec{a_1}} + e^{i\vec{k}.\vec{a_2}}) e^{i\vec{k}.\vec{c_3}} A_{\vec{k}}^\dagger B_{\vec{k}} \quad (17)$$

Now, we obtained the contributions of the second neighbour, where its first term is:

$$\sum_{\langle\langle i,j\rangle\rangle} a_i^\dagger a_j = \sum_{\vec{k}} (e^{i\vec{k}.\vec{a_1}} + e^{i\vec{k}.\vec{a_2}} + e^{i\vec{k}.(\vec{a_1}-\vec{a_2})}) A_i^\dagger A_j \quad (18)$$

And the second term of the contribution of the second neighbour is:

$$\sum_{\langle\langle i,j\rangle\rangle} b_i^\dagger b_j = \sum_{\vec{k}} (e^{i\vec{k}.\vec{a_1}} + e^{i\vec{k}.\vec{a_2}} + e^{i\vec{k}.(\vec{a_1}-\vec{a_2})}) B_i^\dagger B_j \quad (19)$$

Actually, the Hamiltonian in the Tight Binding approach can be expressed as follows:

$$H = -t \sum_{\vec{k}} ((1 + e^{i\vec{k}.\vec{a_1}} + e^{i\vec{k}.\vec{a_2}}) e^{i\vec{k}.\vec{c_3}} A_{\vec{k}}^\dagger B_{\vec{k}} + h.c.)$$

$$-t' \sum_{\vec{k}} ((e^{i\vec{k}.\vec{a_1}} + e^{i\vec{k}.\vec{a_2}} + e^{i\vec{k}.(\vec{a_1}-\vec{a_2})})(A_{\vec{k}}^\dagger A_{\vec{k}} + B_{\vec{k}}^\dagger B_{\vec{k}}) + c.h.) \quad (20)$$

This Hamiltonian can be written in the matrix way as follows:

$$H = \sum_{\vec{k}} \begin{pmatrix} A_{\vec{k'}}^\dagger & B_{\vec{k}}^\dagger \end{pmatrix} \begin{pmatrix} C & D \\ D^* & C \end{pmatrix} \begin{pmatrix} A_{\vec{k'}}^\dagger \\ B_{\vec{k}}^\dagger \end{pmatrix} \quad (21)$$

Where C and D are defined as follows:

$$C = -t'((e^{i\vec{k}.\vec{a_1}} + e^{-i\vec{k}.\vec{a_1}}) + (e^{i\vec{k}.\vec{a_2}} + e^{-i\vec{k}.\vec{a_2}})$$
$$+(e^{i\vec{k}.(\vec{a_1}-\vec{a_2})} + e^{-i\vec{k}.(\vec{a_1}-\vec{a_2})}))$$
$$D = -t(1 + e^{i\vec{k}.\vec{a_1}} + e^{-i\vec{k}.\vec{a_2}})$$

Diagonalizing this new Hamiltonian, the energy bands turn out to be:

$$E(k) = C \pm D \quad (22)$$

If we define the following function *f(k)* as:

$$f(k) = 3 + 4cos(\tfrac{3k_1 a}{2}) cos(\tfrac{\sqrt{3}}{2} k_2 a) + 2cos(\sqrt{3} k_2 a)$$

The energy is expressed as follows:

$$E_\pm(k) = \pm t\sqrt{f(k)} - t'[f(k) - 3] \quad (23)$$

IOP Conf. Series: Journal of Physics: Conf. Series **1143** (2018) 012022 doi:10.1088/1742-6596/1143/1/012022

3.2 Tight Binding model applied to bilayer Graphene

The graphene bilayer consists of 2 parallel graphene monolayers [7,8,9,10]. In this case there are 4 atoms in the unit cell with unequivalent sites A1, B1 and A2, B2 on the bottom and top grapheme sheets, respectively [2] as shown in Figure 2. Each site B1 of the lower layer is directly below A2 [2]. The unit vectors are the same for bilayer and monolayer graphene.

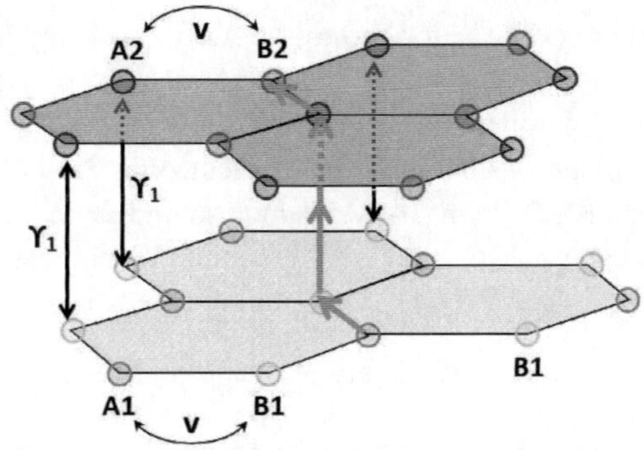

Figure 2. Structure bilayer graphene. The orange shown path goes from A1 to B2 giving the weak coupling $\gamma A1B2$ and is defined as γ_3.

The Tight Binding model of the graph was used adapting the parametrization of Slonczewski-Weiss-McClure [2,9,10] of relevant couplings in order to model the graphene bilayers. The jump in the plane is parametrized by the coupling $\gamma A1B1 \equiv \gamma A2B2 \equiv \gamma_0$, with a velocity $v = (\frac{\sqrt{3}}{2})a\gamma_0/\hbar$,where a is the constant parameter. Additionally, we take into account the strong coupling between layers $\gamma A2B1 \equiv \gamma_1$, within between the pairs of A2B1 orbitals that are directly below and above the other. As the strong coupling produces dimmers of those pairs of orbitals A2B1 leaving the formation of high band energies. In addition, the weak coupling A1-B2, $\gamma A1B2 \equiv \gamma_3$, was included, which leads to an effective velocity $v_3 = (\frac{\sqrt{3}}{2})a\gamma_3/\hbar$, where $v_3 \ll v$. The Hamiltonian is written near the centres of the valleys in a base corresponding to the wave functions:

$$\Psi = (\psi_{A1}, \psi_{B2}, \psi_{A2}, \psi_{B1}) \text{ en el valle } \mathbf{k} \quad (24)$$
$$\Psi = (\psi_{B2}, \psi_{A1}, \psi_{B1}, \psi_{A2}) \text{ en el valle } \tilde{k}$$

$$H = \xi \begin{pmatrix} -\frac{1}{2}\Delta & v_3\pi & 0 & v\pi^\dagger \\ v_3\pi & \frac{1}{2}\Delta & v\pi & \xi\gamma_1 \\ 0 & v\pi^\dagger & \frac{1}{2}\Delta & 0 \\ v\pi & 0 & \xi\gamma_1 & -\frac{1}{2}\Delta \end{pmatrix} \quad (25)$$

Where $\pi^- = p_x + ip_y$, $\pi^\dagger = p_x - ip_y$, $\boldsymbol{P} = (p_x, p_y)$ is the momentum measured with respect to k point. When the magnetic field is zero, H has four degenerate bands valley $s\epsilon_\pm^{(\propto)2}(\boldsymbol{P})$, $\propto = 1,2$

$$\epsilon^{(\alpha)2} = \frac{\gamma_1^2}{2} + \frac{\Delta^2}{4} + (v^2 + \frac{v_3^2}{2})p^2 +$$
$$(-1)^2 \left[\frac{(\gamma_1^2 - v_3^2p^2)^2}{4} + v^2p^2\left[\gamma_1^2 + \Delta^2 + v_3^2p^2\right] + 2\xi v_3 v^2 p^3 \cos 3\phi \right]^{\frac{1}{2}} \quad (26)$$

IOP Conf. Series: Journal of Physics: Conf. Series **1143** (2018) 012022 doi:10.1088/1742-6596/1143/1/012022

The dispersion $\epsilon_{\pm}^{(\propto)2}$ shows that there are 2 bands with energies greater than γ_1 and 2 bands with energies less than γ_1, these do not touch the point k. These bands are the result of the strong pairing γ_1 between the upper and lower layers.

4. Conclusions

- The Tight Binding model was applied to the monolayer and bilayer graphene, obtaining the energies of the band structure. The results show a low energy electronic band structure in the case of bilayer graphene.
- In the case of monolayer graphene, it has 2 electronic bands, while the graphene bilayer has 4 electronic bands of which 2 are conduction and 2 are valence.
- Both, the monolayer and the bilayer graphene, have electronic bands with a gap equivalent to zero.

References

[1] A.C. Ferrari, J.C. Meyer, V. Scardci, C. Casiraghi, M. Lazzeri, F. Mauri, S. Piscanec, D. Jiang, K.S. Novoselov, S. Roth, and A.K. Geim. Raman spectrum of graphene and graphene layers. Physical Review Letters, 97 (18), 187401 1 – 4(2006).

[2] E. McCann, D.S.L. Abergel, and Vladimir I. Fal'ko. The low energy electronic band structure of bilayer graphene. The European Physical Journal Special Topics, 148, 91-103 (2007).

[3] Marcin Mucha-Kruczynski. Theory of Bilayer Graphene Spectroscopy. University of Lancaster.UK (2013).

[4] M.A. Omar. Elementary Solid States Physics: Principles and applications. Lowell Technological Institute. United Stated. (1993).

[5] Anthony T Paxon. An Introduction to the Tight Binding Approximation-Implementation by Diagonalisation. John von Neumann Institute for Computing, UK, (2009).

[6] Christina Bena and Gilles Montambaux. Remarks on the tight-binding model of grapheme. New Journal of Physics 11 095003 (2009).

[7] Jean-Noel Fuchs and Mark Oliver Goerbig. Introduction to the Physical Properties of Graphene. Lecture Notes (2008).

[8] R M Ribeiro, Victor M Pereira, N M R Peres, P R Briddon and A H Castro Neto. Strainde Graphene: tight-binding and density functional calculations. New Journal of Physics 11 (2009).

[9] Victor M. Pereira, A H Castro Neto, and N. M. R. Peres. Tight-binding approach to uniaxial strain in graphene. Physics Review B 80 (2009).

[10] L. M. Zhang, Z. Q. Li, D. N. Basov, M. m. Flogler, Z. Hao, and M. C. Martin. Determination of the electronic structure of bilayer graphene from infrared spectroscopy. Physical Review B 78, 235408 (2008).

IOP Conf. Series: Journal of Physics: Conf. Series **1143** (2018) 012023 doi:10.1088/1742-6596/1143/1/012023

Calibration of remote nanothermometers using nanoparticles of $NaYF_4$: Er^{3+}, Yb^{3+}, Nd^{3+}

D Vásquez Mazzotti, L Alvarado, R Puga and H Loro

Science Faculty, Universidad Nacional de Ingeniería, Lima, Peru

E-mail: diana_mazzotti@uni.edu.pe

Abstract. The nanoparticles of $NaYF_4$: Er^{3+}, Yb^{3+}, Nd^{3+} present fluorescent emissions by upconversion. These emissions are associated with the doping ions of Er, Yb and Nd in the $NaYF_4$ nanoparticles. The nanoparticles of $NaYF_4$ doped with Er^{3+} and Yb^{3+} ions produce upconversion radiation when excited by 980 nm wavelength laser. By incorporating the Nd^{3+} ions, the $NaYF_4$: Er^{3+}, Yb^{3+}, Nd^{3+} arrangement is formed, which produces the same excitation with an 880 nm wavelength laser, the product of a cross-transfer process between the Nd, the Yb and the Er. For applications in human tissue it is preferable to use a 880 nm wavelength laser. These nanoparticles of $NaYF_4$: Er^{3+}, Yb^{3+}, Nd^{3+} have been synthesized in our laboratories with Y (79,3%), Yb (18%), Er (2%) and Nd (0,7%). In this work we have studied the fluorescence behavior of these nanoparticles, which correspond to the transitions $^2H_{11/2} \rightarrow {}^4I_{15/2}$ y $^4S_{3/2} \rightarrow {}^4I_{15/2}$, and that can be used in the calibration of a remote nanothermometer, because they have a sensitivity to very small changes in temperature.

1. Introduction

The nanoparticles of $NaYF_4$: Er^{3+}, Yb^{3+}, Nd^{3+} show fluorescent emissions by upconversion. These emissions are associated with the doping ions of Er, Yb and Nd in the $NaYF_4$ nanoparticles. The nanoparticles of $NaYF_4$ doped with Er^{3+} and Yb^{3+} ions produce upconversion radiation when excited by 980 nm wavelength laser. By incorporating the Nd^{3+} ions, the $NaYF_4$: Er^{3+}, Yb^{3+}, Nd^{3+} arrangement is formed, which produces the same excitation with an 880 nm wavelength laser, the product of a cross-transfer process between the Nd, the Yb and the Er. For applications in human tissue it is preferable to use a 880 nm wavelength laser. These nanoparticles of $NaYF_4$: Er^{3+}, Yb^{3+}, Nd^{3+} have been synthesized in our laboratories in the following proportions Y (79,3%), Yb (18%), Er (2%) and Nd (0,7%). In this work we have studied the fluorescence behavior of these nanoparticles, which correspond to the transitions $^2H_{11/2} \rightarrow {}^4I_{15/2}$ y $^4S_{3/2} \rightarrow {}^4I_{15/2}$, and that can be used in the calibration of a remote nanothermometer, because they have a sensitivity to very small changes in temperature.

In the last decade, the interest in the study of nanoparticles doped with lanthanides by upconversion[1][2] has grown, due to its multiple applications in the field of biomedicine[3][4]. These applications use the fact that photons in the near infrared have an effective penetration length in biological tissues and minimal background autofluorescence. So the nanoparticles doped with lanthanides are also used to measure the temperature with optical sensors. To obtain the temperature of the system, different parameters have been measured, according to

IOP Conf. Series: Journal of Physics: Conf. Series **1143** (2018) 012023 doi:10.1088/1742-6596/1143/1/012023

the material used, such as the absolute luminescent intensity, half-life or the relation between the emission bands[5][6][7][8].

Among the nanoparticles used for these applications, the nanocrystals doped with lanthanides stand out, due to its high photonic stability and its efficient anti-Stokes conversion from the infrared to near visible spectrum. The most studied nanocrystals are those doped with Er and Yb, that are used in luminescent thermometry, because they present sensible changes in the ratio of the emission intensities of the transitions $^2H_{11/2} \rightarrow {}^4I_{15/2}$ a $^4S_{3/2} \rightarrow {}^4I_{15/2}$. When changing the host or the dopants, different arrangements are proposed, according to the physiological temperature range. Among these arrangements stand out the phosphor doped with Nd[9][10], whose luminescence is greater than of other materials in the near-infrared region so they can be used as optical heaters or remote thermometers for biological systems[11][12][13].

The advances in the field of nanothermometry and optics and the constant technological development require testing new materials capable of offering the necessary characteristics for new applications. In this work, the production of $NaYF_4$ nanoparticles doped with Er, Yb and Nd is shown, forming the $NaYF_4$: Er^{3+}, Yb^{3+}, Nd^{3+} array and their application as fluorescent markers and remote nanothermometers.

2. Emission by upconversion in $NaYF_4$: Er, Yb, Nd

The nanoparticles doped with lanthanides, which are used for upconversion, are formed by a crystalline host ($NaYF_4$), a sensitizer (Yb^{3+}) and an emitter (Er^{3+}). As a crystalline host, has been used $NaYF_4$, one of the most efficient due to its low photon energy that allows to minimize the voltages in the network and the non-radiative transitions. As a sensitizer, has been used Yb^{3+} ions, the most efficient because they have only one excited state, the $^2F_{5/2}$ state, which is resonant in energy with the $^4I_{11/2}$ state of the Er3 $+Er^{3+}$ ions, which is used as an emitter. The state of Er^{3+} has an extinction coefficient ten times smaller than that of Yb^{3+}, so it is excited to its $^4F_{7/2}$ state via two successive transfers of nearby Yb^{3+} ion energy, which is promoted from its fundamental state $^4I_{15/2}$ to intermediate state $^4I_{11/2}$ and subsequently to the excited state $^4F_{7/2}$.

To obtain a sensitizer with excitation wavelength near 800 nm, Nd^{3+} is incorporated into the first biological window of human tissue, due to its strong absorption close to this wavelength. The figure 1 shows a scheme of the extinction spectrum, taking into account the absorption and dispersion of human tissue. In this spectrum, the absorption bands that allow the definition of the two biological windows can be identified. The first biological window extends from 700 nm to 980 nm and corresponds to the spectral range defined between the visible absorption band of hemoglobin and the characteristic water absorption band. In this spectral region there is a decrease in the absorption of light, however an important optical extinction is still observed due to the residual dispersion, whose relevance decreases for longer wavelengths. The second biological window extends from 1000 nm to 1400 nm, both limits correspond to the absorption bands of water. In this spectral window the optical absorption does not completely cancel out, since the absorption coefficient of water is close to 0.5 cm^{-1}. However, the optical dispersion is lower compared to that of the first biological window due to the longer wavelengths.

The Nd^{3+} ions are those that allow emitting in wavelengths of 980 nm. This allows the excitation of the Yb^{3+} ions in order to produce upconversion between the ions of Yb^{3+} and Er^{3+}. The figure 2 shows the scheme of the energy levels required to produce upconversion in the nanoparticles of $NaYF_4$: Er^{3+}, Yb^{3+}, Nd^{3+}. To produce upconversion in these nanoparticles we have an initial pump wavelength close to 800 nm, which produces a transfer of energy close to 980 nm to Yb^{3+}. The emissions of the Yb^{3+} ion produce a second energy transfer to the Er^{3+} ion, whose emission spectrum in the visible range consists of three emissions attributed to the transitions $^2H_{11/2} \rightarrow {}^4I_{15/2}$ and $^4S_{3/2} \rightarrow {}^4I_{15/2}$ corresponding to wavelengths of 525 nm and 545 nm respectively, both in green, and $^4F_{9/2} \rightarrow {}^4I_{15/2}$ corresponding to a wavelength of 650 nm in red.

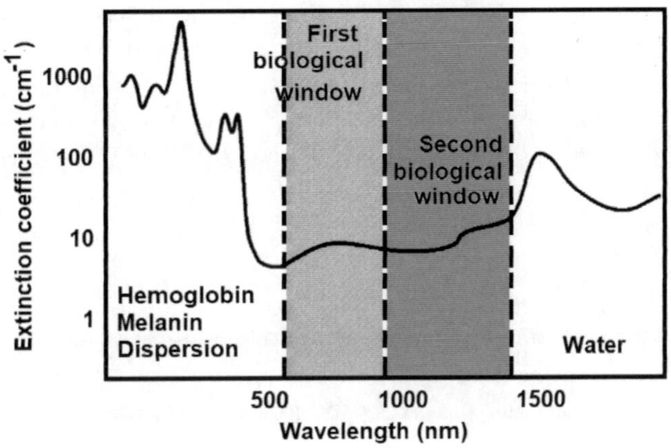

Figure 1. Extinction coefficient as a function of the wavelenght in human tissue where the first and second biological windows stand out

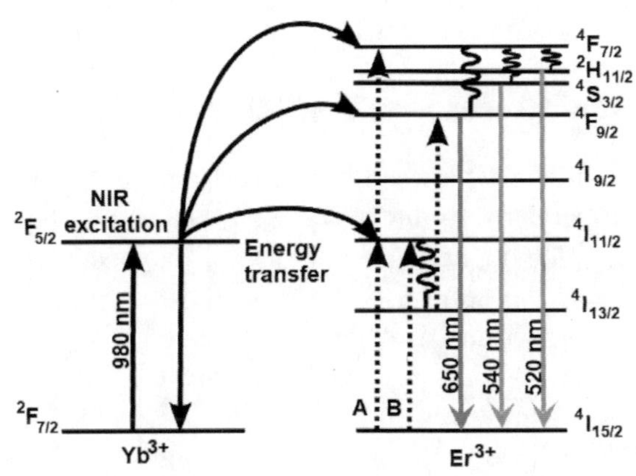

Figure 2. Diagram showing the cross-transfer process in the $NaYF_4$: Er, Yb, Nd. Two possible mechanisms A and B contributing to Er^{3+} emissions are shown. The initial wavelenght is 808 nm

In the transition from Yb^{3+} to Er^{3+}, the excitation energy can be transferred to the energy level $^4I_{11/2}$ of the Er^{3+}, via a cross-relaxation process: $^2F_{5/2} \rightarrow {}^2F^{7/2}$ (Yb^{3+}) $^4I_{15/2} \rightarrow {}^4I_{11/2}$ (Er^{3+}), marked as A in the figure 2. A second cross-relaxation process from Yb^{3+} to an Er^{3+} ion, previously excited, gives rise to the excitation of the Er^{3+} ion at higher energy levels via a new transfer process: $^2F_{5/2} \rightarrow {}^2F_{7/2}$ (Yb^{3+}) $^4I_{11/2} \rightarrow {}^4F_{7/2}$ (Er^{3+}) that allows green emission. Thermal processes or the presence of nearby defects produce non-radiative decays at the levels

IOP Conf. Series: Journal of Physics: Conf. Series **1143** (2018) 012023 doi:10.1088/1742-6596/1143/1/012023

$^4F_{9/2}$, $^4S_{3/2}$ and $^2H_{11/2}$ that give rise to the transitions $^4F_{9/2} \rightarrow {}^4I_{15/2}$ of 650 nm, $^4S_{3/2} \rightarrow {}^4I_{15/2}$ of 540 nm and $^2H_{11/2} \rightarrow {}^4I_{15/2}$ of 520 nm in the Er^{3+}, which leads to a competition between red and green emissions. Another probable mechanism of the red emulsion is shown as B in the figure 2.

3. Relation between temperature and probabilities of transition

The emission in the green of the Er^{3+} ion, product of the energy transfer of the Yb^{3+} ion, consists of two bands between 515 nm and 535 nm, centered at 525 nm and between 535 nm and 570 nm, centered at 545 nm, which come from transitions from the excited states, $^2H_{11/2}$ and $^4S_{3/2}$ respectively to the ground state. These two states are very close, separated only by hundreds of wave numbers, so they reach a thermal equilibrium governed by the Boltzman factor:

$$\frac{I_{525}}{I_{545}} = Ae^{-\Delta E/k_B T} \tag{1}$$

where A is a constant, I_{525} and I_{545} are the integrated intensities of the transitions $^2H_{11/2} \rightarrow {}^4I_{15/2}$ and $^2S_{3/2} \rightarrow {}^4I_{15/2}$ respectively, ΔE is the energy gap that separates the two excited states, k_B is the Boltzmann constant and T is the absolute temperature.

This leads to changes in the emission intensities of these bands as a function of temperature. Thus it is possible to obtain a thermometric scale that gives information on the temperature of the nanoparticles and their environment. From this equation is found a linear relationship between $ln(\frac{I_{525}}{I_{545}})$ and $\frac{1}{T}$, given by the equation 2 shown below:

$$ln\left(\frac{I_{525}}{I_{545}}\right) = \frac{-\Delta E}{k_B T} + B \tag{2}$$

where B is a constant.

Calculating the energy difference between the transitions of interest we have that $\Delta E = 1.389 \times 10^{-20}$ J, at room temperature we have that $k_B T$ is approximately 0.414×10^{-20} J, that is, ΔE is of the order of $k_B T$. The dependence of $ln\left(\frac{I_{525}}{I_{545}}\right)$ has been measured with temperature at temperatures close to room temperature. And with these results the value of the slope $\frac{\Delta E}{k_B}$ is estimated in the previous equation. The calculated value is 1.006 86 K, close to the experimentally found value.

4. Experimental results

The nanoparticles of $NaYF_4$: Er^{3+}, Yb^{3+}, Nd^{3+} have been synthesized by the solvothermal method [14]. The percentages used are: Y (79.3%), Yb (18%), Er (2%), Nd (0.7%). Initially, 3.6 mmol of NaCl (Sigma Aldrich), 1.427 mmol of $YCl_3 \cdot 6H_2O$ (Sigma Aldrich), 0.036 mmol of $ErCl_3 \cdot 6H_2O$ (Sigma Aldrich), 0.324 mmol of $YbCl_3 \cdot 6H_2O$ and 0.013 mmol of $NdCl_3$ (Sigma Aldrich) were dissolved, in a solution of 27 ml of ethylene glycol containing 0.45 g of branched polyethylene (Sigma Aldrich). The mixture was stirred for 60 minutes using a magnetic stirrer. Subsequently, a solution of 17 ml of ethylene glycol with 7.2 mmol of NH_4F (Sigma Aldrich) was added to the initial solution containing the chlorides and stirred for 30 minutes. The resulting clear solution was heated with an 85 ml teflon flask in an oven at 170 °C for 5 hours. Finally, he performed a centrifugation process at 6,000 RPM using a mixture of distilled water and ethanol for 15 minutes in order to wash them. This process was repeated several times. The resulting product was dried for 12 hours at 80 °C. This sample was subjected to a thermal treatment for a further 5 hours at 500 °C. The final product obtained was a fine off-white powder, like the one shown in the figure 3.

Figure 3. $NaYF_4$: Er, Yb, Nd nanoparticle fabricated at the laboratory

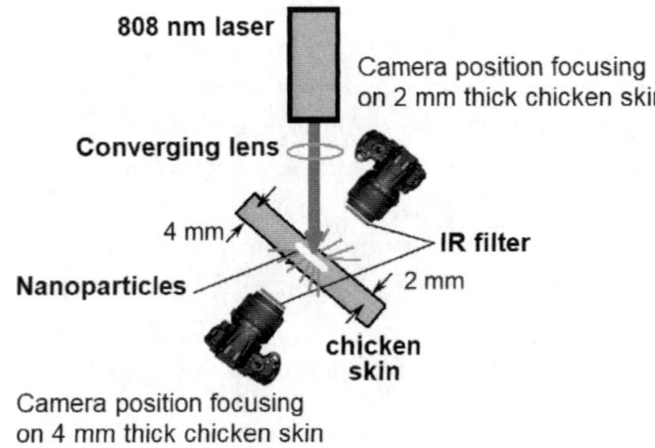

Figure 4. Disposition of the equipment to record the images obtained by the emission of the nanoparticles inside the chicken skin

These nanoparticles convert 808 nm radiation into shorter wavelength radiation such as green and red. The green emissions correspond to the $^2H_{11/2} \rightarrow {}^4I_{15/2}$ and $^2S_{3/2} \rightarrow {}^4I_{15/2}$ transitions. In order to observe this emission under a biological tissue, an experimental arrangement was mounted, schematized in the figure 4. For this experiment a laser diode of 2 W of power was used, a Nikon camera model D3300 with an IR filter, to register the observed images, eliminating the source of excitation and attenuating the emission in red. The nanoparticles were placed inside the chicken tissue. The figures 5 and 6 show the emission of the nanoparticles when passing through a chicken tissue of approximately 2 mm and 4 mm respectively. They show that the efficiency of the emissions is much greater when the particles are covered by a 2 mm thick chicken tissue.

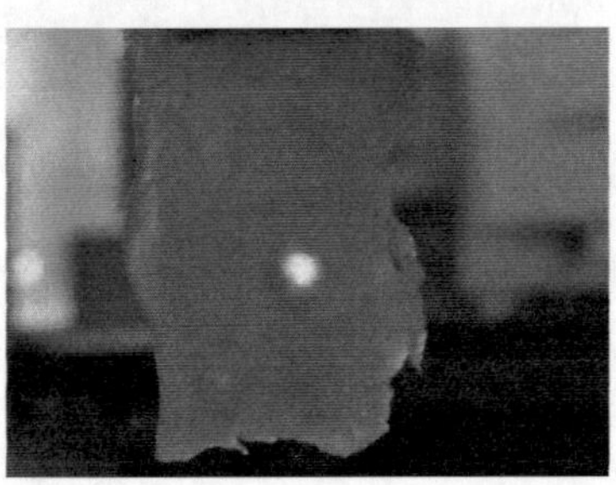

Figure 5. Emission of the nanoparticles inside 2 mm thick of chicken tissue

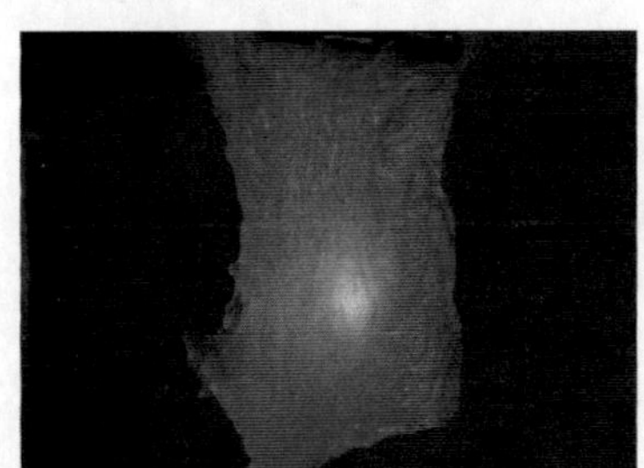

Figure 6. Emission of the nanoparticles inside 4 mm thick of chicken tissue

To observe the changes in the intensities of the emission bands $^2H_{11/2} \rightarrow {}^4I_{15/2}$ and $^2S_{3/2} \rightarrow {}^4I_{15/2}$ from Er^{3+} in nanoparticles of $NaYF_4$: Er^{3+}, Yb^{3+}, Nd^{3+} when the temperature changed, the emission spectra of these nanoparticles were analyzed for different temperatures close to physiological temperatures. The assembly of this experiment is shown in the figure 7 of the

IOP Conf. Series: Journal of Physics: Conf. Series **1143** (2018) 012023 doi:10.1088/1742-6596/1143/1/012023

experiment setup to obtain the emission spectra. For this, a laser diode of 808 nm and up to 2 W of output power was used, a B&W Tek Model BTC-110S spectrometer connected to an optical fiber and controlled by a PC where the data obtained is also stored, a filter IR to eliminate the excitation source of the nanoparticles. The observed emissions correspond to a tablet of $NaYF_4$: Er^{3+}, Yb^{3+}, Nd^{3+}, and are shown in the figure 8. This tablet of 0.02 g of mass was obtained after compressing the powder sample to form a tablet of 0.2 mm in thickness and 6 mm in radius. This tablet was placed on a copper rod 22 mm in diameter and 23 mm in height. The copper bar, wrapped by coated nichrome wire, was connected to a variable voltage source between 0 V and 5 V, in order to vary the temperature of the nanoparticles tablet. To record the temperature and to be able to calibrate the emission spectra, a type J thermocouple was used, which made contact with the tablet. Likewise, converging lenses were used to focus both the incident radiation and the one emitted by the nanoparticles.

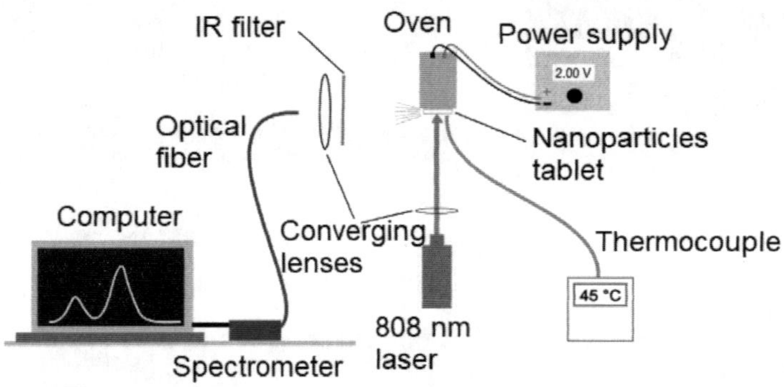

Figure 7. Experiment setup to obtain the emission spectra

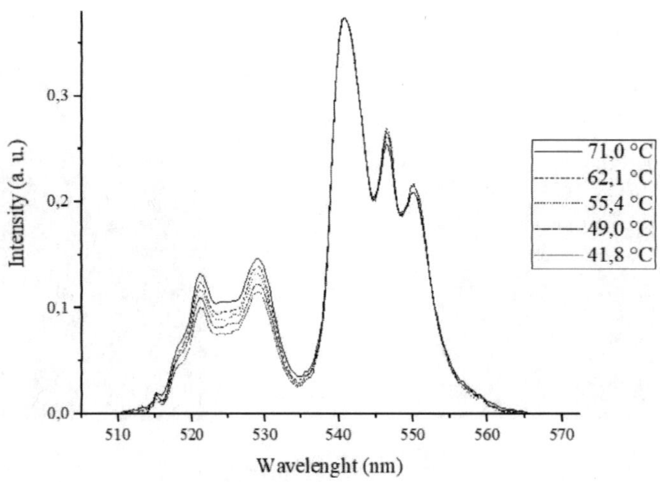

Figure 8. Emission spectra of NaYF4: Er^{3+}, Yb^{3+}, Nd^{3+} with excitation wavelenght of 808 nm obtained at 41.8 °C, 49.0 °C, 55.0 °C, 62.1 °C and 71.0 °C temperatures

The emission spectra obtained consist of two bands between 510 nm and 565 nm. When the temperature of the nanoparticles varies, it is observed that the relative intensity between

IOP Conf. Series: Journal of Physics: Conf. Series **1143** (2018) 012023 doi:10.1088/1742-6596/1143/1/012023

these bands changes, registering a relative growth of the band between 510 nm and 535 nm, corresponding to the transition $^2S_{3/2} \rightarrow {}^4I_{15/2}$, maintaining the same band shapes. The result obtained shows the hypersensitivity of the relation between the intensities of these bands with temperature. This ratio can be used as a parameter to measure the temperatures of the nanoparticles. In order to study the dependence of this ratio of intensities, the integrated intensities of the emission spectra obtained were analyzed. For this the area covered by the first band (I_{525}) was integrated, as well as the area covered by the second band (I_{545}). The figure 9 shows the graph of $ln\left(\frac{I_{525}}{I_{545}}\right)$ as a function of $\frac{1}{T}$. A linear dependence is observed where the slope corresponds to $\frac{\Delta E}{k_B}$.

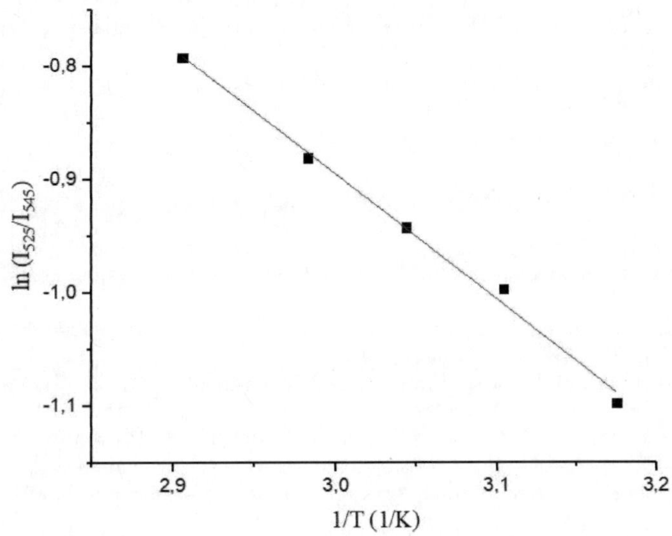

Figure 9. Nanothermometer calibration curve for a $NaYF_4$: Er^{3+}, Yb^{3+}, Nd^{3+} tablet when excited with a laser of 808 nm and 2 W power. The points correspond to the experimental data and the line represents the best fit

By adjusting the experimental data to the expression of $ln\frac{I_{525}}{I_{545}}$ it was found that the slope of the line is $-1,1 \times 10^{-20}$, which is in accordance with the result of $-1,0 \times 10^{-20}$ that was previously calculated. The difference can be attributed to the fact that the bands are not really centered at 525 nm and 545 nm. The adjustment of the curve has a R^2 of 0.99478, which shows how precise the fit is. This line allows calibrating the nanometer. The linear behavior observed in the figure reveals that a precise determination of the relationship between the luminescent intensities of these hypersensitive lines allows to determine the temperature of the nanoparticles of $NaYF_4$: Er^{3+}, Yb^{3+}, Nd^{3+} for temperatures close to room temperature.

5. Conclusions

(i) The 525 nm and 545 nm emissions of $NaYF_4$: Er^{3+}, Yb^{3+}, Nd^{3+} nanoparticles within chicken meat have been used to obtain the temperature of the nanoparticles and that of their environment.

(ii) The same luminescent emissions have been observed when placing the nanoparticles of $NaYF_4$: Er^{3+}, Yb^{3+}, Nd^{3+}, between two pieces of chicken 2 mm and 4 mm thick.

Acknowledgments
This work has been financed by the FONDECYT-CONCYTEC (agreements No. 167 and No. 168) and developed in the Photonics Laboratory of the Faculty of Sciences of the Universidad Nacional de Ingeniería.

References

[1] Tian L. Xu Z. Zhao S. Cui Y. Liang Z. Zhang J. and Xu X. The upconversion luminiscence of $er^{3+}/yb^{3+}/nd^{3+}$ triply-doped β-nayf$_4$ nanocrystals under 808-nm excitation. *Materials*, 7:7289–7303, 2014.
[2] Jaque D. Martínez Maestro L. Del Rosal B. Haro-Gonzalez P. Benayas A. Plaza J. L. Martín Rodríguez E. and García Solé J. Nanoparticles for photothermal therapies. *Nanoscale*, 2014.
[3] Peng H. Stich M. Yu J. Sun L. Fischer L. and Wolfeis O. Luminescent europium (iii) nanoparticles for sensing and imaging of temperature in the physiological range. *Adv. Mater.*, 2010.
[4] Zhou J. Liu Q. Feng W. Sun Y. and Li F. Upconversion luminescent materials: Advances and applications. *Chem. Rev.*, 2015.
[5] Ye F. Wu C. Jin Y. Chan Y. Zhang X. and Chiu D. Ratiometric temperature sensing with semiconducting polymer dots. *J. Am. Chem. Soc.*, 2011.
[6] Kim M. and Yoda M. Infrared quantum dots for liquid-phase thermometry in silicon. *Meas. Sci. Technol.*, 2011.
[7] Zhou J. Xu S. Zhang J. and Qui J. Upconversion luminescence behavior of single nanoparticles. *Nanoscale*, 2015.
[8] Debasu M. Ananias D. Pastoriza-Santos I. Liz-Marzán L. Rocha J. and Carlos L. All-in-one optical heater-thermometer nanoplatform operative from 300 to 2000 k based on er^{3+} emission and blackbody radiation. *Adv. Mater.*, 2013.
[9] Xue F. and Wang H. The stable and water-soluble neodymium-doped lanthanide fluoride nanoparticles for near infrared probing of copper ion. *Talanta*, 2012.
[10] Bednarkiewicz A. Wawrzynczyk D. Nyk M. and Strek W. Optically stimulated heating using nd^{3+} doped nayf$_4$ colloidal near infrared nanophosphors. *Appl. Phys. B*, 2011.
[11] M. Rocha U. Kumar K. Jacinto C. Navarro E. Martín Rodríguez E. Pedroni M. Speghini A. Hirata G. Martín I. Del Rosal B. Pérez-Delgado A. Misiak M. Bednarkiewicz A. Vanetsev A. Orlovskii Y. Jovanović D. Dramićanin and Jaque D. Neodymium-doped nanoparticles for infrared fluorescence bioimaging: The role of the host. *Appl. Phys.*, 2015.
[12] L. Bednarkiewicz A. Stefanski M. Tomala R. Hreniak D. Marciniak and Strek W. Near infrared absorbing near infrared emitting highly-sensitive luminescent nanothermometer based on nd^{3+} to yb^{3+} energy transfer. *Phys. Chem. Chem. Phys.*, 2015.
[13] Wawrzynczyk D. Bednarkiewicz A. Nyk M. Strek W. and Samoc M. Neodymium (iii) doped fluoride nanoparticles as non-contact optical temperature sensors. *Nanoscale*, 2012.
[14] Puga R. Palacios E. Velasco M. Hernández J. and Loro H. Síntesis de partículas α-nayf$_4$: Er, yb por el método solvotermal para florescencia por upconversion. *REVCIUNI*, 2014.

Synthesis of $NaYF_4$: Yb, Tm Nanoparticles by Solvothermal Method and Characterization by Upconversion Visible Radiation Fluorescence

Rubén Mallqui Ch†[1], Rubén Puga[1], Dwight Acosta[2], José M. Hernandez[2], and Héctor Loro[1]

[1] Faculty of Sciences, Universidad Nacional de Ingeniería (UNI), Lima, Peru

[2] Institute of Physics at UNAM, Mexico

E-mail: † ruben.mallqui20@gmail.com

Abstract. We report the synthesis of $NaYF_4$ nanoparticles doped with rare earth ions Yb^{3+}, Tm^{3+}, using the solvotermal method. These nanoparticles were synthesized varying the concentration of Tm^{3+}, considering 0.5%; 1.0%; 1.5% and 2.0% of Tm^{3+} instead of the Y ions in $NaYF_4$ in order to study variations in their emissions around 480 nm. The results show a better performance of 1% Tm^{3+} doped nanoparticles. We report emission spectroscopy results using a 980 nm diode laser.

1. Introduction

Lanthanides are a special group of chemical element of the periodic table, because of the particular characteristics of their 4f orbitals. These inorganic materials with interesting fluorescent properties, when combined with a host, as in the case of Tm^{3+} and Yb^{3+} in $NaYF_4$, present the phenomenon known as upconversion [1] where an Yb transition is transferred to increase the energy of an excited electron in Tm^{3+} atom. This array of doping ions in a host is known as Upconversion nanoparticles (UCNPs).

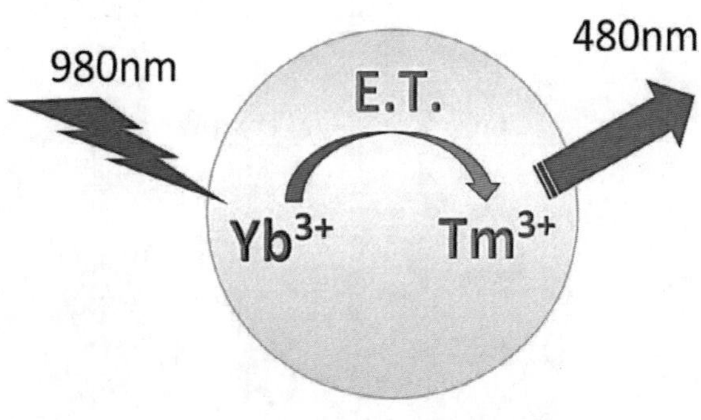

Figure 1. Schematic of the Upconversion process as in $NaYF_4$ matrix with its doping ions Yb^{3+}, Tm^{3+}.

In these UCNPs one of the doping Ln^{3+} ions acts as a sensitizer, absorbing light. The sensitizer emits light, which is absorbed by the other Ln^{3+} doping ion (activator). This activator emits light at a wavelength greater than the absorbed light [2]. In addition, the inorganic matrix $NaYF_4$ is useful because of its low vibrational energy and high refractive index [3].

The $NaYF_4$ matrix has two polymorphic crystalline structure, the α-phase that is a cubic structure and the β-phase that is a hexagonal structure; these appear depending on the synthesis method and conditions [4]. Previous research showed that the β-phase presents a better crystalline for the different doping ions, in addition, the formation of the $NaYF_4$ matrix can be done with various synthesis methods such as coprecipitation, hydrothermal and solvothermal, the latter is developed when a chemical compound in liquid mixed placed in a closed container and heated above its boiling point, generating a pressure above atmospheric, with the objective of a better dissolution between the mixed chemical compound [5]. Finally, the characterization of these nanoparticles is carried out with the luminescence of Upconversion, using diode laser of 980 nm for its good affinity to be absorbed, and analyzing its fluorescence spectrophotometer.

2. Method of synthesis and characterization $NaYF_4$ nanoparticles

The method used for the synthesis of the nanoparticle $NaYF_4$ was solvothermal method. The chemical compounds were initially separated into three beakers of 100ml. The 1st beaker with 3.6 mmol of NaCl (Sigma Aldrich), 1.431 mmol of $YCl_3.6H_2O$ (Sigma Aldrich), 0.0324 mmol of $YbCl_3.H_2O$ (Sigma Aldrich) and 0.036 mmol of $TmCl_3.6H_2O$ (Sigma Aldrich), the 2nd beaker with 27 ml of polyethyleneimine and 17 ml of ethylene glycol and the 3rd beaker with 0.027g NH4F and 17ml of ethylene glycol, then homogenous mixing was carried out with the aid of a magnetic stirrer [Figures 2a and 2b]. The mixed sample was poured in a Teflon and sealed with an autoclave for its respective solvothermal process, for this we heated it to 170°C for 6 hours in an oven and the temperature was controlled by a Eurotherm 2408f [Figures 2c and 2d].

After this process, the heated sample in centrifuge tubes was placed in a centrifuge equipment (universal centrifuge PLC-012E) for its respective precipitation. After this process 3ml of residual liquid was to remove, then 2ml of ultrapure water and 2ml of ethanol were aggregated to clean the sample and it was centrifuged again. This process was repeated three times [Figure 2e]. After this, the sample was separated in a 25ml crucible and taken to an electric oven to perform the drying of the sample [Figure 2f], the dry sample was again taken to the oven to carry out a heat treatment. This consisted of taking the $NaYF_4$ sample at 400°C with two different time periods of 2 and 3 hours, this was in order to study dependence of the characterization of the sample with the periods of heat treatment. [Figure 2g].

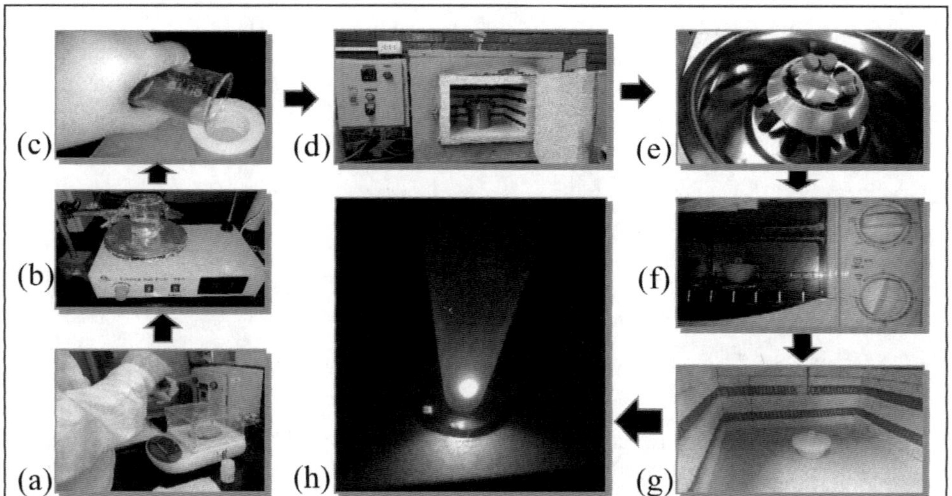

Figure 2. Synthesis of $NaYF_4$: Yb, Tm nanoparticles by the solvothermal method. (a) and (b) sample preparation, (c) and (d) sample heating, (e) and (f) washing and drying, (g) heat treatment and (h) light emission using a diode laser of 980nm as the excitation source.

Finally, the sample was taken to an experimental arrange in order to visualize the luminescence it presented, and to carry out its characterization. This arrange included a 980nm laser diode, 2 lenses placed between the sample, a ramp to position the sample and a filter positioned in such a way that it received the emission of the sample. This emission was taken to a spectrophotometer (Science-surplus), through its software the sample emissions peaks were analyzed [Figures 2h and 3].

Figure 3. Characterization by emission spectrometry. On the left: the encapsulated sample, diode laser, spectrometer, lenses, and the optical fiber used in the experiment. On the right: the sample emission spectrum registered by the computer.

3. Interpretation and Discussion

The samples were placed in two groups. Each group had samples with different molar concentrations of thulium dopant ions at 0.5%, 1.0%, 1.5% and 2.0%. A different thermal treatment was applied to each group of samples.
Emission spectra are shown in Figure 4 for both groups of samples. Spectra shown in Figure 4.1 have three prominent peaks (with a highest intensity of nearly 25 000 u.a.) at wavelengths of 450nm, 477nm and 645nm with the most prominent peak being at 477nm which corresponds to the $^1G_4 \rightarrow {}^3H_6$ transition.

In Figure 4.2, the highest intensity is approximately 21 000 u.a., this also presented three prominent peaks and it is observed again that the molar concentration at 1.0% Tm has better intensity than the other concentrations.

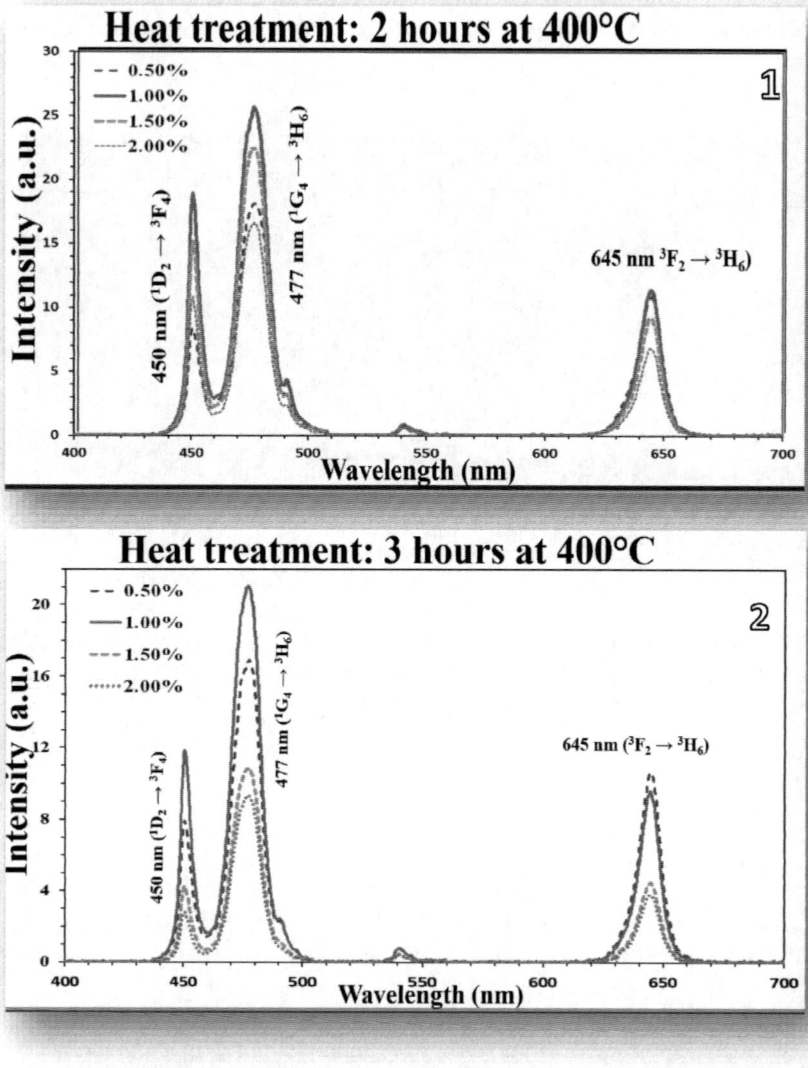

Figure 4. Emission spectra of the 4 Tm samples (0.5%: 1%: 1.5%: 2%). **(1)** Emission spectra with heat treatment 2 hours at 400°C. **(2)** Emission spectra with heat treatment 3 hours at 400 °C.

An X-ray diffraction (XRD) spectrum was carried out to the sample of 1% Tm molar concentration with two hours of heat treatment. Figure 5 shows the XRD spectra of this sample. Spectra shows the presence of α-phase (in red) and the β-phase (in blue). In this Figure, in the small graph, we present the diffraction pattern obtained from the literature for $NaYF_4$ α-phase and β-phase.

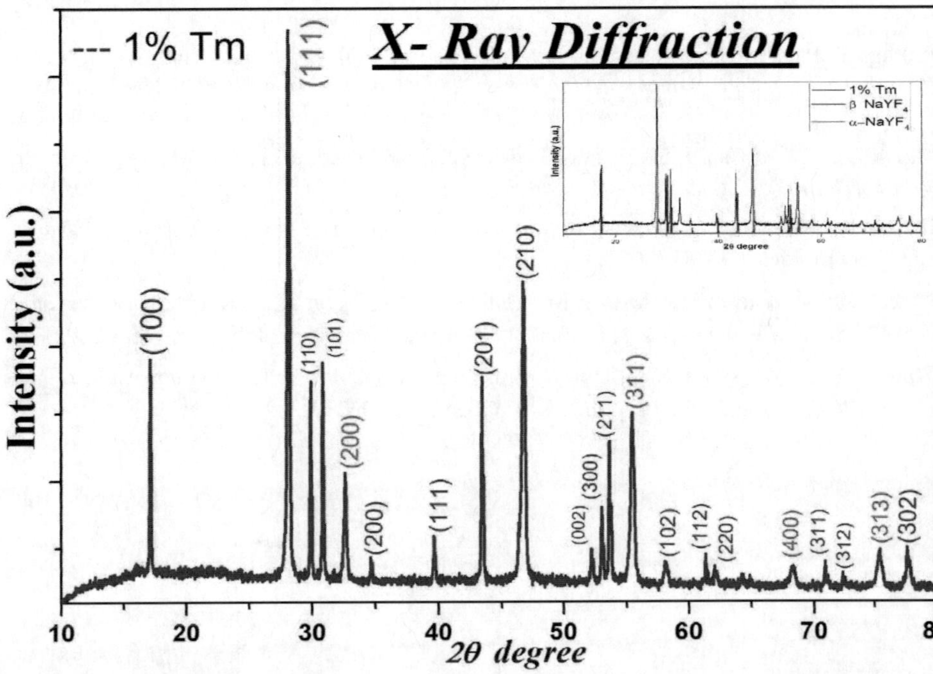

Figure 5. X-ray diffraction spectrum of 1% Tm molar concentration of NaYF4: Yb, Tm, and its miller index of diffraction patterns, small graph shows the comparison between x-ray diffraction spectrum and diffraction patterns of inorganic matrix $NaYF_4$ α and β-phase taken from references [6] y [7].

4. Conclusions

- Using the Solvothermal method it was possible to synthetize $NaYF_4$ nanoparticles doped with Yb^{3+}, Tm^{3+} ions and blue luminescent emission could be observed.
- For both heat treatments (by 2 and 3 hours at 400 °C), 1% Tm concentration samples shows to be the most efficient emission ones.
- Heat treatment at 400°C by 2 hours let us get a better contrast between the observed emissions.
- X-ray diffraction spectra for the 1% Tm molar concentration with 2 hours of heat treatment shows a good correspondence with X-ray patterns of β-NaYF4 and α-NaYF4 matrix.
- Future work with these nanoparticles will be performed using 1% Tm samples.

5. References

[1] P. V. Delgado. *Caracterización del NaYF4 impurificado con Yb3+ y Tm 3+ y sus potenciales aplicaciones en biomedicina.* Universidad de Cantabria, 4 – 47 (2011 – 2012).

[2] V. Romero. *Estudio de las propiedades luminiscentes de oxidos nanocristalinos dopados con Eu, Tb y Tm.* 1- 42 (2012).

[3] R. Puga, E. Palacios, M. Velasco, J. Hernández, H. Loro. *Síntesis de partículas α– $NaYF_4$: Er, Yb por el método Solvotermal para fluorescencia por upconversión. REVCI*UNI 17(1) 39 – 41 (2014).

[4] M. Ding, C. Lu, L. Cao, J. Song, Y. Ni and Z. Xu. *Facile synthesis of β– $NaYF_4$: Ln^{3+} (Ln=Eu,Tb,Yb/Er, Yb/Tm) microcrystals with down- and up-conversion luminescence.* (2013).

[5] R. Zanella. *Metodologías para las síntesis de nanopartículas: controlando forma y tamaño.* Mundo Nano Vol 5(1) 69 – 81 (2012).

[6] Grzechinik A., Bouvier P., Mezouar M., Mathews M.D., Tyagi A.K., Koehler J., *Hexagonal Na(1.5) Y(1.5) F6 at high pressures*, Journal of Solid State Chemistry, 165, 159 – 164 (2002).

[7] Zhurova E.A., Maksimov B.A., Hull S., Wilson S.S., Sobolev B. P., Simonov V. I., *Defect structure of Na0.39 Y0.61 F2.22 crystals*, Kristallografiya, 42, 277-282 (1997).

Synthesis of Bismuth Ferrite $BiFeO_3$ by solution combustion method

J Peñalva[1] and A Lazo[2]

[1] National University of Engineering, Faculty of Sciences, Lima 25, Peru

[2] National University of Engineering, Faculty of Geological, Mining and Metallurgical Engineering, Av. Túpac Amaru, Lima 25, Peru

Email: jpenalvas@uni.pe

Abstract: In this work we describe the synthesis of bismuth ferrite $BiFeO_3$ (BFO) by the method of combustion in solution. Using the techniques of X-ray diffraction (XRD) together with its refinement Rietveld reveals the existence of a distorted rhombohedral perovskite structure (Space group-R3c). The average size of the BFO crystallite was estimated from the half-height width of the X-ray diffraction peaks using the Scherrer's equation. The morphology of the prepared sample was determined by scanning electron microscopy (SEM) and stoichiometry ratio of BFO was determined by energy dispersive spectroscopy (EDS)

Keywords: Combustion synthesis, BFO, Multiferroics, X-Ray Diffraction (XRD)

1. Introduction

Multiferroic materials, with simultaneous coexistence of electrical and magnetic order, are currently the subject of intense research not only due to promising applications of multifunctional devices, but also to the interesting physics taking place in this class of materials.

Multiferroic material $BiFeO_3$ (BFO) has ABO_3 distorted perovskite structure and it is known as magneto-electric material. Due to this magnetoelectric coupling at room temperature, BFO can be used to develop novel applications in the field of information storage, spintronics, sensors and photovoltaic devices [1,2,3]. In addition to potential magnetoelectric applications, the photovoltaic activity of BFO has attracted considerable attention because of its small band gap which is used to the absorption between visible light ranges, making it a promising candidate for a visible light photocatalyst [4]. Bismuth ferrite exhibits simultaneous ferroelectric order at Curie temperature ($T_C \approx 830$ °C) and antiferromagnetic at Neel temperature ($T_N \approx 370$ °C) [5]. However, antiferromagnetic order shows modulated spin spiral structure of periodicity 62nm whereby it show very low magnetization in BFO [6].

BFO has been synthesized by different methods: nitrate combustion methods with several fuel as Glycine [7,8], Tartaric Acid [8,9] and Sucarose [10]; Sol-Gel method [11] and Pechini method [12]. However, each of these methods has its own advantages and limitations. The development of simpler and efficient process to obtain BFO nanoparticles with regular morphology and a homogeneous chemical composition is still an active area of research. The combustion method is found to be a rapid process, simple and effective to get very fine and homogeneous nanopowders. This method is one of the most-convenient methods for the preparation of oxides materials, which allows effective synthesis of a variety of nanosize materials.

In this work we have synthesized the BFO nanoparticles prepared by the combustion method using metallic nitrates with mixture fuel between glycine and urea, in which we have used the ratio of fuels recommended by Ianos et al [13]. The structure of the crystal and morphology are studied by X-ray diffraction (XRD) and scanning electron microscopy (SEM), stoichiometry ratio of BFO was determined by energy dispersive spectroscopy (EDS).

2. Experimental procedure

The reagents used in this synthesis were bismuth nitrate pentahydrate Bi $(NO_3)_3 .5H_2O$ (Aldrich> 99.99%), iron nitrate nanohydrate Fe $(NO_3)_3 .9H_2O$ (Merck> 99.99%), diluted nitric acid HNO_3, glycine $C_2H_5NO_2$ and urea CH_4N_2O. The mixture of the total solution was prepared by mixing the solution of Bi $(NO_3)_3.5H_2O$ (dissolved in deionized water and in nitric acid HNO_3) and Fe $(NO_3)_3 .9H_2O$ (dissolved in deionized water) under magnetic stirring at room temperature until a clear solution was obtained, then the glycine and urea fuels were added to the mixture in the required stoichiometry:

$$6\,Bi(NO_3)_3\,.5H_2O + 6\,Fe(NO_3)_3.9H_2O + 10C_2H_5NO_2 + 15\,CH_4N_2O \rightarrow 6\,BiFeO_3 + 35CO_2 + 38\,N_2 + 139\,H_2O \quad (1)$$

The total mixture in a porcelain dish was taken to a heating mantle where the temperature was gradually raised to 400 °C. The mixture was heated until all the solvent was evaporated, then a combustion reaction was observed with a large amount of gas released. The brown powder obtained was taken to a muffle furnace for a thermal treatment at a temperature of 500 ºC (1h) with a ramp of 2 ºC / min. The powder obtained with thermal treatment was taken to its respective characterization.

The crystal structure of $BiFeO_3$ crystal was determined by and X-ray diffractometer with a Co source (Kα, λ = 1.7889Å), its morphology was studied by a scanning electron microscopy (SEM) of model EVO MA10 and stoichiometry ratio of BFO was determined by energy dispersive spectroscopy (EDS).

3. Results and discussions

3.1. X-ray powder diffraction

The powder X-ray pattern (XRD) of the $BiFeO_3$ obtained using a mixture of urea and glycine is shown in Figure 1. In Figure 1a, the powder obtained from the synthesis presents an amorphous phase which can be due to incomplete combustion and that there is probably deficient oxygen in BFO due to the release of gases in the combustion [8]. When the sample was heated to 500 °C for one hour, the peaks corresponding to the BFO phase appear in the DRX pattern in Figure 1b. The result of the DRX is in accordance with PDF # 01-071-2494 and reveals that the sample has a rhombohedral perovskite crystal structure, with space group R3c.

The average crystallite size of BFO were estimated using Debye-Scherrer equation:

$$D = \frac{0.9\lambda}{\beta\cos\theta} \quad (2)$$

where D is the average the crystallite size, λ = 1, 7889 $\dot{A}$ (X-ray wavelength) and β is width of diffraction peak at half maximum for the angle 2θ. The average crystallite size of BFO was 27nm calculated by the half-height width of the peaks (012) and (024).

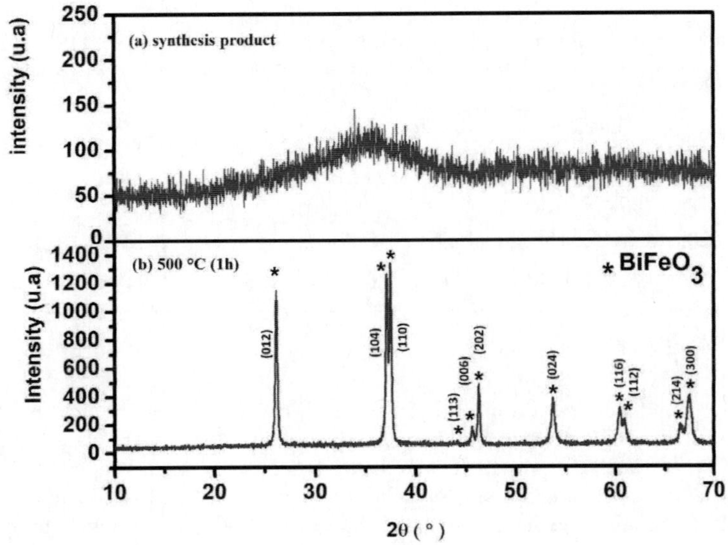

Figure 1. (a) Diffraction pattern of BFO without heat treatment and (b) with thermal treatment at 500°C for 1h.

The X-ray pattern of the BFO has been refined by the Free Full Suit software [14] and is shown in Figure 2. The diffraction XRD has been used in the refinement by the Rietveld method [15]. This method determines structural parameters and atomic positions of the BFO from the construction of a theoretical model that is able to predict the experimental diffraction pattern. The BFO DRX pattern shows fit with the rhombohedral phase model (group space R3c). Structural, atomic position and R-factor parameters obtained from the refinement are observed in Table 1.

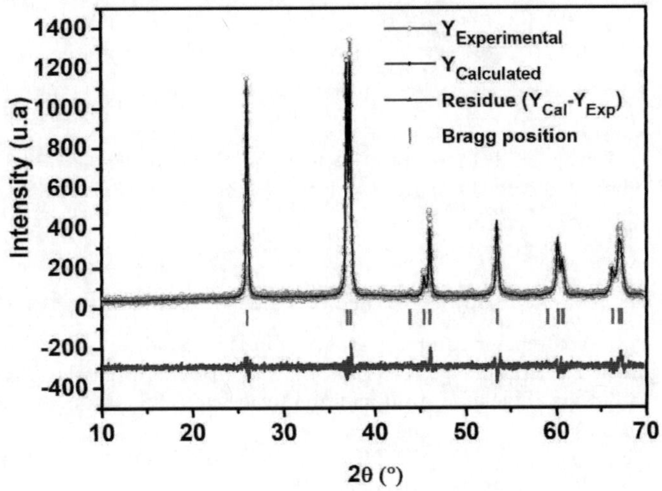

Figure 2. XRD pattern after structural refinement procedure using Rietveld's method.

Table 1. Structural parameters, atomic position and R factor of the DRX refinement for the BFO.

Sample	Lattices parameters(A)	Atomic position	X	Y	Z	R factor
	a=b=5.5972	Bi 6a	0.0000	0.0000	0.0000	R_{wp}=18.7%
$BiFeO_3$	c=13.9155	Fe 6a	0.0000	0.0000	0.2219	R_{exp}=16.5%
	v=377.5540	O 18b	0.4481	0.0171	0.9488	χ^2=1.29

The structure simulation based on XRD pattern was carried out on Rietveld crystal structure refinement. The data of the lattice parameters and atomic positions were placed in the input file of the VESTA software [16], which allowed us to visualize the crystal structure of the $BiFeO_3$ in the hexagonal cell unit including six formula units and rhombohedral cell unit including two formula unit. In addition, we have visualized the Fe^{3+} cations located inside an octahedron of oxygen atoms (FeO_6).

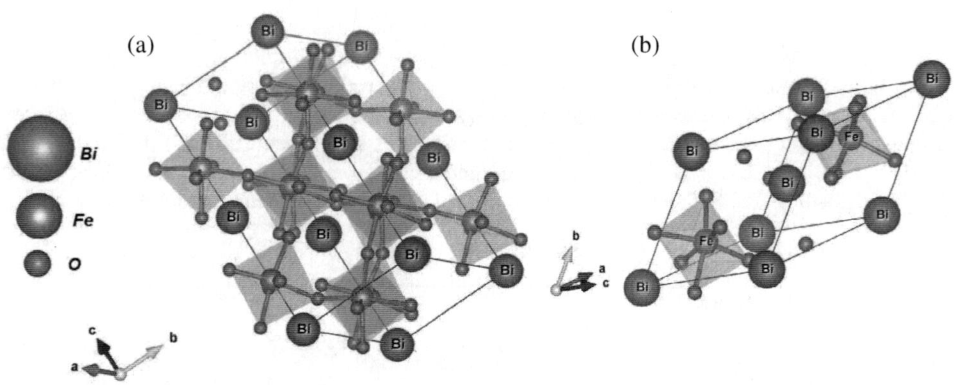

Figure 3. (a) Hexagonal cell unit of BFO with lattice parameters a=b=5.5972 y c=13.9155 Å, (b) Rhombohedral cell unit of BFO with parameters a=b=c=5.5972 Å.

3.2. Scanning electron microscopy (SEM) and energy dispersive spectroscopy (EDS)

The morphology of the prepared sample was determined by SEM at different magnification and is shown in figure 4. The SEM images clearly reveals that BFO powder show pores or voids with different size and shape (Figure 4 a), in addition to the presence of an agglomerate and non-homogenous particles (Figure 4b). The presence of pores in the sample are attributed by the release of a large amount of gases (NO, NO_2, CO, NH_3 and H_2O) during the reaction between fuels and nitrates in the combustion process [8,13].

IOP Conf. Series: Journal of Physics: Conf. Series **1143** (2018) 012025 doi:10.1088/1742-6596/1143/1/012025

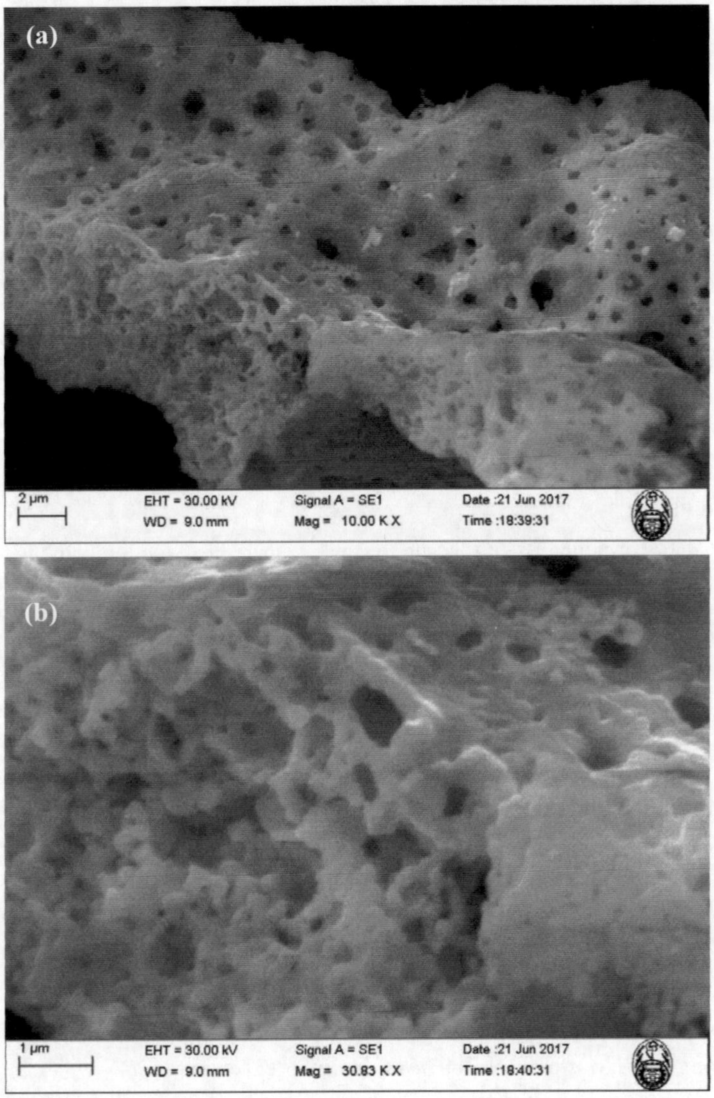

Figure 4. SEM image of the BFO powder with (a) a scale of 2μm and 10kx magnifications, (b) at a scale of 1μm and 30.83kx magnifications.

The stoichiometry ratio of BFO was determined by the SEM-EDS technique. The theoretical values for the atomic percentage between Bi, Fe and O in $BiFeO_3$ are 20, 20 and 60, respectively. The values obtained for Bi, Fe, O were 18.1%, 22.9%, 59%, these values obtained experimentally by SEM-EDS are close to the theoretical.

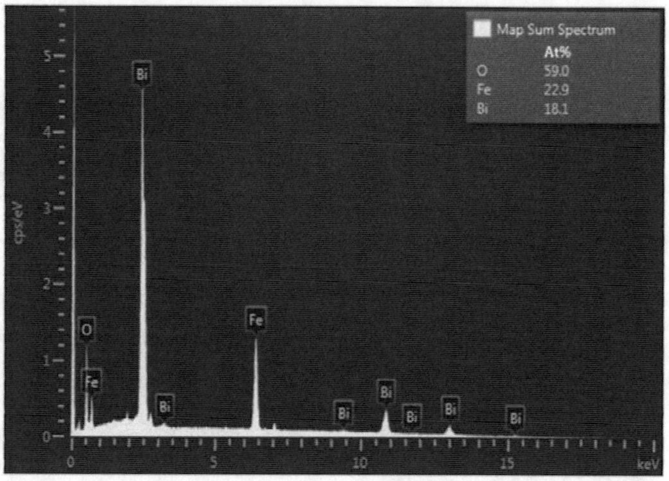

Figure 5. EDS analysis of BFO powders.

4. Conclusions

In this study, nanocrystalline $BiFeO_3$ is successfully prepared by solution combustion synthesis. The BFO phase was formed after heat treatment at 500°C for only 1h, in which the diffraction peaks are highly compatible with the presence of the BFO phase with a rhombohedral structure and space group R3C. The average crystallite size of about 27 nm by Scherrer's equation. Scanning electron microscopy analysis confirmed the presence of pores or avoids and agglomerate of particles. Dispersive energy spectroscopy (EDS) reveals the proportion of Bi, Fe, O are close to the nominal 1: 1: 3 ratio for $BiFeO_3$.

Acknowledgments

This work was possible thanks to *CONCyTEC* within the Master's program in Physical Sciences of the National University of Engineering (UNI). We thank the company *Ingenieros SAC* for the measures of X-ray diffraction and the Physical Sciences laboratory for the measurements with their Scanning Electron Microscope.

References

[1] Cheong S and Mostovoy M 2007 *Nat. Mater.* **6** 13-20.

[2] Anar S, Vibhav P, Kotnala R and Dhananjai P 2008 *Phys. Rev. Lett.* **101** 247602.

[3] Wang J, Neaton J , Zheng H, Nagarajan V, Ogale S, Liu B, Viehland D, Vaithyanathan V, Schlom D, Waghmare U, Spaldin A, Rabe K, Wuttig M and Ramesh R 2003 *Science* **209** 1719-22.

[4] Feng G, Xinyi C, Kuibo Y, Shuai D, Zhifeng R, F Y, Tao Y, Zhigang Z and Jun-Min L 2007 *Adv. Mater.* **19** 2889-92.

[5] Erenstein W, Mathur N and Scott J 2006 *Nature* **442** 759-65.

[6] Sosnowska L, Peterlin-Neumaier T and Steichele E 1982 *J. Phys. C: Solid State Phys.* **15** 4835-46.

[7] Vijayanad S, Potdar H and Pattayil J 2009 *Appl. Phys. Lett.* **94** 182507.

[8] José O, David D, Inti Z, Humberto A, Israel B, Patricia S and Noel N 2013 *Inorg. Chem.* **52** 10306-10317.

[9] Xuelian Y and Xiaoqiang A 2009 *Solid Stat. Comm.* **149** 711.
[10] Farhadi S, Zaidi M and J Mol 2009 *Catal. A-Chem.* **299** 18.
[11] Ting L, Yebin X and Jingyuan Z 2010 *J. Am. Ceram. Soc.* **93**, 3637-3641.
[12] Elidia A, Vetter F, Ivair A, Eduardo R, Rafaelle B and Emerson G 2008 *J. Braz. Chem. Soc.* **19** 1153-57.
[13] Robert I, Radu L, Silvana B and Roxana B 2015 *J. Mater. Sci.* **50** 6382-87.
[14] Rodriguez C 1993 *Phys. B.* **192** 55-69.
[15] Hugo R 1969 *J. Appl. Cryst.* **2** 65-71.
[16] Koichi M and Fujio I 2008 *J. Appl. Cryst.* **41** 653-58.

Synthesis of LaF_3 inorganic nanoparticles: Er^{3+}, Yb^{3+} by the coprecipitation method and characterization by optical spectroscopy and electron microscopy

Luis Acosta A[1], Rubén Puga[1], Hector Loro[1], Karla Santacruz Gómez [2], José Manuel Hernández[3] and Dwight Acosta[3]

[1]Faculty of Sciences, National University of Engineering, Lima, Peru
[2]Department of Physics, University of Sonora, Mexico
[3]Institute of Physics, UNAM, Mexico

E-mail: luisomega_21@hotmail.com

Abstract. It has been possible to synthesize LaF_3: Er^{3+}, Yb^{3+} nanoparticles, which are light emitters by upconversion, using the co-precipitation method followed by a thermal treatment at temperatures of 400°C, 500°C and 600°C. In these nanoparticles, luminescence is attributed to upconversion process, where green and red radiative transitions emissions are observed under NIR irradiation of 980nm. The upconversion results reveals the emission dependence with Er^{3+} and Yb^{3+} concentrations in the synthesis of these nanoparticles, as well as their dependence on temperature as a result of the applied thermal treatment , being in all cases the most intense at 525nm, 546nm and 668nm. By means of diffuse reflectance spectra, the absorption bands of Er^{3+} ions were obtained. On the other hand photoluminescence measurements were made for the identification of emission bands in Er^{3+}, in the visible range. Furthermore, data and additional information about crystalline nature of the material were obtained using the electron microscopy technique, in order to confirm the presence of LaF_3, the obtained samples showed a good correspondence with the hexagonal phase. Thus, from the electron microscopy and optical spectroscopy results, it was verified that temperature increment in the thermal treatment of the nanoparticles, significantly influenced their morphology and luminescent intensity.

Keywords: Nanoparticles, wet chemistry, upconversion, nonradiative transitions

1. Introduction

Recently, there is great interest in certain luminescent inorganic multifunctional materials such as nanoparticles doped with lanthanide ions (Ln^{3+}), which have been extensively investigated for their potential optical applications [1], [2]. The interest in this type of nanoparticles is due to the fact that, under excitation of 980nm, it presents emissions in the visible range due to the phenomenon of upconversion. The benefit of using inorganic matrices of LaF_3 lies in the structural flexibility, an important factor that allows the assembly to easily host various types of dopants, especially other rare- earth ions [3]. Also the advantage of this host is that their low phonon energy could reduce the probability of non-radiative transitions, achieving a good luminescent efficiency [4]. These nanoparticles have a great potential in biological studies due to their good dispersibility in aqueous solutions [5]. This is not the case of conventional nanomaterials such as luminescent quantum dots (QD) applied in biological analyzes, These could have certain disadvantages, due to the use of short wave (high energy) as well, for example, if the cells were subjected to high doses of light, especially at short wavelengths, these would suffer a process called phototoxicity which can lead to cell death, these disadvantages include low depth of penetration and possible damage of living organisms [6]; so that, if we

IOP Conf. Series: Journal of Physics: Conf. Series **1143** (2018) 012026 doi:10.1088/1742-6596/1143/1/012026

compare the luminescent quantum dots with the nanoparticles based on lanthanides by upconversion, the lanthanide nanoparticles have several remarkable advantages, such as sharp emission bands, long photoluminescent lifetimes, and large anti-Stokes changes, as well as high resistance to photobleaching [7]. Considering their attractive characteristics, the luminescent nanoparticles of LaF_3 are a promising class of luminescent markers, thus, they have been investigated for applications in bioimages [8]. Therefore, the study of the optical properties of nanoparticles with optically active ions of rare earths, would allow a better knowledge of these materials, in order to be able to use it in applications that include visualization devices.

2. Experimental procedure

For the preparation of $LaF_3:Er^{3+},Yb^{3+}$ nanoparticles we use the following chemicals. Lanthanum Chloride $LaCl_3$ (99.9%), ammonium fluoride NH_4F (99.9%), erbium chloride hexahydrate $ErCl_3$. $6H_2O$ (99.9%) and ytterbium chloride hexahydrate $YbCl_3$. $6H_2O$ (99.9%) were obtained from Sigma Aldrich.

2.1 Synthesis of nanoparticles LaF_3: Er^{3+}, Yb^{3+}

Initially for the first solution, 0.312mmol (76.520g) of $LaCl_3$ was mixed with 0.011mmol (4.229g) of $ErCl_3.H_2O$ and 0.111mmol (43.007g) of $YbCl_3.H_2O$ in 50ml of distilled water with ethanol. In another beaker, 3mmol (111.108 g) of NH_4F was dissolved in 5ml of distilled water, then the first solution was in fast magnetic stirring for about 12 minutes. Subsequently the second solution is added, by a slow process (drop by drop) and immediately the mixture was heated, maintained at 65°C for 4 hours. After this procedure was completed, white precipitated powders were formed in the beaker, which were placed for 8 minutes at centrifugation (5500 rpm), then it was washed with distilled water and ethanol several times; the obtained substance was dried at about 62°C. Then, thermal treatment was carried out, the material obtained was heated in the oven at about 600°C for 2 hours, the resulting product was extracted from the oven where a total of 0.211g was obtained. For its spectroscopic test the nanopowders were bombarded with IR laser (980nm) where a high fluorescence was observed. Following the previous procedures, they were synthetized several more times but the temperature of the thermal treatment was modified, heating to 500°C, 400°C and 200°C.

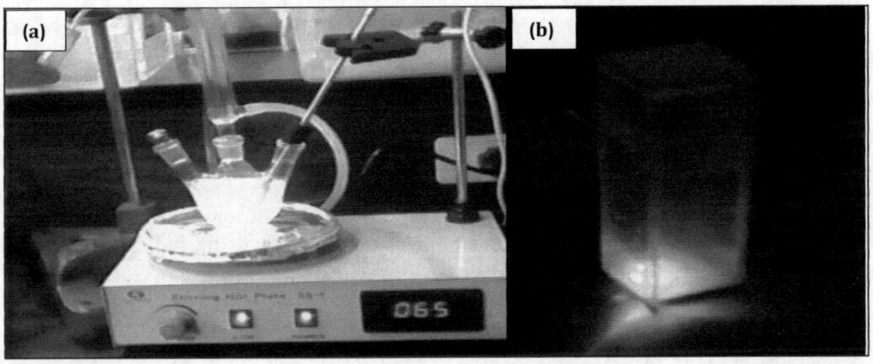

Figure 1. (a) Experimental assembly of the components for the synthesis of LaF_3 nanoparticles doped with Er^{3+}; Yb^{3+}. **(b)** Fluorescence of LaF_3 nanoparticles: Er^{3+}; Yb^{3+} dissolved in ultra-pure water, under 980nm laser excitation.

2.2 Characterization

For the measurements of the spectra by upconversion on the samples synthesized, a Science Surplus spectrophotometer equipped with a linear CCD sensor and a 980nm IR laser diode with power of 400mW was used. In the measurements of the absorption spectra, a UV-Vis-NIR Cary 5000 series Spectrophotometer was used, which was analyzed within the UV-visible, visible and part of the near-infrared UV range; While for the measurements of the fluorescence spectra a Perkin Elmer Fluorimeter was used model LS-55. TEM and HRTEM electron microscopy images were measured with an E1 JEOL 2010 FEG-TEM microscope.

IOP Conf. Series: Journal of Physics: Conf. Series **1143** (2018) 012026 doi:10.1088/1742-6596/1143/1/012026

3. Obtained results

3.1 Spectroscopy of emission by upconversion

In order to study upconversion emissions, LaF_3 nanoparticles doped with 20% mol Yb^{3+} and 2% mol Er^{3+} were excited with a 980nm laser diode. This was done for the samples with thermal treatment of 600°C, 500°C and 400°C, registering the emission spectra obtained by upconversion. To study the effect of the heat treatment on the synthesized samples, we compared the emission spectra, where we observed significant emission differences, stands out a higher luminescent intensity at a higher thermal treatment temperature.

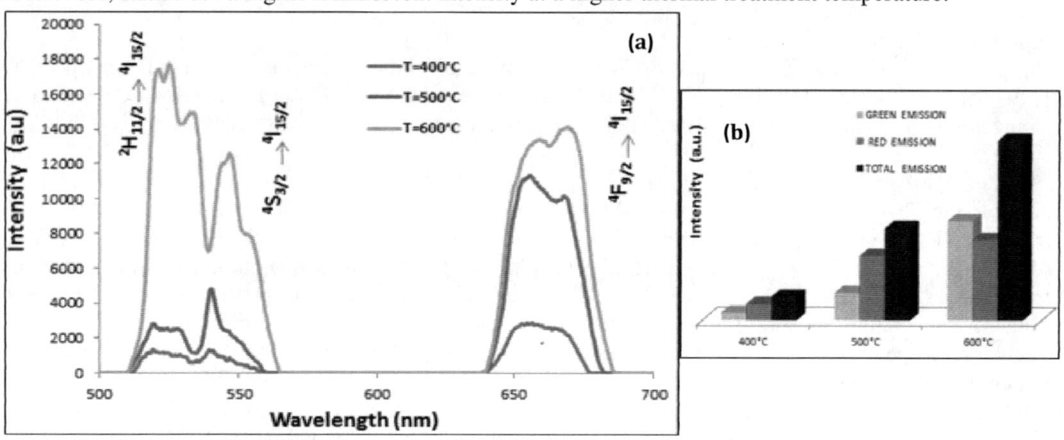

Figure 2. (a) Emission spectra produced by upconversion (excited with a laser diode of 980nm with 400mW) of LaF_3 nanoparticles doped with 20% mol Yb^{3+} and 2% mol Er^{3+}, with thermal treatment of 400°C, 500°C and 600°C. **(b)** Diagram showing the comparison of the total integral intensities of the LaF_3 nanoparticles: Yb^{3+}, Er^{3+} (20/2)% mol at different temperatures of 400°C, 500°C and 600°C.

In figure 2, the presence of bands centered on green and red are observed in the three emission spectra. These bands present a series of peaks associated with the electronic transitions of excited states of the Er^{3+} which are: $^2H_{11/2} \rightarrow\ ^4I_{15/2}$; $^4S_{3/2} \rightarrow ^4I_{15/2}$ visible green luminescence and $^4F_{9/2} \rightarrow\ ^4I_{15/2}$ visible red luminescence with wavelengths centered at 524.9nm, 545.8nm and 668.6nm respectively.

In the sample with heat treatment at 600°C, we see that the band formed by the radiative transition in the multiplets $^2H_{11/2} \rightarrow ^4I_{15/2}$ of Er^{3+} has a higher luminescent intensity compared to the band formed by the transitions $^4S_{3/2} \rightarrow ^4I_{15/2}$. On the contrary , for the samples with treatment at 500°C and 400°C, the transitions $^4S_{3/2} \rightarrow ^4I_{15/2}$ are more efficient than transitions $^2H_{11/2} \rightarrow ^4I_{15/2}$.

In the luminescence spectra for upconversion of the LaF_3 nanocrystals: 20% Yb, 2% Er given by figure 3, the formation of bands coming from the strong emissions in the visible green corresponding to the electronic transitions can be seen $^2H_{11/2} \rightarrow\ ^4I_{15/2}$ and $^4S_{3/2} \rightarrow ^4I_{15/2}$ which are centered at 525.3nm and 545.9nm respectively, while the strong emission in the visible red is assigned to the transition $^4F_{9/2} \rightarrow\ ^4I_{15/2}$ centered at 667.5nm.

In the same way for the LaF_3 nanocrystals: 12% Yb, 3% Er, the bands formed presented the same characteristics as the concentration Yb, Er (20/2). As both samples were treated at 600°C, then the high temperature and the heating time favored the luminescent intensity, however, the most remarkable fact is that the LaF_3 nanocrystals with a concentration of 2% Er, 20% Yb, showed a greater intensity of emission both in their bands with visible green emission and in the band in the visible red, compared to the samples with concentration 3% Er, 12% Yb.

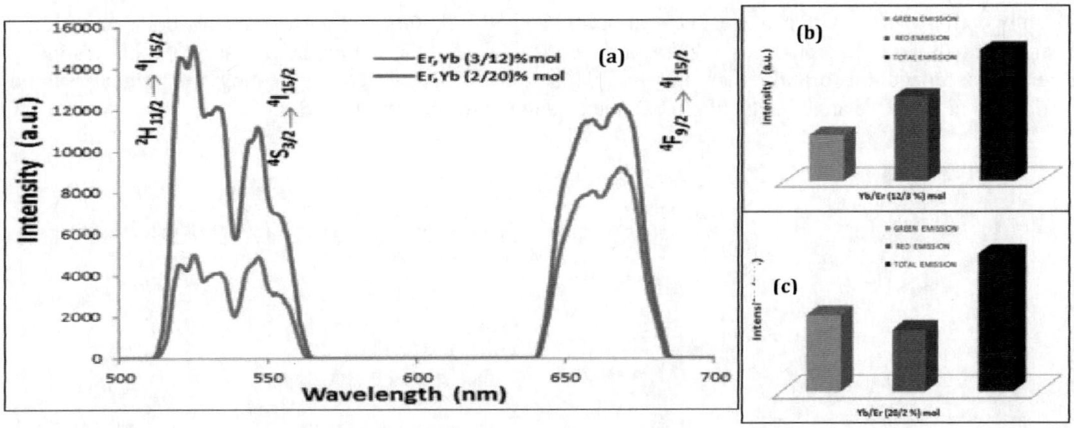

Figure 3. (a) Emission spectra produced by upconversion (excited with laser diode of 980nm with power of 400mW) of LaF_3: Yb^{3+}, Er^{3+} (20/2)% mol, (red line) and LaF_3: Yb^{3+}; Er^{3+} (12/3)% mol (blue line) both with heat treatment of 600°C. **(b)** Comparative diagrams of the total intensities, corresponding to the samples of LaF_3: Yb^{3+}; Er^{3+} (12/3)% mol and **(c)** Comparative diagrams of the total intensities of LaF_3: Yb^{3+}; Er^{3+} (20/2)% mol.

3.2 Diffuse reflectance spectroscopy

By this technique, the absorption spectra were determined. These spectra obtained allow the identification of the excited energy levels of the optically active Er^{3+} ion inside the LaF_3 nanocrystals.

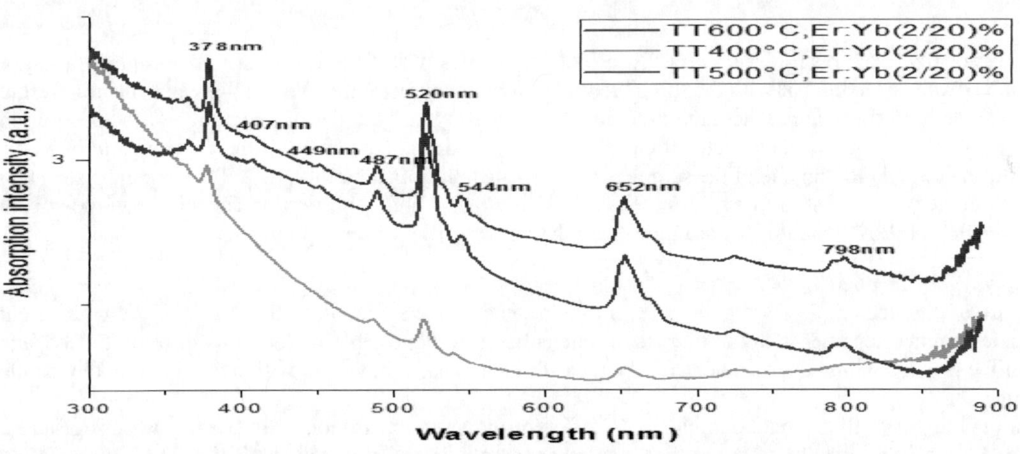

Figure 4. Absorption spectra of LaF_3 nanoparticles doped with 20% mol Yb^{3+} and 2% mol Er^{3+}, at different thermal treatments of 400°C, 500°C and 600°C. The measurement conditions were the same in each one of the samples.

In figure 4, a comparison between the absorption spectra of three LaF_3 samples is observed: Er^{3+}, Yb^{3+} (2/20)% mol at different heat treatment temperature, where it is indicated for each band the centered wavelengths belonging to the excited states of the Er^{3+}. The spectra were taken between the ranges of 300 nm to 900 nm.

By comparing the absorption spectrum of the sample at 400°C with the samples at 500°C and 600°C, we observed a significant decay of the intensity of absorption, as well as the decrease in the number of sharp peaks in certain regions of the spectrum, due to there is no formation of the bands corresponding to the multiplets $^2H_{9/2}$, $^4F_{5/2}$, $^4I_{9/2}$, of Er^{3+}.

In the absorption spectra with heat treatment at 600°C and 500°C, the number of peaks for both samples are the same, also the forms adopted by the bands are identical, as well the sample treated at 600°C has a slightly greater magnitude of absorption than the sample at 500°C, therefore, the heating temperature would significantly favor the relative intensity of absorbance and the formation of bands.

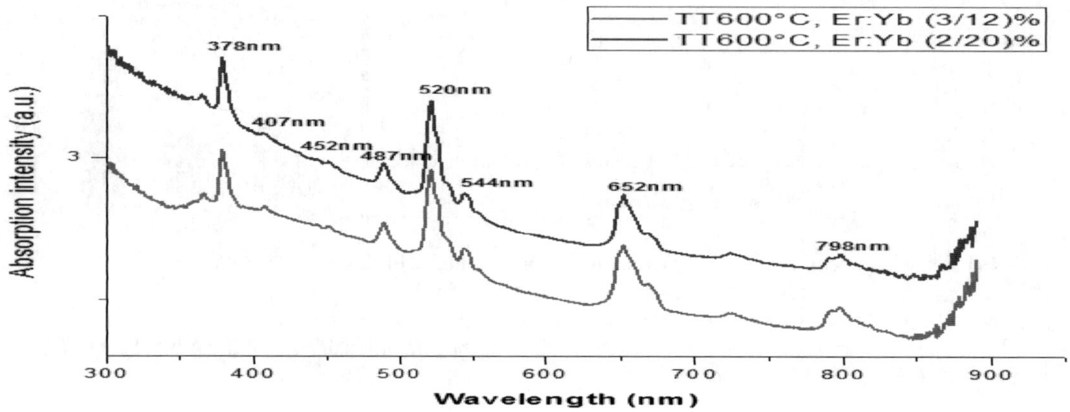

Figure 5. Spectra of absorption in the visible range of LaF_3 nanoparticles doped in molar ratio of Yb^{3+}, Er^{3+} (20/2)% mol, and Yb^{3+}, Er^{3+} (12/3)% mol, with temperature of thermal treatment at 600°C . The measurement conditions were the same in each one of the samples.

In the case of figure 5, absorption spectra given by two samples of LaF_3 at different molar concentration of Er^{3+} and Yb^{3+} are observed, but with the same thermal treatment temperature. We see that both spectra present identical acute band formation, where the variation occurs mainly in the order of magnitude of the absorption intensities. As the well-defined characteristic peaks, come from the excited levels of the Er^{3+} ion. Although the concentration of (3%) Er^{3+} in one of the samples is greater than the concentration (2%) Er^{3+}, it can be seen that the LaF_3 sample with dopant ratio Yb^{3+}, Er^{3+} (20/2)% mol, it has a higher absorption capacity than the dopant ratio Yb^{3+}, Er^{3+} (12/3)% mol, this according to the relative intensities of absorption.

3.3 Fluorescence spectroscopy

By means of this technique, excitation and emission spectra were recorded, thus by analyzing the light frequencies emitted together with their emission intensities, it was possible to determine different electronic states of the Er^{3+} ion within the host of LaF_3, the measurements made they were within the range of the visible spectrum.

Spectra of LaF_3 crystallized nanoparticles with different molar concentration of impurities were registered, one with doping 20% mol Yb^{3+}: 2% mol Er^{3+}, the other sample were doped with 12% mol Yb^{3+}: 3% mol Er^{3+} both treated at 600°C, these synthesized samples were excited at 378nm.

From figure 6, it was observed peaks distributed in five groups of bands, centered at 456nm, 487nm, 520nm, 548nm and 660nm, which comes from the transitions $^4F_{5/2} \rightarrow {}^4I_{15/2}$; $^4F_{7/2} \rightarrow {}^4I_{15/2}$; $^2H_{11/2}$, $^4S_{3/2} \rightarrow {}^4I_{15/2}$ y $^4F_{9/2} \rightarrow {}^4I_{15/2}$ respectively.

The results indicate that the luminescent intensity is better in the dopant relationship Yb^{3+}, Er^{3+} (12/ 3) % than Yb^{3+}, Er^{3+} (20/2) %, this is associated with the higher concentration of optically active ions of Er^{3+}. If compare the bands with the highest emission, we can see that the group of bands with peaks at 660nm and 674nm visible red band, is relatively more efficient than the emission in the visible green band with peaks at 520nm and 548nm.

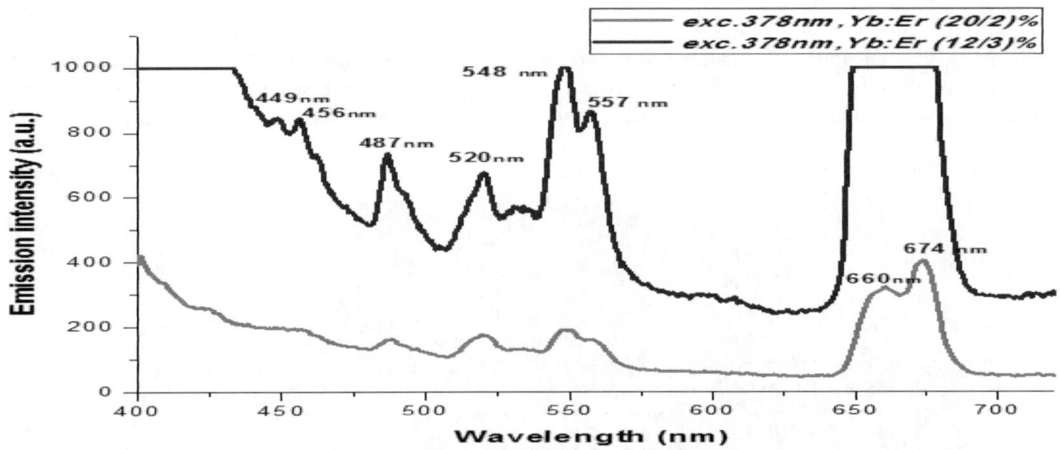

Figure 6. Emission spectra produced by photoluminescence of LaF_3 nanoparticles doped with 20% mol Yb^{3+}: 2% mol Er^{3+} and 12% mol Yb^{3+}: 3% mol Er^{3+}, with thermal treatment at 600°C. The samples were excited at $\lambda = 378$nm.

3.*4 Electron microscopy*

In order to observe crystalline planes, the shape and size of the synthesized nanoparticles, it shows the results of the high resolution images obtained by transmission electron microscopy (TEM). The image of figure 7 allows observing the crystalline planes of the sample studied. In this region, the crystallographic parameters given in figure 8 were calculated, so the Fourier transform (FFT) in the selected area shows a hexagonal crystalline system for LaF_3, where a good crystallinity is observed.

In figure 7, a region of the studied sample is shown. This figure shows the reciprocal spaces for a certain region with values 8.13nm^{-1}, 7.71nm^{-1} y 7.73nm^{-1}. Likewise, the FFT pattern is shown. The interplanar spacing of 0.33nm corresponds to the crystalline plane (111) of LaF_3; also the nanocrystals with interplanar spacing of 0.37nm correspond to the crystal plane (002) of LaF_3 given by reference [9].

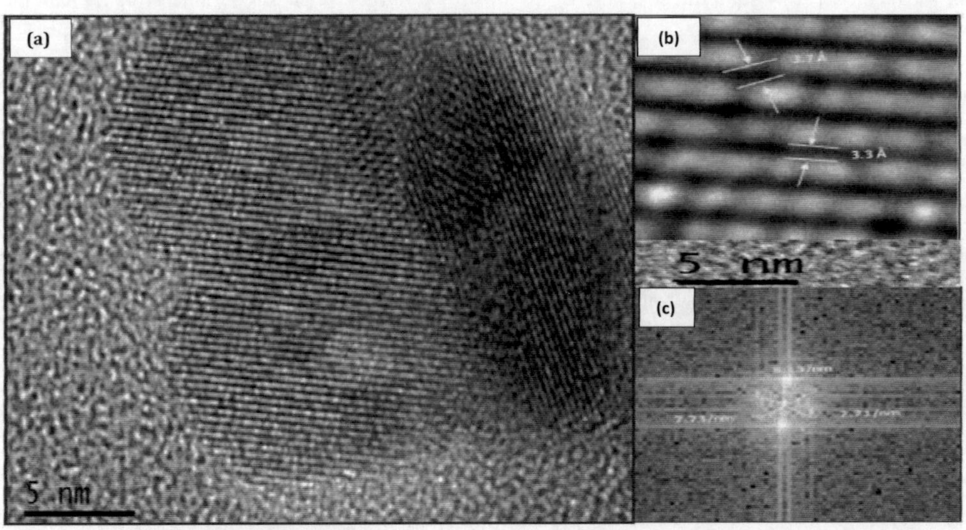

Figure 7. (a) Image (TEM) with unit of measurement at 5nm of nanoparticles LaF_3: Yb^{3+}; Er^{3+} (20/2)% mol, with a heat treatment of 600°C. **(b)** Approach the region marked in figure (a). **(c)** The FFT pattern of the region indicated in (a) is shown.

IOP Conf. Series: Journal of Physics: Conf. Series **1143** (2018) 012026 doi:10.1088/1742-6596/1143/1/012026

The results of the high resolution images obtained by transmission electron microscopy (TEM), In figure 8, HR-TEM images of the LaF_3: Er^{3+} nanoparticles are shown; Yb^{3+} thermally treated at 200°C with units of measurement of 100nm and 50nm. These images show a greater contrast, which allows to see the agglomeration of the nanoparticles, despite that a certain distribution of particles. From this image, is observed a particle size of approximately 15 nm.

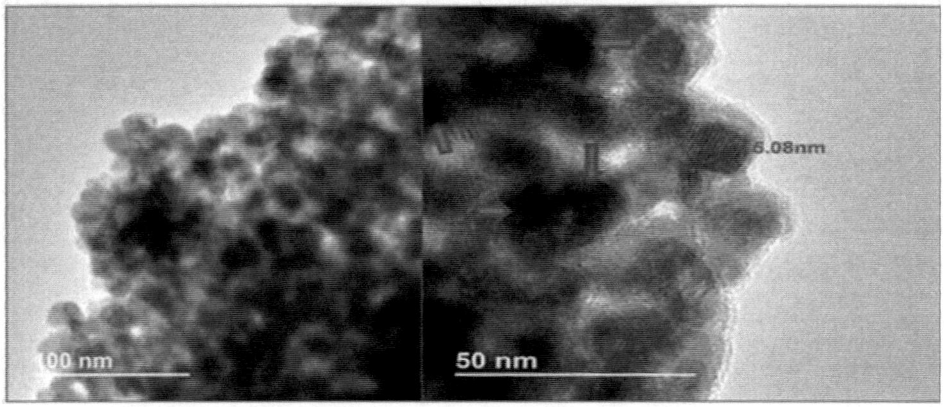

Figure 8. Micrographs obtained by HR-TEM with unit of measurement 100nm and 50nm of the LaF_3 nanoparticles: Yb^{3+}; Er^{3+} (20/2)% mol, with a heat treatment of 200°C.

Figure 9 is a HR-TEM image, where the crystal planes can be observed. The crystalline system corresponds to the hexagonal structure, with well-defined crystallinity. Its structure corresponds to Lanthanum trifluoride (LaF_3) given by the reference [9], this can be corroborated with the pattern of the Fourier transform (FFT), which is located in the zone axis [1,-2, 1,-3]. Some crystal lattice planes are shown where the interplanar spaces are labeled in the image, which gave distances of 3.2 Å and 3.5 Å, which corresponding to the families of planes (111) and (110) respectively, with reciprocal spaces of 6.421nm^{-1}, 6.451nm^{-1} and 5.751nm^{-1}. The doping of the trivalent ions Yb, Er, are within the structure of the host material, because the parameter of the crystal lattice of LaF_3 is not altered.

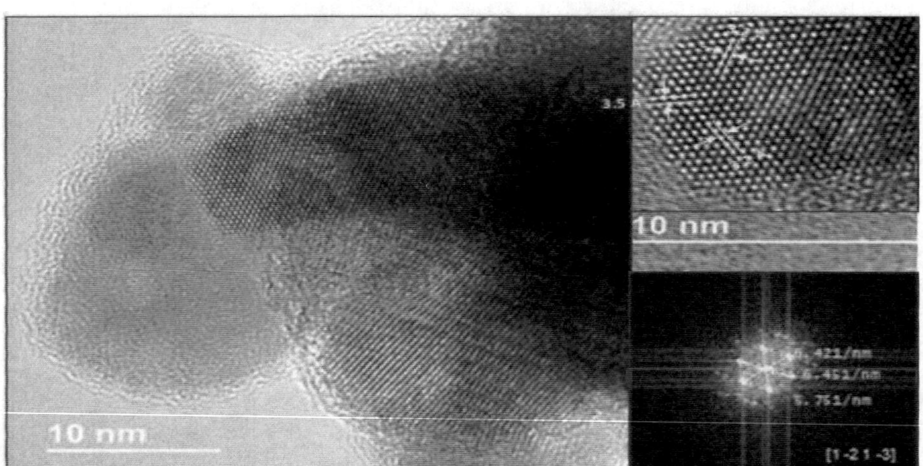

Figure 9. High resolution image (HRTEM) of the LaF_3 nanoparticles: Er^{3+}; Yb^{3+} (20/2)% mol, with thermal treatment at 200°C and unit of measurement 10nm in a certain region. The FFT pattern is indicated as well its interplanar distances of a single nanocrystal.

4. Conclusions

In the work presented in this paper, we were able to synthesize inorganic nanoparticles LaF_3: Er^{3+} Yb^{3+}, effectively implementing a relatively practical chemical method.

- The emission spectra obtained by upconversion with dopant concentration Yb^{3+}, Er^{3+} of 20%, 2% have greater luminescent efficiency with respect to the sample with concentration Yb^{3+}, Er^{3+} of 12%, 2% respectively. In addition, the efficiency of the upconversion of nanoparticles is favored by increasing the heat treatment temperature.

- It has been determined that the most outstanding peaks in the absorption spectra are centered at 387nm, 521nm and 652nm. The formation of these bands presents relatively sharp lines.

- From TEM images, the existence of LaF_3 nanoparticles was verified: Er^{3+}; Yb^{3+}. The crystalline system corresponds to the hexagonal structure.

- The information obtained from the HRTEM images allowed to verify the estimated average grain size, which is around 15nm, belonging to the structure of LaF_3.

- In the future it is expected to be able to functionalize the nanoparticles by means of a nanostructured design from core to layer, covering the LaF_3: Er^{3+}, Yb^{3+}, in order to provide it with characteristics for possible applications in biomedicine, where these nanoparticles would produce emissions when placed inside cells.

Acknowledgements
This research was possible thanks to the support of CONCYTEC- Peru, agreement No. 167 FONDECYT-UNI and the Faculty of Sciences of the National University of Engineering, Peru. The spectra of absorption, fluorescence and electron microscopy were obtained thanks to the collaboration with the IFUNAM of Mexico. The high resolution electronic microscopy images were obtained thanks to the collaboration with University of Sonora, Mexico.

References

[1] Yang Wei, Fengqi Lu y Xinrong Zhang 2007 *Mater. Lett.* **61**, 1337-1340.

[2] He Hu, Zhigang Chen, Tianye Cao, Qiang Zhang, Mengxiao Yu, Fuyou Li, Tao Yi y Chunhui Huang 2008 *Nanotech.*, **19**, 375702.

[3] Navadeep Srivastava, L U Khan, J M Vargas, Carlos Ospina , J A Q Coaquira , Giorgio Zoppellaro , H F Brito, Yasir Javed , D K Shukla and M C F C Felinto 2017 *Phys. chem .Chem. Phys.***19**, 18660-18670.

[4] Xiaoyong Huang 2016 *J. Mate. Sci.* **51**, 3490-3499.

[5] Min Hua Chen, Yi Jhen Jenh , Sheng Kai Wu , Yo-Shen Chen, Nobutaka Hanagata and Feng Huei Lin 2017 *Nanoscale , Res. Lett.*, **12** 62.

[6] Pedro Villanueva y Rafael Valiente, *Caracterización del $NaYF_4$ impurificado con Yb^{3+} y Tm^{3+} y sus potenciales aplicaciones en biomedicina,* University of Cantabria Faculty of Sciences, Santander Spain 2012.

[7] Peng Li, Qing Peng y Yadong Li 2009 *Adv. Mater.* **21,** 1945-1948.

[8] Huaxiang Shen,Feng Wang, Xianping Fan and Minquan Wang 2007 *J. Experimental Nanoscience* **2**,303-311.

[9] Wright A F, Fuess H ,Fender B E F ,Cheetham A K 1976 *Acta Crystallographica B (24,1968-38,1982)***32**, 94 – 97.

IOP Conf. Series: Journal of Physics: Conf. Series **1143** (2018) 012027 doi:10.1088/1742-6596/1143/1/012027

Influence of substrate temperature on the properties of Ag-C films produced by DC Magnetron Sputtering

N Z Calderon, J L Ampuero, A La Rosa-Toro and B R Pujada

National University of Engineering (UNI), Department of Physics, Av. Tupac Amaru s/n, Rimac, Lima, Peru

E-mail: zully1410@gmail.com

Abstract. Silver-carbon films (Ag-C), rich in silver, deposited by DC magnetron sputtering from a silver target (Ag) into silicon substrate (100) in atmosphere of argon/acetylene plasma, have been studied in function of the substrate temperature (as-deposited, 200°C and 300°C). The Ag-C films were obtained using an acetylene gas flow of 0.3 sccm and dc cathode current of 180mA. Ag-C films were characterized by X-ray diffraction, Raman spectroscopy and scanning electron microscopy. It has been observed the decrease of graphite-like bonds in the Ag-C films with the substrate temperature. The SEM revealed that the substrate temperature promotes changes on the surface morphology. These changes on the chemistry and surface morphology are discussed in terms of atoms diffusion on the surface and the chemical composition of the Ag-C films.

Key Words: Thin films, Sputtering, Ag-C.

1. Introduction

Films of Ag-C (silver-carbon) and Ag-DLC (diamond-like silver-carbon), which are composed of nanoparticles of silver and carbon, have attracted increasing scientific and industrial interest due in part to their interesting properties such as low coefficient of friction, low chemical reaction and antibacterial properties [1-16]. These films can be prepared by various deposition techniques, including Physical Vapor Deposition (PVD) [17-18]. Due to the need to gather as much information about morphology, phase composition and microstructure as possible, advanced deposition and characterization techniques for Ag-C and Ag-DLC films are necessary to obtain a deeper understanding of the relationship between the conditions of the sputtering process and the properties of the film. During the last years, a lot of work has been devoted to the study and understanding of the properties of the Ag-C and Ag-DLC films. These works have shown that the microstructure, mechanical and antibacterial properties of these materials depend to a large extent on the deposition conditions. Choi et al., using a hybrid system of ion beam deposition, studied the composition, microstructure and mechanical properties of Ag-C coatings deposited on Si (100), as a function of the argon/acetylene fraction [1]. They found that by increasing the fraction of argon in the plasma, the concentration of silver increases. In addition, the hardness and comprehension effort of the Ag-DLC films decrease with the content of silver. The authors related their results to the variation of the chemical composition in the film and to the sp^2/sp^3binding ratio. Garcia-Zarco et al., studied the influence of the magnetron power supply and the distance of the target to substrate, on the structure and composition of the Ag-DLC films deposited by PVD magnetron sputtering [7]. They found a nanostructure of silver crystalline clusters uniformly distributed in the amorphous carbon matrix with a cathode power of 40 W and the target to substrate distance of 28 mm. In the present paper we report and analyze surface morphology and chemical composition of Ag-C films at different substrate temperatures using a dc magnetron sputtering.

2. Experimental details

Ag-C films were produced in a home-made sputter deposition system. The samples were deposited by dc magnetron sputtering of Ag target in an argon/acetylene plasma. The films were deposited on silicon wafers (100) during 90 minutes. Before deposition the chamber was evacuated to $1x10^{-5}$ mbar. The diameter of the target is 5.04 cm and the target-substrate distance 6 cm. The dc current applied to the target was of 180 mA. The argon gas flow was maintained constant at 6 sccm whereas the acetylene gas flow at 0.3 sccm. In order to study the effect of the substrate temperature on the properties of the Ag-C films, we deposited the samples at three different temperatures, as-deposited (60 °C), 200°C and 300°C. The structure of the Ag-C films was determinate by x-ray diffraction (XRD) using a Bruker D8 advanced diffractometer. The surface morphology of Ag-C films was studied using scanning electron microscopy (JEOL, JSM-6610LV)). Raman spectra were used to characterize the structural arrangement of C sites in the films. Raman spectra were obtained using a Xplora Horiba spectrometer.

3. Results and Discussions

Figure 1 shows the XRD patterns of pure Ag film and Ag-C films deposited at acetylene flow of 0.3 sccm and substrate temperatures of as-deposited (60 °C) and 200 °C. From Fig. 1-(a), it is clear the presence of peaks arising out from the (111), (200), (220) and (331) planes of Ag crystalline phase. In Ag-C films (Fig. 1, b and c), a slight broadening of the peaks is observed. This effect could be caused by the structural disorder in silver grains due to the increase of carbon content in the films such as observed by Auger spectroscopy (not shown here).

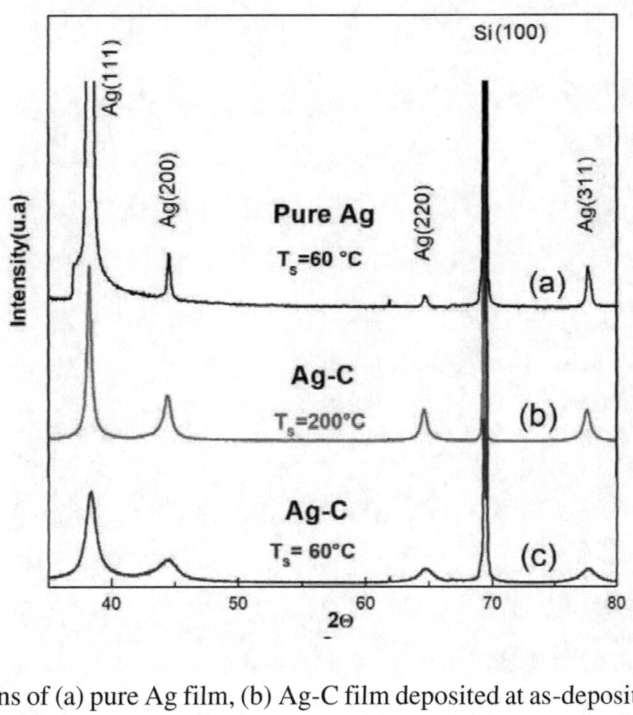

Figure 1. XRD patterns of (a) pure Ag film, (b) Ag-C film deposited at as-deposited (60 °C) and (c)Ag-C film deposited at 200°C.

IOP Conf. Series: Journal of Physics: Conf. Series **1143** (2018) 012027 doi:10.1088/1742-6596/1143/1/012027

From SEM micrographs of the Ag-C films, it can be seen that the surface morphology changes with the substrate temperature. The surface of Ag-C film deposited at a substrate temperature of as-deposited consist of a smooth, granular structure with a small grain diameter. At higher substrate temperatures, the average grain sizes increase; these Ag-C films are characterized by a rich silver content such as observed by XRD. Due to the physics of the sputtering process, these changes are closely related with changes in the atoms diffusion on the surface, producing an atomic rearrangement of the silver and carbon atoms in the structure and consequently on the surface morphology of Ag-C films. At higher substrate temperature, the grain size increases and Figure 3 shows the Raman spectra for Ag-C films deposited at different substrate temperatures. The double peak Raman spectra of the Ag-C films were deconvoluted into D and G peaks commonly named as the disorder and graphite lines, respectively. The intensity ratio of D peak to G peak (I_D/I_G) was also determined. For the Ag-C film deposited at substrate temperature of as-deposited (60 °C), the Raman spectra shows the D and G peaks at 1393,26 cm^{-1} and 1570,58 cm^{-1}, respectively. For Ag-C film deposited at substrate temperature of 300 °C, the Raman spectra exhibits the D and G peaks at 1360,44 cm^{-1} and 1583,81 cm^{-1}, respectively.

Figure 2. SEM images of amorphous Ag-C films deposited at substrate temperatures of (a) as-deposited (60°C) (b) 200°C (c) 300°C.

The I_D/I_G ratio decreases with the increases of the substrate temperature. Choi at al., in their studies of Ag-incorporated DLC films deposited by a hybrid ion beam system, found the same dependence of the I_D/I_G ratio with the incorporation of Ag [1], indicating therefore the decrease of sp^2-bonded clusters in the Ag-C film with the substrate temperature.

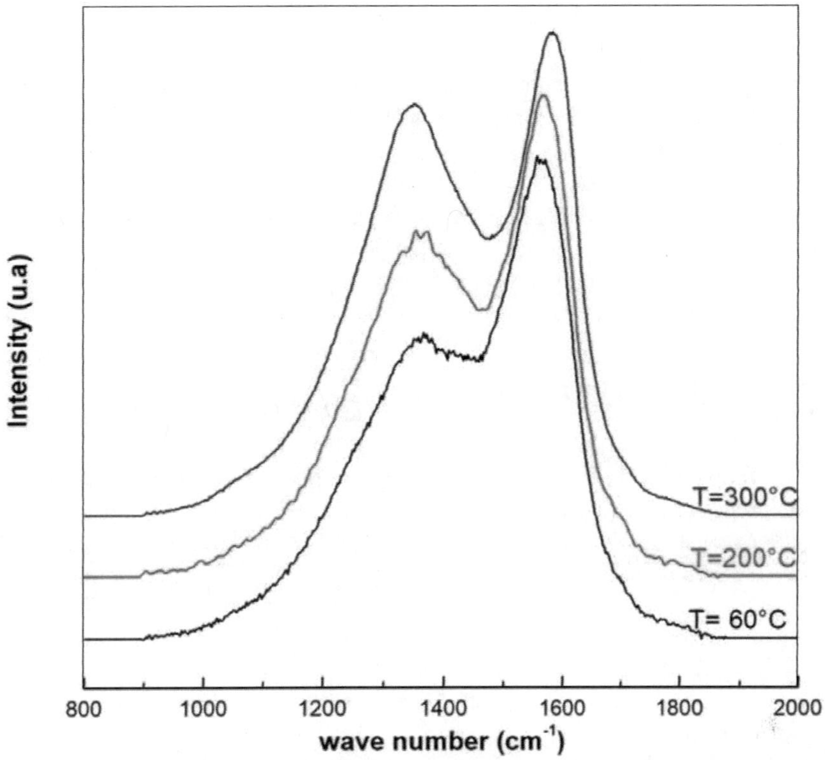

Figure 3. Raman spectra of thin films Ag-C at substrate temperatures of (a) as-deposited (60°C), (b) 200°C, and (c) 300 °C.

4. Conclusions

Ag-C films produced by dc magnetron sputtering were characterized for their chemistry and surface morphology. The results showed that the surface morphology, chemical composition and structure can be altered by changing the substrate temperature. The Ag-C films exhibited a structure composed of silver rich predominant composition, increasing with the substrate temperature. The XRD patterns indicated that the addition of acetylene gas produces slight broadening of the silver peaks, increasing at substrate temperature of as-deposited. The decreases of the I_D/I_G ratio of the Ag-C films with the substrate temperature have been attributed to the decreases of sp^2-bonded clusters in the Ag-C film. Finally, the results are useful in understanding of how process parameters affect the deposition of Ag-C films and consequently its properties.

Acknowledgements

Calderon would like to express her thanks to the Fondecyt project number 167-2015 (Concytec, Perú). Pujada express his thanks to the project number FC-F-3-2017, Vicerrectorado de Investigación (UNI-Perú), and also to the Universidad de Chile and the Fondecyt Project 1110148 (Conicyt, Chile).

IOP Conf. Series: Journal of Physics: Conf. Series **1143** (2018) 012027 doi:10.1088/1742-6596/1143/1/012027

References

[1] H.W. Choi, J.H. Choi, K.R. Lee, J.P.Ahn and K.H Oh,"Structure and mechanical properties of Ag-incorporated DLC films prepared by a hybrid ion beam deposition system". Thin Solid Films 516 (2007)248-251

[2] F.R. Marciano, L.F. Bonetti, R.S. Pessoa, J.S. Marcuzzo, M. Massi, L.V. Santos and V.J. Trava-Airoldi,"The improvement of DLC film lifetime using silver nanoparticles for use on space devices". Diam. Relat. Mater 17 (2008) 1674-1679

[3] H.-S. Zhang, J.L. Endrino and A. Anders,"Comparative surface and nano-tribological characteristics of nanocomposite diamond-like carbon thin films doped with Silver". Appl. Surf. Sci. 255 (2008) 2551-2556.

[4] H.W. Choi, R.H. Dauskardt, S.C. Lee, K.R Lee and K.H. Oh,"Characteristic of silver doped 48 DLC films on surface properties and protein adsorption". Appl. Surf. Sci.Diamond Relat. Mater. 17 (2008) 252-257

[5] C. Wang, X. Yu and M. Hua,"Microstructure and mechanical properties of Ag- containing diamond-like carbon films in mid-frequency dual-magnetron sputtering". Appl. Surf. Sci. 256 (2009) 1431-1435

[6] S.F. Ahmed, M.W. Moon and K.R. Lee,"Effect of silver doping on optical property of diamond like carbon films". Thin Solid Films. 517 (2009) 4035-4038

[7] O. Garcia-Zarco, S.E. Rodil and M.A. Camacho-Lpez, "Deposition of amorphous carbon-silver composites". Thin Solid Films. 518 (2009) 1493-1497

[8] V. Ionescu, C.P. Lungu, M. Osiac and V. Ciupina, "Silver Containing Carbon Amorphous Nanocomposite

Films Deposited by Termionic Vacuum Arc Technique". Rom. Journ. Phys. 55 (2010) 119-12

[9] F. P. Schwarz, I.Hauser-Gerpach, T. Waltimo and B. Stritzker, "Antibacterial properties of silver containing diamond like carbon coatings produced by ion induced polymer densification". Surf. Coat. Technol. 205 (2011) 4850-4854

[10] K. Kato, H. Omoto and A. Takamatsu, "Microstructure Analyses of Ag Films Deposited by Low-voltage

Sputtering". Thin Solid Films. 520 (2012) 4139-4143

[11] A.A. Onoprienko and M.I. Danylenko, "Annealing effects in Ag-doped amorphous carbon films deposited by dc magnetron sputtering". Surf. Coat. Technol 206 (2012) 3450-3453

[12] K. Baba, R. Hatada, S. Flege, W. Ensinger, Y. Shibata, J. Nakashima, T. Sawase and T. Morimura, "Preparation and antibacterial properties of Ag-containing diamond- like carbon films prepared by a combination of magnetron sputtering and plasma source ion implantation". Vacuum 89 (2013) 179-184

[13] X. Yu, Y. Qin, C.B. Wang, Y.Q. Yang and X.C. Ma, "Effects of nanocrystalline silver incorporation on sliding tribological properties of Ag-containing diamond-like carbon films in multi-49 ion beam assisted deposition". Vacuum 89 (2013) 82-85

[14] N.M. Chekan, N.M. Beliauski, V.V. Akulich, L.V. Pozdniak, E.K. Sergeeva, A.N. Chernov, V.V. Kazbanov and V.A. Kulchitsky, "Biological Activity of Silver-doped DLC Films". Diamond Relat. Mater 18 (2009) 1006-1009

[15] W.C. Lan, S.F. Ou, M.H. Lin, K.L. Ou and M.Y. Tsai, "Development of silver- containing diamond-like carbon for biomedical applications. Part I: Microstructure characteristics, mechanical properties and antibacterial mechanisms". Ceram. Intern 39 (2013) 499-4104

[16] C.A. Love, R.B. Cook, T.J. Harvey, P.A. Dearnley, R.J.K. Wood, "Diamond like carbon coatings for potential aplication in biological implants". a review, Tribology International 63 (2013) 141-150

[17] B. Chapman,John Wiley y Sons "Glow Discharge Processes: Sputtering and Plasma Etching". Inc. (1980)
[19] M.Ohring "The Materials Science of Thin Films: Deposition and Structure". Academic Press (1992)

[18] W. D. Westwood "Sputter Deposition". AVS Education Committee Book Series, Vol. 2 (2003)

IOP Conf. Series: Journal of Physics: Conf. Series **1143** (2018) 012028 doi:10.1088/1742-6596/1143/1/012028

Identification of superconductivity mechanisms and prediction of new materials using Density Functional Theory (DFT) calculations

J A Alarco, P C Talbot and I D R Mackinnon

Institute for Future Environments and Science and Engineering Faculty, Queensland University of Technology (QUT), Brisbane, Queensland, Australia 4001

E-mail: jose.alarco@qut.edu.au

Abstract. Superconductivity at room temperature or higher is considered a fundamentally viable goal, which may bring significant benefits to society, if achieved. However, methods to predict and design new superconductors remain largely empirical, without extensive guidance from computational quantum chemistry techniques, such as Density Functional Theory (DFT). This paper presents our progress, using DFT, towards a predictive tool for superconductivity from knowledge of the crystal structure. Excellent correlations between predicted and experimentally determined values have been demonstrated for MgB_2 under a wide range of external conditions, including isotopic forms, metal substitutions, pressure and temperature effects. Model ideas have recently been shown to work equally well for hydrogen sulphide (H_3S), which is the current record holder for the highest T_c (at 200K), albeit requiring very high pressures (~160 GPa). Consistent patterns across families of superconductors, as determined from DFT calculations, are emerging that suggest a method to predict T_c seems possible.

1. Introduction

Superconductivity at room temperature or higher is considered achievable by world experts [1] and remains an ultimate goal that may revolutionize the transport and storage of electricity with tremendous environmental benefits. However, approaches to design new superconducting materials remain largely empirical. Current theories, including those that make use of Density Functional Theory (DFT), predominantly explain properties *post facto* using additional formulae and corrections to match predictions to experimental values. Thus, the *ab initio* merit of the DFT computation is largely destroyed.

The magnitude of the superconducting transition temperature, T_c, not only contributes to determine the practical applicability of the material (together with the critical current and the critical field), but also can help gain insight to the mechanisms and underlying interactions from which superconductivity emerges [2]. Based on the fact that the McMillan equation reasonably successfully determines T_c from normal state parameters [3], we are led to believe that standard DFT may be capable of providing all the necessary superconducting characteristics of a material from knowledge of the crystal structure.

In this work, we present a summary of our progress towards the development of a predictive tool for superconductivity, as determined for hexagonal AlB_2-type structures, from the identification of anomalies in DFT calculated phonon dispersions (PDs). Highlights of our consistent extension of the approach to a different cubic structure case, that of hydrogen sulphide (H_3S), a superconductor at about

200K -- the temperature of a cold Antarctic day [4], are also presented.

2. Computational methods

All calculations are performed within the framework of first-principles DFT using the CASTEP module of Materials Studio 2017. LDA and GGA approximations are used as checks for consistency and a source for estimates of calculation errors [5-7]. Norm-conserving pseudo-potentials and density mixing schemes are used in the calculations. The electronic wave functions are expanded in terms of a plane-wave basis set with an energy cut-off of 990 eV. Other parameters are chosen to guarantee that the total energy convergence is within 1.0×10^{-6} eV atom^{-1}, maximum force at 0.01 eV Å^{-1}, maximum stress at 0.02 GPa and maximum atom displacement at 5×10^{-4} Å. The separation of k-points (Δk) is generally set to $\leq$ 0.03 Å^{-1}. Fine grids, increasing in fineness with the complexity of the structure, are required to resolve Fermi surface details and for reliable correlation between predicted and experimentally determined values within experimental error.

We measure the thermal energy δ given by the height or depth of the PD anomaly and calculate a corresponding temperature, T_δ, of the E_{2g} phonon(s) near the $\mathbf{\Gamma}$ zone centre, for each external condition, based on the following equation (1) (see Fig. 1) [5-7]:

$$\delta = (nN/Z)\cdot(k_B T_\delta/2) \tag{1}$$

where δ is the phonon anomaly (in cm^{-1}), n is the degrees of freedom per atom, N is the number of atoms per unit cell, Z is the number of formula units per unit cell, k_B is Boltzmann's constant, and $k_B T_\delta/2$ is the well-known relationship between thermal energy and degrees of freedom.

3. Modelling results

Most results presented in summarized form here, have been collected over an extended period, and have been published in more detail in separate publications. Only highlights are given below.

3.1. Application to MgB_2 and other AlB_2-type structures

Using this approach, we have consistently correlated the PD anomalies associated with the E_{2g} phonon modes to the superconducting properties for MgB_2 and other related AlB_2-type structures [8], for the different isotopic forms of MgB_2 [9], for a series of substituted MgB_2 compositions with Al [5] and with transition metals [6] and for the pressure dependence of superconductivity in MgB_2 [7]. We have also used these *ab-initio* calculations to predict potential new superconductor materials with enhanced T_c [5-6]. Figure 1 below shows the definition of the PD anomaly depth or height, δ. Table 1 below summarizes the excellent correlation obtained between the calculated and published experimentally determined values of T_c, for which there is good synthesis and materials and property characterization work. The list would significantly increase if isotopic results, all other intermediate substitutions and determined values for the pressure dependence are tabled.

3.2. Temperature effects

More recently, temperature effects, simulated via frozen displaced atomic positions, have provided invaluable insight on the actual superconducting mechanism, give a potential explanation for the superconducting gap and confirm the correlation between PD anomalies and T_c [10]. The calculations for frozen, displaced atoms clearly show that the fraction of the highly covalent bond charge (or electron density), that responds to the atomic core attractions and repulsions as the atoms vibrate, produce the change in force constants responsible for the abrupt change in E_{2g} frequency. The depth of the PD anomaly corresponds to a critical amplitude of interatomic vibration before the energy of the E_{2g} mode (red) goes above the energy of the B_{1g} mode (blue) (see Fig. 1). These effects on the PDs coincide with the Fermi surface undergoing a topological transition. Topological transitions have more frequently been discussed in terms of variable composition [11]. However, according to our results, it would appear that concepts of analogous topological transitions may well apply to the influence of temperature effects.

Table 1. Comparison of DFT calculated and published experimentally determined T_c's

Compound	Predicted T_c This work (K)$^+$	Experimental T_c By Others (K)$^+$	References
MgB_2	42.0 (3.0)	39.2 – 40.2	[8-9]** [15]
$(Mg_{0.8}Al_{0.2})B_2$	31.3 (3.4)	25 (2)–33.0 (0.1)	[5-6]** [16-17]
$(Mg_{0.66}Al_{0.34})B_2$	16.2 (3.3)	13.5 (0.1)	[5-6]** [16-17]
$(Mg_{0.5}Al_{0.5})B_2$	4.6 (2.5)	4.0 – 13.5	[5-6]** [16-17]
$(Mg_{0.86}Sc_{0.14})B_2$	31.5 (4.4)	23.0 (2.5)*	[6]** [18-19]
$(Mg_{0.75}Sc_{0.25})B_2$	11.4 (1.0)	~8.2 (5.0)*	[6]** [18-19]
MgB_2(P=-5to 20 GPa)	49 – 19 (3.0)	49 - 21	[7]** [20-21]
AlB_2	0.0	0.0	[5-6]** [22]
$BaSi_2$	9.3 (0.5)	8.9	[5-6]** [23]
$Ca(Al_{0.5}Si_{0.5})_2$	7.5 (0.5)	7.8	[5-6]** [23]

* T_c for Sc substituted compositions were reported as the peak position of the derivative of the surface resistance, and since the transitions are about 8-10K broad, they shift the experimental T_c by about 4-5K to lower values.
**References therein.
+ Values in parentheses are one standard deviation.

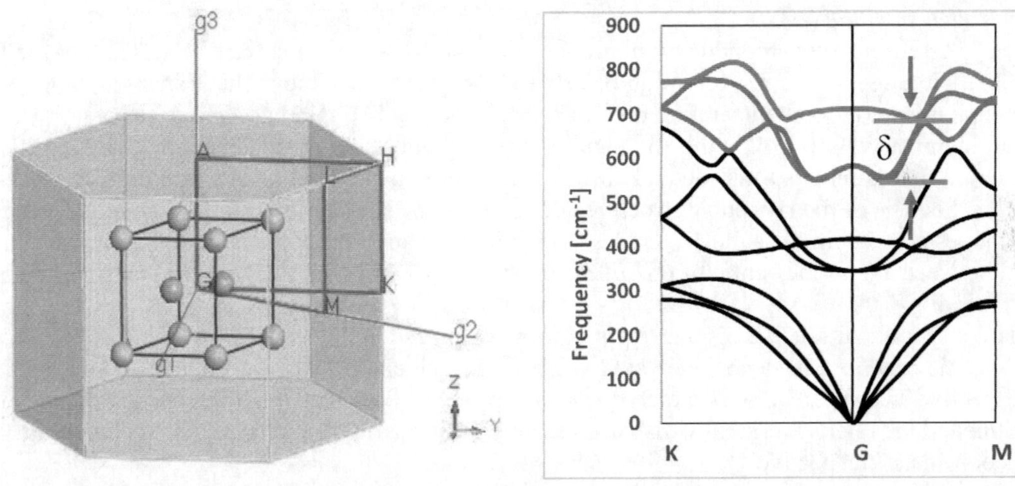

Figure 1. (a) Schematic of the MgB_3 structure. Mg and B are green and pink/gray, respectively. Reciprocal space directions are labelled; (b) PD anomaly for MgB_2 at atmospheric pressure where the definition of δ is indicated [5-7].

The Fermi surface of MgB_2 consists of approximately parallel warped cylindrical sections [5,7]. As the atoms are displaced along the movement patterns of the E_{2g} modes, at a critical relative displacement of ~ 0.6% (0.006 in relative coordinates), the light effective mass σ Fermi surface cylinder collapses and can no longer participate in purely in-plane transfer of electrons and phonon generation with the heavy effective mass (that is without involvement of substantial k_z components). The difference in Fermi energies at the equilibrium position and this critically displaced position correlates with the superconducting gap (see Fig. 2 below). MgB_2 offers a more straightforward correlation for this phenomenon, because charge transfers and atom displacements remain essentially in the same plane and directions up to and including the topological transition.

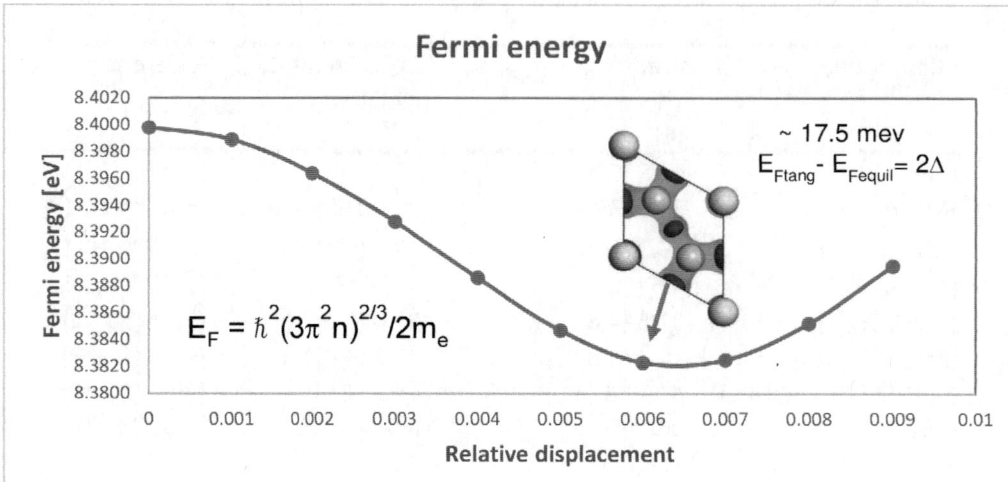

Figure 2. Variation of the Fermi energy of MgB_2 as function of the relative displacement of the B-B atom positions along the E_{2g} mode, which stretches and contracts the inter-bond distance. Notice that the difference in Fermi energies at the equilibrium position and at the critical displacement of about 0.6% (0.006 in relative coordinates) is about twice the superconducting gap energy for MgB_2.

3.3. Hydrogen sulphide (H_3S)

The recently discovered superconductor, hydrogen sulphide (H_3S), is the current record holder for T_c at about 200K, although it requires very high pressure (~ 160 GPa) to achieve this transition temperature which is equivalent to room temperature on a cold Antarctic day [4]. H_3S has a cubic structure with Im-3m group symmetry [12]. Following the same modelling approach used for MgB_2, we identify the phonon mode that is most strongly coupled to the electronic bands and/or electron density. Figure 3 displays a schematic of the H_3S cubic structure at 200 GPa. The strongly electronic coupled, degenerate phonon modes are identified as the T_{1u} modes, which involve stretching and contracting movements of the bond between a H in the centre of one of the cube faces (light blue) and S in the centre of the cube (yellow) (see Fig. 3).

Remarkably, critical relative displacements for the displaced atoms, which make the light effective mass electronic band lose its degeneracy at **G** and become tangential to the Fermi level, is in proportion to the respective T_c's, as are approximately the ratios of their differences in Fermi energies between the equilibrium and the relative, critically displaced atomic positions (0.03 and 0.006 for H_3S and MgB_2, in relative coordinates, respectively).

4. Discussion

A significant part of the work presented here follows the early lead from An and Pickett [13] on deformation potentials created by the displacement of boron atoms along the E_{2g} modes in MgB_2. The methodology displays significant potential for explaining mechanisms associated with superconductivity. However, earlier work failed to recognize that electronic bands tangential to the Fermi level is the point at which superconductivity appears to be destroyed. It is at this critical displacement that the Fermi surface of MgB_2 undergoes a topological transition, from connected approximately parallel warped tubes to separate paraboloids facing in opposite directions. Between the equilibrium position and this critical displacement, important correlations between PD anomalies, differences in Fermi energies, the superconducting transition temperatures (T_c) and superconducting

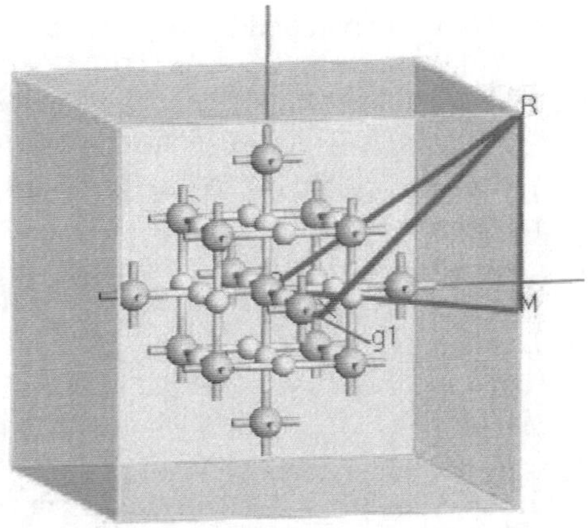

Figure 3. Schematic of the H_3S structure. H and S are light blue and yellow, respectively.

energy gaps are established. It is thus the interplay between the heavy and light effective mass Fermi surfaces (and the volume in between), which appears to control T_c.

The E_{2g} modes are strongly coupled to the electron density and are the most affected by variation in external conditions such as pressure, isotopic effects and elemental substitutions. Similarly, a degenerate T_{1u} mode in H_3S has been identified as strongly coupled to the electron density and the most responsive to variations in external conditions. Regular, consistent patterns in the characteristics of superconductors, as determined from standard DFT calculations of electronic bands, phonon dispersions and the effects of frozen displaced atomic positions, particularly if they follow key phonon modes, are becoming understood.

Given the growing list of examples [5-10, 14], which are producing excellent matches between calculated values of T_c and well determined experimental values, we posit that a method to determine T_c from a knowledge of the crystal structure is now emerging.

5. Conclusions

Consistent examples of accurate determinations of T_c for MgB_2 and other related AlB_2-type structures under a range of external conditions have been provided. This approach has also been validated for other crystal structures. Thus, standard DFT calculations can provide direct information about superconductivity of compounds from a knowledge of the crystal structure without requiring post calculation corrections or parameter adjustments. Identification of consistent patterns of properties across families of superconductors, as determined from DFT calculations, indicate that a potential tool for prediction of superconductivity is emerging.

Acknowledgments

Assistance from Ashley Wright and Abdul Sharif from the High-Performance Computing (HPC) facilities at QUT is gratefully acknowledged. No specific grants have been received to carry out this research.

References

[1] Chu C W et al 2015, “Epilogue: Superconducting materials past, present and future”, Physica C, vol 514, pp 437-443.
[2] Geballe T H, Hammond R H and Wu P M 2015, “What T_c tells”, Physica C, vol 514, pp 9-16.
[3] Carbotte J P 1990, "Properties of boson exchange superconductors," Rev. Modern Phys., vol 62, pp. 1027-1157.

[4] Bianconi A and Jahrlborg T 2015, "Superconductivity above the lowest Earth temperature in pressurized sulfur hydride", EPL, vol 112, pp 37001-1-7.
[5] Alarco J A, Talbot P C and Mackinnon I D R 2015, "Phonon anomalies predict superconducting T_c for AlB_2 type structures", PCCP, vol 17, pp 25090-25099.
[6] Mackinnon I D R, Talbot P C and Alarco J A 2017, "Phonon dispersion anomalies and superconductivity in metal substituted MgB_2", COMMAT, vol 130, pp 191-203.
[7] Alarco J A, Talbot P C and Mackinnon I D R 2017, "Phonon dispersion models for MgB_2 with application of pressure", Physica C, vol 536, pp 11-17.
[8] Alarco J A, Talbot P C and Mackinnon I D R 2014, "Phonon modes of MgB_2: super-lattice structures and spectral response", PCCP, vol 16, pp 24443-24456.
[9] Alarco J A, Talbot P C and Mackinnon I D R 2014, "Coherent phonon decay and the boron isotope effect for MgB_2", PCCP, vol 16, pp 25386-25392.
[10] Alarco J A, Talbot P C and Mackinnon I D R 2017, "Electron density response to the phonon dynamics in MgB_2", submitted.
[11] Bianconi A 2005, "Feshbach shape resonance in multiband superconductivity in heterostructures", J. Supercond. Nov. Magn. vol 18(5/6),pp 626–636.
[12] Papaconstantopoulos D A, Klein B M, Mehl M J and Pickett W E 2015, "Cubic H_3S around 200 GPa: An atomic hydrogen superconductor stabilized by sulfur", Phys. Rev. B, vol 91, pp 184511-1-5.
[13] An J M and Pickett W E (2001) "Superconductivity of MgB2: covalent bonds driven metallic". Phys. Rev. Lett. Vol 86, pp 4366–4369.
[14] Alarco J A, Talbot P C and Mackinnon I D R 2018, "A Complete and Accurate Description of Superconductivity of AlB_2-Type Structures from Phonon Dispersion Calculations", J. Supercond. Nov. Magn. Vol 31, pp 727-731 (https://doi.org/10.1007/s10948-017-4328-6).
[15] Nagamatsu J, Nakagawa N, Muranaka T, Zenitani Y and Akimitsu J (2001) "Superconductivity at 39 K in magnesium diboride", Nature vol 410, pp 63-64.
[16] Karpinski J, Zhigadlo N D, Katrych S, Puzniak R, Rogacki K and Gonnelli R (2007), "Single crystals of MgB_2: synthesis, substitutions and properties", Physica C 456, pp 3-13.
[17] Sevastyanova L G, Gulish O K, Stupnikov V A, Genchel V K, Kravchenko O V, Bulychev B M, Lunin R A, and Tarasov V P (2012), "Structure and properties of solid solutions in the Mg-Al-B system", Cent. Eur. J. Phys. Vol 10 (1), pp 189-196.
[18] Agrestini S, Metallo C, Filippi M, Campi G, Sanipoli C, De Negri S, Giovannini M, Saccone A, Latini A and Bianconi A (2004), "Sc doping of MgB_2: the structural and electronic properties of $Mg_{1-x}Sc_xB_2$", J. Phys. Chem. Solids vol 65 (8–9) pp 1479-1484.
[19] Agrestini S et al. (2004), "Substitution of Sc for Mg in MgB_2: effects on transition temperature and Kohn anomaly", Phys. Rev. B vol 70 (13) pp 134514-1-5.
[20] Tang J, Qin L -C, Matsushita A, Takano Y, Togano K, Kito H and Ihara H (2001), "Lattice parameter and T_c dependence of sintered MgB_2 superconductor on hydrostatic pressure", Phys. Rev. B vol 64, pp 132509-1-4.
[21] Deemyad S, Tomita T, Hamlin J, Beckett B, Schilling J, Hinks D, Jorgensen J, Lee S and Tajima S (2003), "Dependence of the superconducting transition temperature of single and polycrystalline MgB_2 on hydrostatic pressure", Physica C: Superconductivity vol 385, pp 105-116.
[22] Awana V P S, Vajpayee A, Mudgel M and H. Kishan H (2009), "Superconductivity of various borides and the role of carbon in their high performance", Supercond. Sci. Technol. vol 22, pp 034015.
[23] Flores-Livas J A, Debord R, Botti S, San Miguel A, Pailhes S, Marques M A L (2011) "Superconductivity in layered binary silicides: a density functional theory study", Phys. Rev. B vol 84, pp 184503-1-7.

IOP Conf. Series: Journal of Physics: Conf. Series **1143** (2018) 012029 doi:10.1088/1742-6596/1143/1/012029

Synthesis of YBCO superconductor by the method of combustion reaction in solution

J Rodriguez[1] **and A Lazo**[2]

[1]Faculty of Science, National University of Engineering, Lima, Peru

[2]Faculty of Geological, Mining and Metallurgical Engineering, National University of Engineering, Lima, Peru

E-mail: jeancarlosr1505@gmail.com

Abstract. Solutions of metallic nitrate were prepared to obtain the $Y_1Ba_2Cu_3O_{7-\delta}$ and $Y_3Ba_5Cu_8O_{18-\delta}$ phases in different mixtures of urea and glycine balanced stoichiometrically according to the principle of propellant chemistry to be synthesized by the method of combustion reaction in solution using a thermal blanket type furnace. The obtained powders required a post-reaction heat treatment at 920 ° C for 2h in air. The X-ray diffraction (XRD) characterization showed that the synthesized powders have a predominant superconducting $Y_1Ba_2Cu_3O_{7-\delta}$ phase with an orthorhombic crystal structure and a complementary phase $Y_3Ba_8Cu_5O_{18-\delta}$. The $Y_3Ba_5Cu_8O_{18-\delta}$ phase was not found in any of the cases. The results of the transition electron microscopy (TEM) showed most grains with an average size of 300 nm and confirmed the Y123 phase. In addition, it was observed grains with an average diametrer of 1.6 μm. The measurement of electrical transport gave a $T_c \sim 91$ K by the 4-probe method and by SQUID magnetometry it was obtained a $T_c \sim 92$ K for zero field cooled (ZFC) and field cooled (FC) magnetizations.

1. Introduction

Superconductor materials are applied for different purposes such as the producction of very large magnetic fields and designing train prototypes based on magnetic l evitation. Additionally, the latest investigations report a critical current density J_c within the YBCO grains or film of about $1MAcm^{-2}$ at 77 K. Moreover, the YBCO superconductor has an advantage over other superconductor based of Bi related to its applications in relation to its high critical current density J_c. Therefore, this advantage is suitable for applications in electrical industry.

The YBCO superconductor was synthesized for first time by C. Chu *et al.* [1] in 1987 by the solid state reaction method with transition temperature of $T_c \sim 92$, however, this method has some disadvantages due to its long sample preparation procedure and a long synthesis time, around 16 h at 100 h in a procedure that it is repeated up to 3 times in a controlled atmosphere of oxygen. As a result, different methods have been proposed to obtain this superconductor in order to optimize the amount of material obtained or to simplify the procedures. Hence, Hsu *et al.* [2] in 1993 synthesized this material by spontaous reaction requiring for the best obtained sample an heat treatment for 5 h to $P_{O_2} = 1$ atm with the aim of decreasing the additional phase as $Y_2Ba_1Cu_1O_5$ y $BaCO_3$ found. In 1997, Kato *et al.* [3] achieved to synthesize the pure phase of $Y_1Ba_2Cu_3O_{7-\delta}$ using metal oxides by domestic microwave oven without post-heat treatment.

IOP Conf. Series: Journal of Physics: Conf. Series **1143** (2018) 012029 doi:10.1088/1742-6596/1143/1/012029

Although they managed to reduce the synthesis time to a few minutes, their method requires an laborious process of preparation of the sample previously.

Surprisingly in 2009, the last member of the YBCO family was discovered, which is also a superconductor. A. Aliabadi *et al.* [4] synthesized the superconductor $Y_3Ba_5Cu_8O_{18-\delta}$(Y358) by the traditional method of solid state reaction at 840°C for 12 h. The sample had a $T_c \sim 102$ K and the crystalline structure of this phase is very similar to the crystal structure Y123 with the exception of the number of CuO chains and CuO_2 planes.

Since 2012, a new trend has been developed in using biopolymers to synthesize YBCO as Hall *et al.* [5] that successfully synthesized the $Y_1Ba_2Cu_3O_{7-\delta}$ superconductor by the sol-gel method using biopolymers. Additionally, the advantage of using these biopolymers is due to the formation of texturized materials as nanowires. The powders obtained were dried for one day at 40°C and required a thermal treatment of 920°C in air for 2 h. The best obtained temperature T_c was 83 K with a critical current density J_c of 1.5 $MAcm^{-2}$ at 10 K and 1 T of applied field. This critical current density exceeds the commercially available superconducting powders by almost 2 orders of magnitude (Aldrich 99.9%). Afther that Konne *et al.* [6] continued with this methodology to synthesize for the first time nanowires of the new superconducting phase Y358 obtaining T_c of 89.4 K that would naturally correspond to the Y123 phase. They justified the decrease in T_c is possibly due to morphology effects such as grain size or shape of the sample. Additionally, the critical current density J_c was 7.6 $MAcm^{-2}$ at 10 K and 1 T of applied field.

In this letter, we report the synthesis of superconducting powders of $Y_1Ba_2Cu_3O_{7-\delta}$ by the method of combustion in solution for first time using fuels: glycine and urea. The temperature requires for the synthesis of the phase Y123 or Y358 have to be over to 800 ° C [7] [6] and the previous experience in the synthesis in other compound of similar crystal structure [8] guarantees that mixture of the fuels achieves temperature above 800°C. The obtained powders needed a post-heat treatment at 920°C for 2 h. However, new phase Y358 have not observed in any of cases for this method.

2. Sample preparation and characterization

In order to synthesize the superconducting Y123 and Y358 powders, we have used the method of combustion in solution using nitrates as reactants, $Y(NO_3)_3.6H_2O$, $Ba(NO_3)_2$ y $Cu(NO_3).3H_2O$, and the fuels used in this synthesis were urea (U) and glycine (G). The nitrates and fuels were balanced weighted stoichiometrically according to the principle of propellant chemistry [9] to get 2 g of compound YBCO with a molar ratio of Y:Ba:Cu::1:2:3 and Y:Ba:Cu::3:5:8.

Additionally, other syntheses were made by varying the urea-glicine molar ratio $r = U/G$ of the value obtained by the principel of propellant chemistry , $r = 1.5$. The Y123 samples with molar ratios r were of 1.5 named as $Y123_1$ and 2.02 named as $Y123_2$; and the Y358 samples with molar ratios r were of 1.4 named as $Y358_1$, 1.5 named as $Y358_2$ and 1.72 named as $Y358_3$. The nitrates reactants and the fuels were dissolved in 10 mL of ultrapure water in each case and mixed homogeneously.

The color of solutions of the mixture was turquoise (Figure 1 a) and it was carried to thermal blanket type furnace. The mixture was heated until ignition temperature T_i ~390°C in a ramp of 10° C/min. To that temperature T_i, the fuels exercise their role starding a ignition process reaching high temperature. This ignition process lasted about 1 min, after that, the samples were cooled down rapidly to room temperature. Therefore, the obtained powders for all syntesis were dark brown (Figura 1 c).

However, their respective XRD patterns showed that the powders were not totally ordered then they required heat treatment. The samples was carried out a heat treatment at 920°C in Mufla Yamato FO210CR for each samples in air and cooled down slowly to room temperature.

IOP Conf. Series: Journal of Physics: Conf. Series **1143** (2018) 012029 doi:10.1088/1742-6596/1143/1/012029

(a) Solution of nitrates and fuels mixture (b) the same solution heated near T_i (c) obtained powders

Figure 1. Procedure for obtaining YBCO powders.

X-ray diffraction (XRD)was carried out to the synthesized powders after the heat treatment in order to determine the phases composition of the powders, using a Cu K_α radiation source of Bruker D8-advance powder diffractometer. The position of peaks and intensities were obtained between 20° y 70° with a step size of 0.02. The results of XRD measurement indicate that both synthesis of Y123 samples have predominat superconducting $Y_1Ba_2Cu_3O_{7-\delta}$ phase as shown in Figure 2.

The peaks in $2\theta = 47°$ indicate an orthorhombic crystal structure with simetric group $Pmmm$ according to PDF 038-1433 [10]. Also an secundary phase was observed , $Y_3Ba_8Cu_5O_{18-\delta}$, which is an ordered deformation of $Y_1Ba_3Cu_2O_{6.5+\delta}$ with a simetric group $P4/mmm$ [11] and small peaks corresponding to Y_2BaCuO_5 [12], $BaCuO_2$ (PDF 038-1402) and CuO [13].

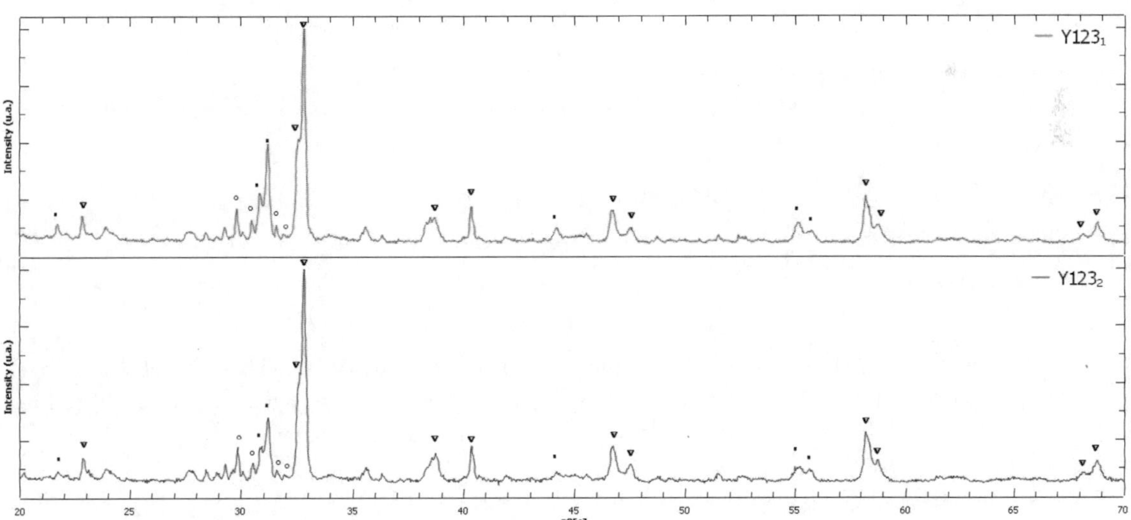

Figure 2. XRD patterns of the Y123 samples. Synthesized powders with $r = 1.5$ (red line) and synthesized powders with $r = 2.02$ (blue line). Symbols $\bigtriangledown$, ■ and ○ indicate peaks due to $Y_1Ba_2Cu_3O_{7-\delta}$, $Y_3Ba_8Cu_5O_{18-\delta}$, Y_2BaCuO_5, respectively. Also, it was observed impurities such as $BaCuO_2$ and CuO.

The presence of these impurity phases is due to the decomposition of the $Y_1Ba_2Cu_3O_{6.5+\delta}$ phase [14] probably given post-reaction heat treatment. The $Y_3Ba_8Cu_5O_{18-\delta}$ phase decrease

IOP Conf. Series: Journal of Physics: Conf. Series **1143** (2018) 012029 doi:10.1088/1742-6596/1143/1/012029

when it was increased the molar ratio of fuels r of 1.5 to 2.02, however, the impurity phases as green phase Y_2BaCuO_5 y $BaCuO_2$ and $BaCuO_2$ increases due to greater decomposition of superconducting YBCO.

Measurement of X-ray diffraction (XRD) for the Y358 samples was carried out using a Co radiation source of Bruker D8-advance powder diffractometer. The position of peaks and intensities were obtained between 20° y 70° with a step size of 0.02. In Figure 3 is shown XRD patterns of a group of Y358 samples which in the samples $Y358_1$ have a predominant phase Y123 with several compounds due to imcomplete reactions given in the synthesis possibly due to the deficit of a urea or excess of glycine. The $Y358_2$ samples have better XRD patterns with two phases: $Y_1Ba_2Cu_3O_{7-\delta}$ and phase green Y_2BaCuO_5, and has only an impurity corresponding to $Y_3Ba_8Cu_5O_{18-\delta}$ phase. This samples were synthesized with $r = 1.5$ according to Fumo. Finally, The $Y358_3$ samples have only two phases, $Y_1Ba_2Cu_3O_{7-\delta}$ and Y_2BaCuO_5.

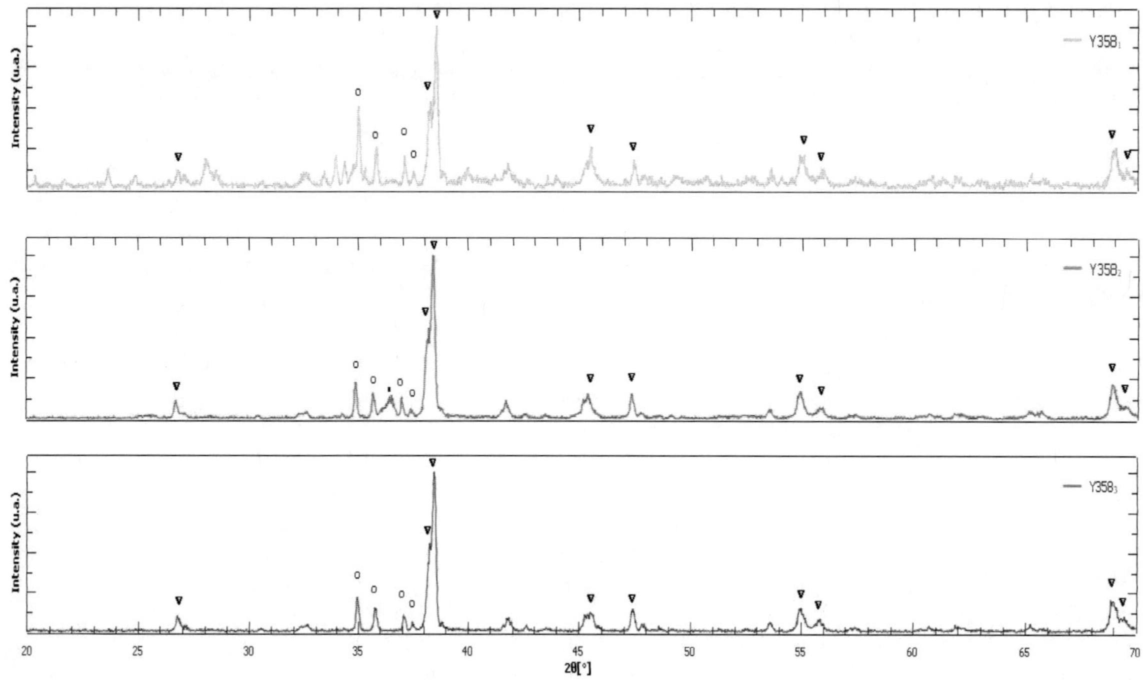

Figure 3. XRD patterns of the Y358 samples. Synthesized powders with $r = 1.4$ (cian line), synthesized powders with $r = 1.5$ (red line) and synthesized powders with $r = 1.72$ (blue line) . Symbols $\bigtriangledown$, ■ and O indicate peaks due to $Y_1Ba_2Cu_3O_{7-\delta}$, $Y_3Ba_8Cu_5O_{18-\delta}$, Y_2BaCuO_5, respectively.

However, it was not observed the new phase Y358 reported by Aliabadi [4] in any of the prepared samples with molar ratio of Y:Ba:Cu::3:5:8. S. Gholipour *et al.*[15] reported the coexistence of phases Y123 and Y358. In addition, it is observed that these phases have similar XRD patterns and it is not easy to distinguish by this measurement.

3. Results and dicussions

The microstructure of the grains were determined by transmission electron microscope (TEM) (MET) TECNAI G^2 F20 equipped with an energy dipersive x-ray (EDX) for analysis. Samples for transmission of electron microscopy were prepared by placing a drop of presonicated-aliquot on a carbon coated Cu grid.

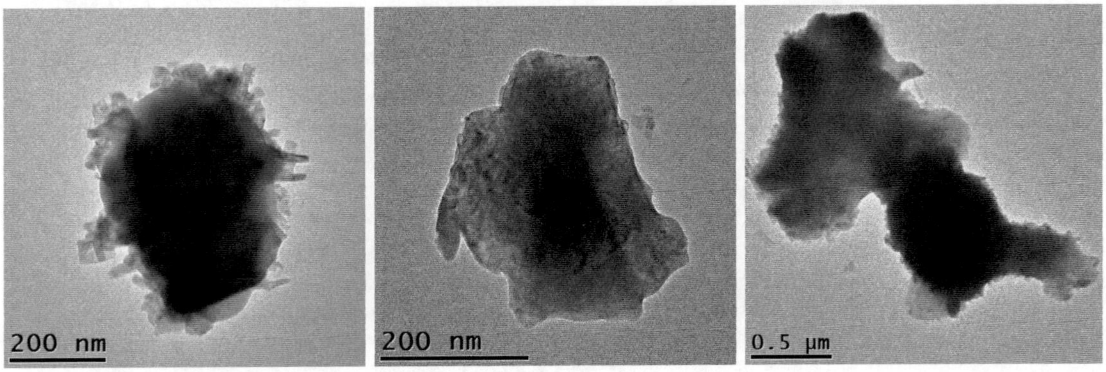

(a) (b) (c)

Figure 4. HR-TEM images of grain of $Y358_2$ sample.

The results of TEM are in Figure 4 it was observed two grain sizes for the Y358 samples: some of 300 nm in diameter on average and others of 1.6 μm. Moreover, the same size distribution was observed in Y123 samples (Figure 5). The grain belongs to the $Y123_1$ sample and its size was $2.05\mu m$x$1.36\mu m$.

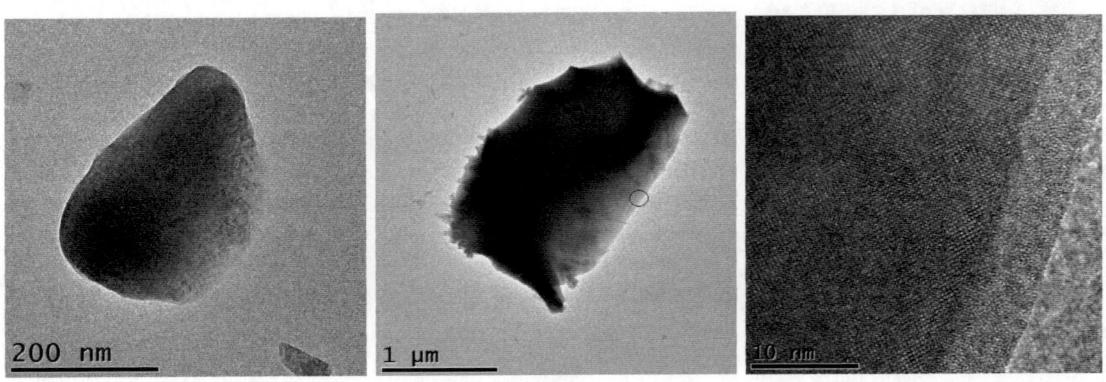

(a) (b) (c)

Figure 5. HR-TEM images of grain of $Y123_1$ sample.

It was carried out high resolution TEM (HR-TEM)in the marked region shown in Figure 5 b. In this figure, it was observed the crystalline planes of the material what shown that the powders have an ordered structure. It is known that Y123 phase has a preferential phase Y123 has a preferential orientation for the plains [0 0 1] and the analysis of fast fourier transform (FFT) (Figure 5 c)) confirms this characteristic of the YBCO. The lattice parameters measured by FFT were $3.84\,\mathring{A}$ y $3.86\,\mathring{A}$, and the lattice parameters of superconducting YBCO according [10] to literature are $a = 3.8185\,\mathring{A}$, $b = 3.8856\,\mathring{A}$ y $c = 11.6804\,\mathring{A}$. Taking into consideration the measurement uncertainty, the measured lattice parameters are very close to the parameters a y b which indicate that the planes are orientated in direction [001] and the measured parameters correspond to an orthorhombic structure.

IOP Conf. Series: Journal of Physics: Conf. Series **1143** (2018) 012029 doi:10.1088/1742-6596/1143/1/012029

SQUID magnetometry was performed on a Quantum Design magnetic property measurement system (MPMS)-5S equipped with a 5 T superconducting magnet. Field cooled (FC) and zero field cooling(ZFC) dc magnetization were measured as a function of temperature under an applied field of H= 100 Oe. Figure 6 shows that $Y123_1$ and $Y123_2$ samples had a diamagnetic behavior in the ZFC and FC measurements. Therefore, SQUID magnetometry showed that both samples were superconducting with $T_{c_{onset}}$ of 91.79 K and 90.98 K, respectively in Figure 6). The variation of obtained T_c in each sample is intimately related to the oxygen content according to M. Kogachi *et al.*[16]. By the obtained T_c above 90 K, the oxygen conten of $Y_1Ba_2Cu_3O_{7-\delta}$ is estimated to be $\sim$ 6.9.

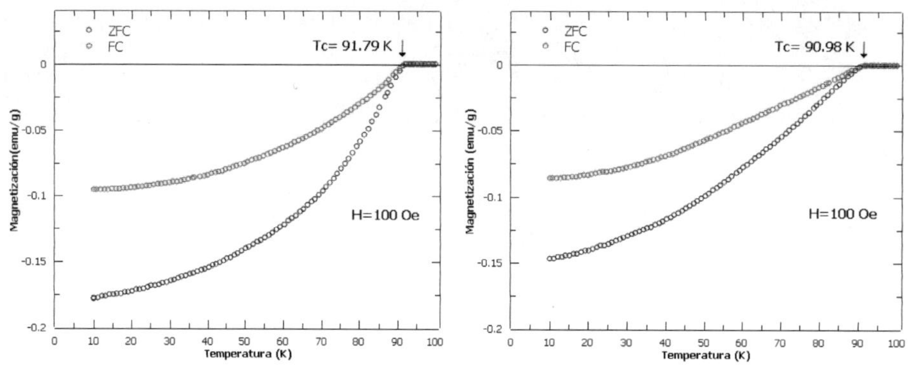

(a) $Y123_1$ sample (b) $Y123_2$ sample

Figure 6. ZFC y FC magnetization of Y123 samples as a function of temperature under $H = 100$ Oe.

It was considered the influence of other secondary phases found by XRD in the magnetization or magnetic moments measured in all sample volume, however, those phases are not superconducting. Therefore, they do not contribute significantly to the measured magnetic moments, except the $BaCuO_2$ phase which has been reported that decrease diamagnetic signal. For that reason, it was observed that magnetization curves are greatest slope and have greater magnetization for below temperature of T_c when synthesis was $r = 1.5$ ($Y123_1$) due to it has less concentration of $BaCuO_2$.

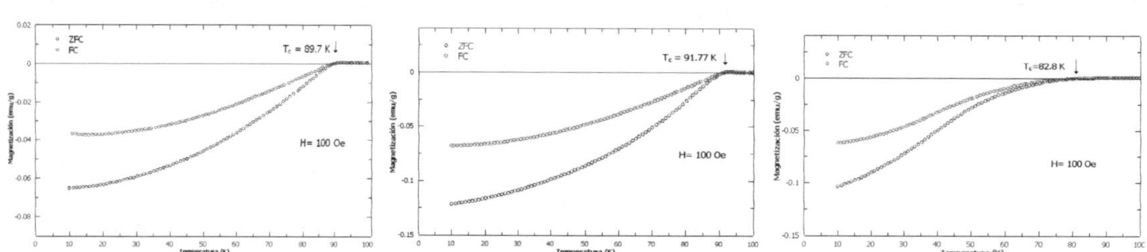

(a) $Y358_1$ sample (b) $Y358_2$ sample (c) $Y358_3$ sample

Figure 7. ZFC and FC magnetization of Y358 samples as function temparature under $H = 100$ Oe.

Additionally, the ZFC and FC measurements of Y358 were shown in Figure 7. The best sample has a $T_{c_{onset}}$ of 91.77 K which was synthesized by molar ratio $r = 1.5$. However, the contribution of i ncomplete reaction was observed decresing diamagnetic signal i n the magnetic measurement f or the $Y358_1$, although, i t has a good critical temperatura T_c ~89.7 K On the other hand, magnetic measurements of $Y358_3$ sample has a poor oxygen content on samples. That is the reason why the samples have a low critical temperature T_c ~82.8 K

Konne *et al.* [6] synthesized nanowires of YBCO which identified as new Y358 phase, however their samples has low critical temperature $T_c \sim 90$ K. They justified the difference between the literature onset T_c values for Y358 is likely due to morphology effects such. In spite of the fact that most grains observed by TEM have an average size of 300 nm, these powders do not suffer the effects of morphology and maintain their critical high temperatures $T_c \sim 92$ K, Figure 6-7.

The electrical transport mesurement was carried out measuring the resisitivity as a function of temperature by the dc four-point method on sample $Y123_1$ inside a Dewar with liquid nitrogen and helium. In order to perform this measurement it was needed to compact the synthezed powders in order to get a pellet and it was applied a heat treatment again at 920°C for 2 h in air. Then the pellet was cut to get a slab of $5.8x2x1.22mm^3$. Figure 9 shows the electrical resistivity dependence on temperature, it was observed that the $Y123_1$ sample has a superconducting transition with a prolonged slope. The superconduting transition temperature $T_{c_{onset}}$ was 90.7 K . The resistivity at room temperature was 6.44 Ωmm which was too big and tended to increase until T_c temperature. This contrasts with other reports where they got lower values of resistivity.

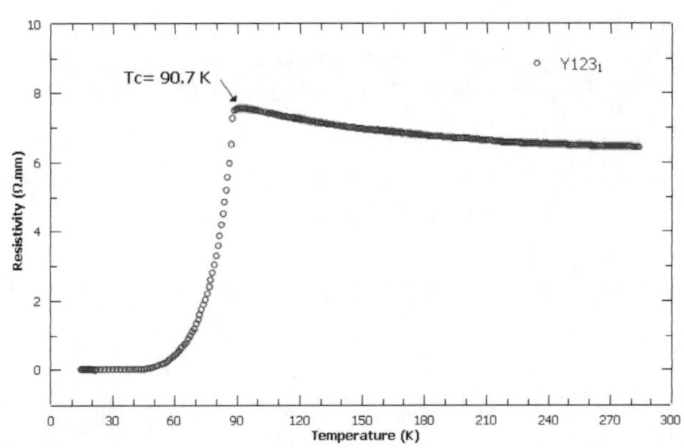

Figure 8. Measurement of electrical transport by the dc four-point method on the $Y123_1$ sample.

The great value of mesurement of resistivity and the prolonged slope could be explained by the bad percolation of current inside the sample as a consecuence of grain colescence or the impurities presence. This slab was oxidized by atmosphere of O_2 at 460 ° C for 6 h and the mesurement of electrical transport showed that the effect of oxigenation atmostphere did not have a significant contribution in sample.

4. Conclusions

We have succesfully demostrated that using this modified combustion in solution method using mixture urea-glycine for fisrt time, it is possible to prepare fine powder of $Y_1Ba_2Cu_3O_{7-\delta}$ phase at temperatures as low 920 ° C. We have showed that process of YBCO phase formation is highly dependent of the molar ratio urea/glycine. The measurements of ZFC and FC were obtained $T_c \sim$ 92 K when the reaction was with molar ratio r = 1.5 for the $Y123$ and $Y358$ samples. We have not observed the new Y358 phase in any of prepared samples. The measurement of

IOP Conf. Series: Journal of Physics: Conf. Series **1143** (2018) 012029 doi:10.1088/1742-6596/1143/1/012029

electrical transport was found $T_c \sim 91$ K. The obtained T_c above to 90 K is in according to oxygen content of 6.9. Although most grains have an average size of 300 nm, these powders do not suffer the effects of morphology and maintain their high critical temperatures $T_c \sim 92$ K.

Acknowledgments
This work was made possible thanks to the programe of master's degree UNI-CONCYTEC, agreement Nro. 167. The authors would like to thank the laboratory of low temperatures of Centro Atómico Bariloche, especially to Dr. J. Guimpel, Dr. G. Nieva and Dr. A. Tolley for electrical, magnetic and TEM measurements.

References
[1] Wu M *et al.* 1987 *Phys. Rev. Lett.* **58**, 908
[2] Hsu S *et al.* 1993 *Physica C.* **207**, 159
[3] Kato M *et al.* 1997 *Jpn. J. Appl. Phys.* **36**, 1291
[4] Aliabadi A *et al.* 2009 *Physica C* **469**, 2012
[5] Hall S *et al.* 2012 *Supercond. Sci. Technol.* **25**, 035009
[6] Konne J *et al.* 2012 *Supercond. Sci. Technol.* **25**, 115005
[7] Schnepp Z *et al.* 2008 *Adv. Mater.* **12**, 1782
[8] Ianos R *et al.* 2015 *J. Master Sci.* **50**, 6382
[9] Fumo D *et al.* 1997 *Mater. Res. Bull.* **32**, 1459-1470
[10] Wong-Ng W *et al.* 1987 Powder Diffraction **2**, 192
[11] De Leeuw W *et al.* 1988 *Physica B+C.* **152**, 39
[12] Mechel C and Raveau B, 1982 *Journal of Solid State Chemistry.* **43**,72
[13] Langford J and Louer D 1991 *J. Appl. Crystallogr.* **424**, 149
[14] Murakami M 1992 Melt Processed High-Temperature Superconductors. Chapter 2 page 15. Ed World Scientific Publishing Co. Pte. Ltda
[15] Gholipour S *et al.* 2012 *Supercond. Nov. Mag.* **25**, 2253
[16] Kogachi M *et al.* 1989 *Jpn. J. Appl. Phys.* **28**, 609
[17] Funaki S *et al.* 2016 *Jpn. J. Appl. Phys.* **55**, 04EJ13

IOP Conf. Series: Journal of Physics: Conf. Series **1143** (2018) 012030 doi:10.1088/1742-6596/1143/1/012030

Some aspects of electromagnetic interference, electromagnetic compatibility and geomagnetic induced currents

Ener Salinas

ABB Corporate Research, Electrotechnology and Applied Physics Team,
Forskargränd 7, 721 78 Västerås, Sweden

E-mail: ener.salinas@se.abb.com

Abstract. Electromagnetic compatibility is studied in special environments represented by renewable energy systems of two types, namely wind power farms and large photovoltaic parks. First, the sources of interference are identified which can be internal or external to these systems. Then suggestions of how to mitigate these effects are proposed. In addition, geomagnetic induced currents are also described as rare events but relevant sources of interference for both systems. The focus is on immunity of these systems to internal and external sources, but emissions and their effects on external equipment/installations will also be mentioned.

1. Introduction

The future of a modern society is highly dependent on the efficiency and reliability of the power network. In this respect, the inclusion of renewable energies has brought several improvements improvements. Yet, at the same time new problems have emerged. For example in wind power (WP) and solar photovoltaic (PV) systems, electromagnetic interference (EMI) can be produced at various frequencies, from 50/60 Hz harmonics up to MHz order and propagate via cables or field emissions. Typical sources of EMI are solid-state converters and connecting cables. In the latter case, wiring in large PV systems (e.g. solar parks of over MW power) can act as antennas injecting disturbances into the power grid or causing serious interference in parts of the PV system itself. Solving such problems by applying computational tools in electromagnetism and performing experimental validation, with further implementation, makes it possible to reach the state of electromagnetic compatibility (EMC) for these systems [1,2].

EMI in power systems can also be generated by natural phenomena, for example lightning, a major source of disturbance, or by coronal mass ejections (CME) from the sun, which under certain conditions are able to create ionospheric intense currents which couple to long wiring ground connections inserting geomagnetic induced currents (or GICs) into large transformers, saturating the core and generating destructive overheating that may, in the worst case, produce a black out. This work is a sample of the investigations of various CIGRE working groups to which the author has contributed [3-7].

2. Electromagnetic compatibility (EMC)

The topic of EMC describes, firstly, the ability of an electronic or electric equipment to operate correctly in its electromagnetic environment, also known as *immunity,* which may relate to another nearby (or external) equipment or, in a wider sense, it may involve natural causes (such as lightning or GICs), all these being *sources of interference*. Secondly, EMC also describes the possibility that the initially considered equipment could cause interference on another equipment, also known as victim, which can be observed in figure 1. For the purpose of illustrating the concepts, a source of unwanted interference and various "victims" have been located for each type of interfering interaction. Thus, we have electromagnetic or radiative type (victim 1), via cables or conductive (victim 2), electric or capacitive (victim 3) and magnetic or inductive (victim 4).

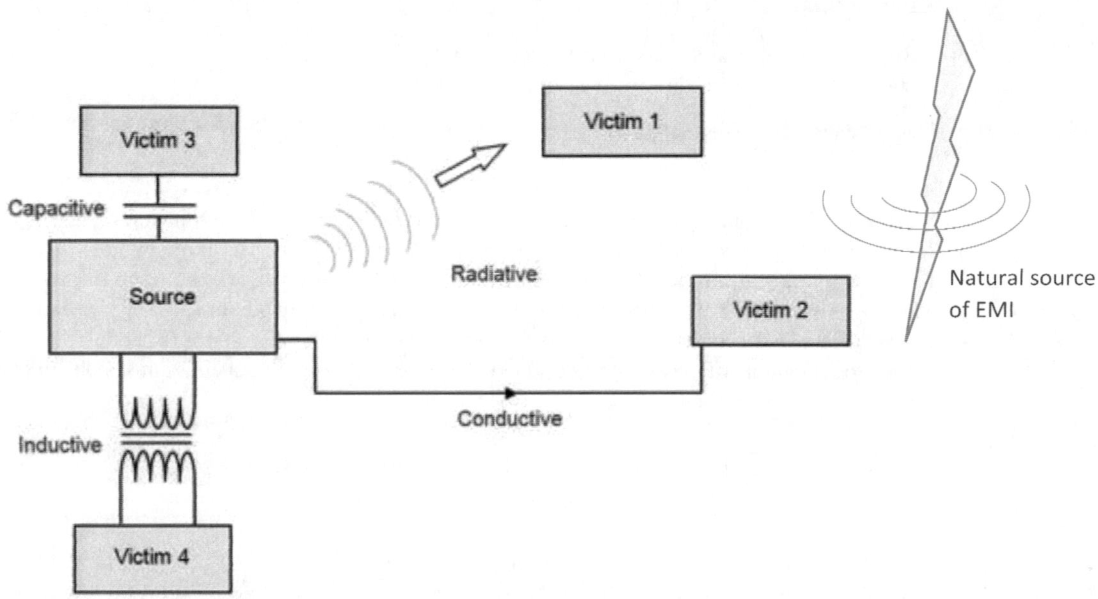

Figure 1. Simplified illustration of various types of EMI produced by an artificial or a natural source (e.g. lightning).

It is important to point out that these interferences are due to natural, stray or unintended emissions from the source. However, intentional electromagnetic interference (IEMI) has recently been subject of intense investigation due to increasing awareness of attacks on cybersecurity, high power microwave (HPM) devices, high altitude electromagnetic pulse (HEMP) attacks and other related risks [8]. Yet, these topics are outside the scope of this paper.

3. EMC in wind energy systems

Wind energy systems, both types, onshore and offshore are increasingly being produced and developed, providing not only energy but also stability to the main grid. The more powerful, the higher and larger the wind turbines are. This fact creates a potential for lightning strikes and interference (figure 2). Lightning protection is designed to "guide" the impulsive incursion through a conductor (located in each blade) then the impulse is transmitted via the nacelle, where sensitive equipment and controls are located, down to the bottom part of the tower where controls are also located. The impulse current is finally distributed to reach the ground in a specially designed conductive mesh (copper) that will avoid any destructive effect on the foundations. The traveling impulse is able to produce serious disturbances in the operation of controls, thus robust design and shielding of such equipment must be carefully applied in order to mitigate these effects.

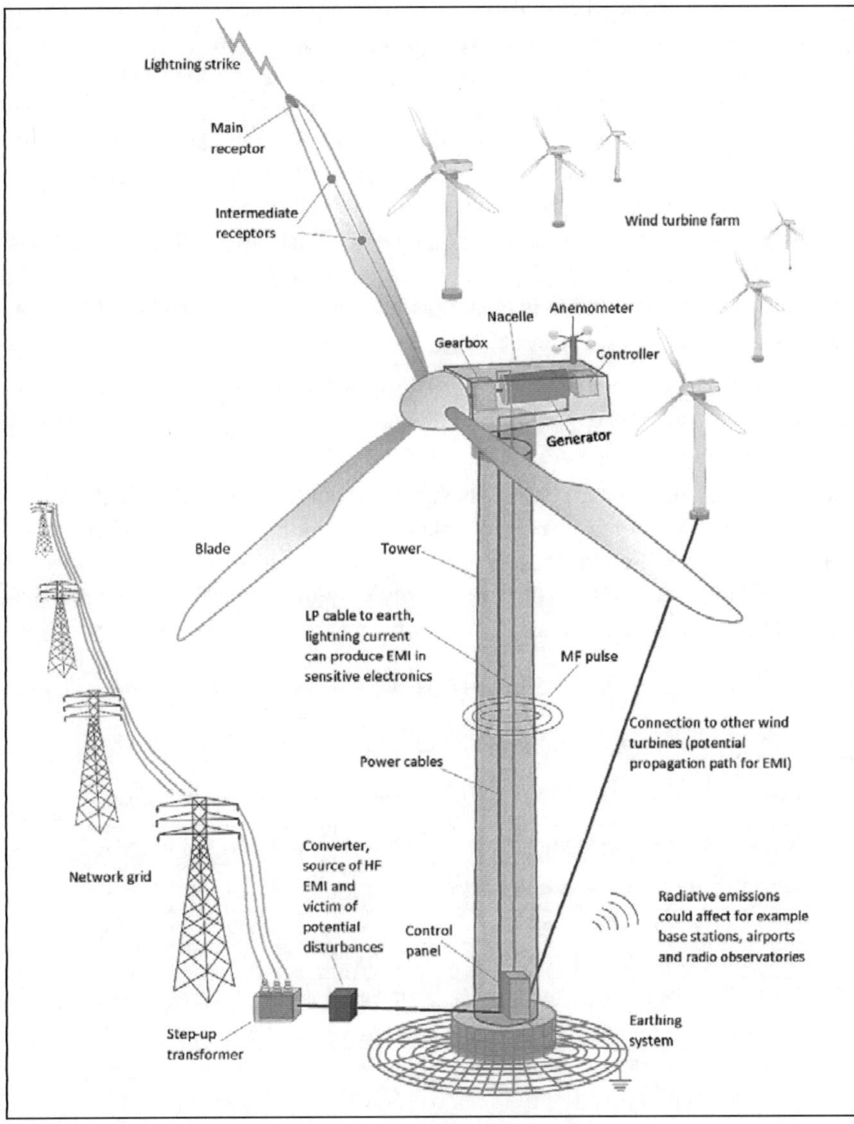

Figure 2. A wind power farm connected to the network grid showing sources of EMI with disturbances propagating inside the turbine, to other turbines, to the grid or to external installations.

The mitigation designs should allow the equipment to comply with standards of lightning protection for wind turbines such as IEC 61400-24 and IEC 62305-4 [5]. Another source of EMI is the high-frequency emissions by converters. The power flow starts with a rotational energy/torque from the blades transformed into AC electric output by an induction generator, which then goes to an AC/DC/AC converter that has two parts: the rotor side converter and the grid-side converter. Both are voltage-source converters that use forced-commutated power electronic devices (IGBT's) to synthesize an AC voltage from a DC voltage source. These fast switching operations tend to emit high frequency (kHz to MHz or even more) disturbances that couple to wires and can be radiated and affect sensitive parts of the system. These emissions can also propagate outside and reach other wind turbines via connecting cables, at the same time they can reach the network grid. Protection techniques such as HF filters and dampers have to be implemented. An IEC EMC standard that can be applied to a WT is IEC 61800-3 as suggested in [5]. Furthermore, radiative emissions can reach the network grid and other sensitive installations/equipment located externally, for example mobile communication base stations, airports, and radio-telescopes, as shown in figure 2.

A source of EMI which is typical for wind energy systems is caused by electrostatic charging of the components of these systems. Specifically large rotor blades, which are made from composites, build up charges while in relative motion with air. There is not an easy way to discharge the blades during that process. Eventually a source of interference is created, which subsequently produces flash-over discharges (ESD) to various parts of the wind turbine, especially those with metallic sharp edges. ESD processes can be damaging to nearby electronics.

4. EMC for large PV systems

In the last couple of decades, large PV systems have shown an exponential growth. Table 1 shows the currently largest solar parks in the world. Yet, the table in a few years may be very different, as every year new members will enter to this table. In fact some forecast gives a total worldwide PV installed capacity of 500 GW by 2020. Therefore, as in the case of wind energy systems, since PV is becoming such a relevant energy source, related EMC considerations are becoming increasingly relevant especially for large PV plants, connected to the grid.

Table 1. Largest solar parks in the world.

Name	Country	Capacity (MW)	Size (km^2)
Tengger Desert Solar Park	China	1,547	43
Datong Solar Power Top Runner Base	China	1,000	30
Kurnool Ultra Mega Solar Park	India	1,000	24
Longyangxia Dam Solar Park	China	850	23
Bhadla Solar Park	India	746	40
Kamuthi Solar Power Project	India	648	10.1
Pavagada Solar Park	India	600	53
Solar Star (I and II)	US	579	13
Topaz Solar Farm	US	550	24.6
Desert Sunlight Solar Farm	US	550	16

IOP Conf. Series: Journal of Physics: Conf. Series **1143** (2018) 012030 doi:10.1088/1742-6596/1143/1/012030

The EMI/EMC issues are a bit different in PV systems (PVs) than in wind turbine systems (WTs). Since the height of PVs is not in the order of ~hundred meters, but just a few meters. Large PVs are not prone to lightning in the same way than WTs. In WTs, the highest top of the blade will attract lightning while in large PVs, lightning "sees" a large mesh of long conductors connecting the large set of modular panels. The lightning protection in WTs can be realized with dedicated receptors in the blades and a robust earthing mesh. In PVs this could be realized with lightning rods (with much simpler earth connections than in the WTs). Still, the emitted impulse field generated when the lightning strike goes through the rod is large and its effect on electronic equipment has to be consider at the design stage. The DC/AC inverters are also a major source of conductive interference being able to inject disturbances into the power grid. The EU directive and a few standards for PVs already exists, such as IEC 62305, while integration into the grid is causing several other standards to be considered and cross referred (e.g. IEC 62561, CLC 50539).

In terms of emission, in large PV parks, long cables act as antennas propagating coupled high frequencies (e.g. RF) from inverters. Thus radiated interference can reach as far as several km distance. Thus airports, radio observatories, base stations and radar sites could be affected by these RF emissions.

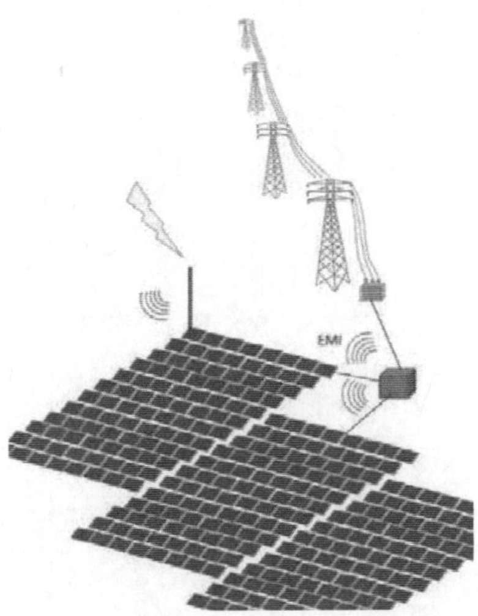

Figure 3. Two major sources of EMI in large PV systems (the system is connected to the grid). One is lightning and the other is the inverter (in red), which is a known source of high frequency stray fields, being able to inject disturbances not only in the PV system but also into the power grid.

5. Geomagnetically induced currents (GIC)

One of the main characteristic of our sun is its activity, which include solar storms related to coronal mass ejections (CMEs) that are huge and highly energetic plasma expulsions from the sun. Figure 4 shows one detected in 2003. From time to time these ejections are able to reach earth going around the magnetosphere and penetrating through the Polar Regions to form in the ionosphere quasistatic currents of hundreds of km long and with an intensity of ~1 million amperes known as electro-jets. These intense currents couple with long conductors in the ground producing GICs which are able to penetrate the grid causing transformer core saturation and serious damage to the power grid in the worst case producing blackout. The most severe CME registered since the use of electricity is the *Carrington event* in Sept 1, 1859 which managed to overheat and destroy some telegraph lines and explode some powering batteries. The second largest event in recent times happened in March 1989 when GICs managed to knock out the power grid causing hundreds of millions in damage.

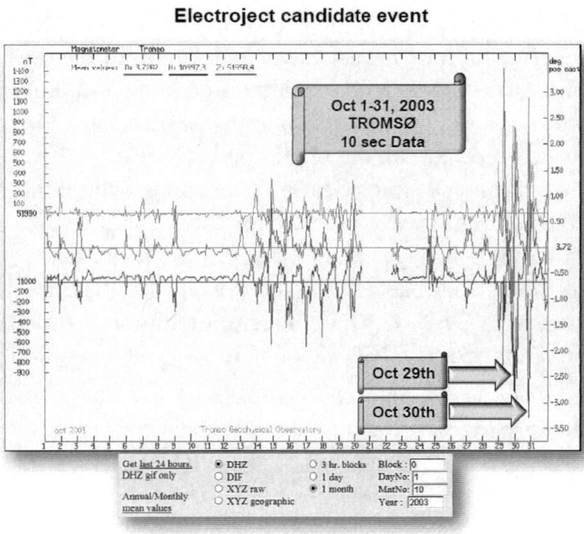

Figure 4. Registration of a CME on October 29-30, 2003 by the geo-magnetic station located in Tromso, Norway. This event is also known as the *Halloween storm*. It caused a power blackout in the city of Malmö in Sweden.

The effects of these natural phenomena can be so severe on the power grid, and therefore affecting our entire society, that serious investigations are currently being performed by various organizations [6, 9].

6. Conclusion

EMC for both, WTs and PVs, present some similarities. Being lightning and the DC/AC converters the most relevant sources of EMI. On the other hand, their electromagnetic environments are rather different, especially in relation with lightning protection. GIC is an extremely relevant topic for the reliability of the power grid and more efforts should be placed for finding mitigation techniques and devices.

Acknowledgements

The author thanks his colleagues in various CIGRE working groups for valuable discussions on the topics presented in this paper.

References

[1] Francesco Lattarulo, "*Electromagnetic Compatibility in Power Systems*", Elsevier Science Ltd. 2006

[2] David Weston, "*EMC – Methods, Analysis, Circuits, and Measurement*", CRC Press, (3rd edition) 2017

[3] CIGRE TB 320, "*Characterization of ELF Magnetic Fields*", 2007

[4] CIGRE TB 535, "*EMC in Power plants and substations*", 2013

[5] CIGRE WG C4.30, "*EMC in Wind Power Systems*", 2017

[6] CIGRE WG C4.32, "Understanding Geomagnetically Induced Currents" 2018

[7] CIGRE WG C4.44, "*EMC in Large Photovoltaic Systems*", stablished in 2017

[8] CIGRE WG C4.206, "*Protection of High Voltage power network control electronics against Intentional Electromagnetic Interference (IEMI)*", 2015

[9] ABB Press Release, "*ABB engineering protects power plant from solar storm*" 20016-10-24 http://www.abb.com/cawp/seitp202/c99eb3b89c85b7b2c12571c6004579aa.aspx

IOP Conf. Series: Journal of Physics: Conf. Series **1143** (2018) 012031 doi:10.1088/1742-6596/1143/1/012031

Improved variational method that solves the energy eigenvalue problem of the hydrogen atom

L M Leon Hilario[1] and E Aldo Arroyo[2]

[1] Facultad de Ciencias, Universidad Nacional de Ingenieria, Apartado 31-139, Lima, Peru
[2] Centro de Ciências Naturais e Humanas, Universidade Federal do ABC, Santo André, 09210-170 São Paulo, SP, Brazil

E-mail: mleon@uni.edu.pe

Abstract. In most quantum mechanics textbooks for graduate studies, the hydrogen atom is studied in an approximate way by means of the variational method. The type of trial functions commonly used are the Gaussian and the Lorenzian. In this paper we consider a natural generalization of the Mei method to improve the trial wave functions [1] applied to hydrogen atom. We propose a sequence of functions as trial wave functions to calculate the eigen-energies of the hydrogen atom. These trial wavefunctions are given in terms of three variational parameters, one of them is fixed by means of the normalization condition and the other two are adjustable parameters. One of these parameters can be chosen as being an integer number, then this parameters will define the sequence of functions. We will show that when the integer parameter approaches to infinity, the ground state, first excited state and second excited state converge to the exact results.

1. Introduction

The hydrogen atom is the simplest atom which consists of a bound state of a proton and an electron. This simpler quantum mechanical system has been used as test-bed for new theories and researches. For instance, transitions in hydrogen are being used to examine in a more accurate way the fine structure constant over a cosmological time scale [2], and more recently, using the variational approach to the determination of the spectral of the hydrogen atom [3], the celebrated Wallis product formula for π has been calculated.

Although the conceptual simplicity of the hydrogen atom to analytically calculate its energy eigenvalues, we need mathematical tools such as partial differential equations in spherical coordinates, the theory of angular momentum, and the introduction of some special functions as the associated Laguerre polynomials. Let us remark that the hydrogen atom is actually one of few quantum mechanical systems whose energy levels can be obtained exactly.

Generally, and not only in quantum mechanics, most of the problems cannot be solved analytically, and hence need to be dealt with numerical or other alternative approximation. In quantum mechanics, the most used methods to approximate solutions are perturbation theory and the variational method. Perturbation theory is useful when there is a small dimensionless parameter in the problem, and the system is exactly solvable when this parameter approaches zero. On the other hand, the variational method allow us to calculate approximations to the lowest energy eigenstate or ground state in a quantum system.

IOP Conf. Series: Journal of Physics: Conf. Series **1143** (2018) 012031 doi:10.1088/1742-6596/1143/1/012031

The variational method consists of choosing a trial wave function that depends on one or more parameters and finding the values of these parameters such that the expectation value of the Hamiltonian operator which defines the energy as the lowest possible [5]. By finding a suitable trial wave function orthogonal to the ground state, an upper bound state on the first excited state can be calculated, then, by computing a set of orthogonal trial wave functions, it is also possible to determine approximately the excited states.

Regarding to the variational method applied to the case of the hydrogen atom, as far as we know, the proposed trial wave functions provide results that have an error close to 10% when compared to the exact answer. For instance, the most popular types of trial wave functions are the gaussians and the lorentziants, which lead to error about of 15%, and 10% respectively for the ground state energy [4, 5, 6].

One way to improve the errors is by adding more parameters to the trial wave functions, such as the Hylleras method for the He atom [7]. In using this approach, calculations soon become very lengthy and tedious. Another and more effective way to improve the trial functions was performed by Mei [1] for the harmonic oscillator problem. He concludes that introducing an integer parameter that determines the asymptotic behavior of the trial function, he is able to show that the energy of the ground state and the first excited state converge to the exact results.

Motivated by the work of Mei, in this article we propose a general trial wave functions for the hydrogen atom that can be indexed by a integer number. This index can be identified as being one of the variational parameters. We are going to show that when this positive integer parameter approaches to infinity, the analytical solutions of the energy eigenvalue problem are recovered.

The paper is organized as follows. In Section 2 we define our trial wave function and the variational parameters. In Section 3 and 4, we apply the variational method to calculate the ground state energy and the ground state wave function respectively. In Section 5 we calculate the energy eigenvalue and wave function for the first exited state. In Section 6 we calculate higher level energies. We end with some concluding remarks summarizing the main results.

2. The ground state trial wave function

Since the energy of the hydrogen atom does not depend on the angular momentum quantum number l, it is possible to use trial wave functions that only depend on the radial coordinate, namely, states with $l = 0$. Furthermore, to take into account the asymptotic behavior of the trial wave function, we are going to choose the following trial wave function for the ground state:

$$\Psi_1 = \frac{C_1}{(A+r)^n}, \tag{1}$$

where $C_1 > 0$, $A > 0$ are real parameters and n is an integer number $n > 0$. From the normalization condition for the wave function, we obtain

$$4\pi C_1^2 \int_0^\infty \frac{r^2}{(A+r)^{2n}}\, dr \quad = \quad 1. \tag{2}$$

To perform the radial integral, we are going to employ the following very useful integral identity:

$$\int_0^\infty \frac{r^M}{(A+r)^N}\, dr = \frac{M!(N-M-2)!A^{M-N+1}}{(N-1)!}, \tag{3}$$

where in order to the integral be a convergent one, we must have $-1 < M < N-1$.

Using equation (3) and equation (2), we have

$$8\pi C_1^2 A^{3-2n} \frac{(2n-4)!}{(2n-1)!} \quad = \quad 1, \tag{4}$$

so that we get

$$C_1^2 = \frac{A^{2n-3}(2n-1)(2n-2)(2n-3)}{8\pi}. \tag{5}$$

This result shows that the parameter C_1 depends on the parameter A. Thus, our trial function will depend only on parameters A and n. We will take A as a variational parameter and n as an index that defines the sequence of functions.

3. The ground state energy by means of a variational calculation

The next step is to calculate the expected value of the Hamiltonian. Using spherical coordinates, the Hamiltonian of the hydrogen atom can be written as [5]

$$H \;=\; H_r - \frac{\hbar}{2\mu}\Big[\frac{1}{r^2\sin\theta}\frac{\partial}{\partial\theta}\left(\sin\theta\frac{\partial}{\partial\theta}\right) + \frac{1}{r^2\sin^2\theta}\frac{\partial^2}{\partial\phi^2}\Big], \tag{6}$$

where μ is the reduced mass and H_r is the radial part of the Hamiltonian given by

$$H_r \equiv -\frac{\hbar}{2\mu r^2}\frac{\partial}{\partial r}\left(r^2\frac{\partial}{\partial r}\right) - \frac{e^2}{4\pi\epsilon_0\, r}. \tag{7}$$

Using the trial wave function given by equation (1) and the expression for C_1 in equation (5), we can compute the energy associated to this state Ψ_1

$$E_1(A) \equiv \langle H\rangle_{\Psi_1} = 4\pi C_1^2 \int_0^\infty \frac{r^2}{(A+r)^n}\, H_r\Big(\frac{1}{(A+r)^n}\Big)\, dr. \tag{8}$$

Performing the integration, we find

$$E_1(A) = 4\pi C_1^2\Big[-\frac{\hbar(n+1)A^{-2n+1}}{2\mu(2n+1)(2n-1)} + \frac{\hbar A^{-2n+1}}{2\mu(2n-1)} - \frac{e^2 A^{-2n+2}}{8\pi\epsilon_0(2n-1)(n-1)}\Big]. \tag{9}$$

Replacing the expression of the parameter C_1 which is given in equation (5) into this equation (9), we obtain

$$E_1(A) = \frac{\hbar n(n-1)(2n-3)}{2\mu(2n+1)\,A^2} - \frac{e^2(2n-3)}{8\pi\epsilon_0\, A}. \tag{10}$$

We now compute the derivative respect to the parameter A to find the minimum value of $E_1(A)$:

$$\frac{dE_1}{dA} = -\frac{\hbar n(n-1)(2n-3)}{\mu(2n+1)\,A^3} + \frac{e^2(2n-3)}{8\pi\epsilon_0\, A^2} = 0. \tag{11}$$

The solution to this equation (11) is given by

$$A_{min} = \frac{8\pi\epsilon_0 \hbar n(n-1)}{e^2\mu(2n+1)}. \tag{12}$$

Therefore, using equation (10), the energy evaluated at this point A_{min} is given by

$$E_1(A_{min}) = -\frac{e^4\mu\,(2n-3)(2n+1)}{128\pi^2\hbar\epsilon_0^2\, n(n-1)}. \tag{13}$$

IOP Conf. Series: Journal of Physics: Conf. Series **1143** (2018) 012031 doi:10.1088/1742-6596/1143/1/012031

Note that taking the limit to $n \to \infty$, the energy of the ground state tends to the exact answer:

$$\lim_{n\to\infty} E_1(A_{min}) = \lim_{n\to\infty} \Big[-\frac{e^4\mu\,(2n-3)(2n+1)}{128\pi^2\hbar^2\epsilon_0^2\, n(n-1)} \Big] = -\frac{e^4\mu}{32\pi^2\hbar^2\epsilon_0^2} = -\frac{e^4\mu}{2(4\pi\epsilon_0)^2\hbar^2}. \tag{14}$$

Just as a matter of comparison, let us write the exact analytical eigenvalue for the energy of the hydrogen atom obtained by solving explicitly the differential equation [5]

$$E_N = -\frac{e^4\mu}{2(4\pi\epsilon_0)^2\hbar^2}\frac{1}{N^2}, \qquad N = 1, 2, 3, \cdots. \tag{15}$$

So in fact, the result (14) corresponds to the energy for the ground state $N = 1$.

4. The ground state wave function

To simplify the expressions, let us write the quantities in terms of the Bohr radius $a = \frac{4\pi\epsilon_0\hbar^2}{\mu e^2}$. For instance, the expression for the A_{min} given in equation (12) in terms of the Bohr radius reduces to

$$A_{min} = \frac{2a(n-1)n}{2n+1}. \tag{16}$$

Now using equation (5), we can obtain the expression for the parameter C_1 evaluated at the minimum value

$$C_{1,min} = \frac{A_{min}^{(2n-3)/2}\sqrt{(2n-1)(2n-2)(2n-3)}}{2\sqrt{2\pi}}. \tag{17}$$

Therefore, the trial wave function (1) evaluated at this minimum value is given by

$$\Psi_{1,min}(r) = \frac{C_{1,min}}{(A_{min}+r)^n} = \sqrt{\frac{(n-1)(2n-3)(2n-1)}{4\pi a^3 n^3}} \left(\frac{2n-2}{2n+1}\right)^{n-\frac{3}{2}} \left(\frac{2n-2}{2n+1} + \frac{r}{an}\right)^{-n}. \tag{18}$$

In order to compare the wave function obtained with the well known analytical expression for the ground state wave function of the hydrogen atom ($\Psi_{100}(r) = \frac{e^{-\frac{r}{a}}}{\sqrt{\pi}a^{3/2}}$) we define the function $\Psi_{1,n}(r) \equiv \frac{r^2}{\sqrt{a}}\Psi_{1,min}(r)$.

Figure 1 shows the behavior of $\frac{r^2}{\sqrt{a}}\Psi_{100}$ and $\Psi_{1,n}$ as a function of r/a. We observed that the function $\Psi_{1,n}(r)$ is approaching to the analytical solution $\frac{r^2}{\sqrt{a}}\Psi_{100}$ (solid line) when we increase the value of n. Remarkably, in the limiting case of $n \to \infty$ both functions are identical.

For completeness, we also have calculated the limit of (18) as $n \to \infty$ analytically:

$$\begin{aligned}\lim_{n\to\infty} \Psi_{1,min}(r) &= \lim_{n\to\infty} \sqrt{\tfrac{(n-1)(2n-3)(2n-1)}{4\pi a^3 n^3}} \left(\tfrac{2n-2}{2n+1}\right)^{n-\frac{3}{2}} \left(\tfrac{2n-2}{2n+1} + \tfrac{r}{an}\right)^{-n} \\ &= \tfrac{1}{a^{3/2}\sqrt{\pi}} e^{-\frac{r}{a}}.\end{aligned} \tag{19}$$

Where we have used the well know limit

$$\lim_{n\to\infty}\left(1+\frac{z}{n}\right)^n = e^z.$$

In that way, the solution converges to the analytical expression for the ground state.

Our next task will be the derivation of the energy for the first excited state, namely with $N = 2$ by means of the variational method and using a clever trial function just like the one we have done in the case of the ground state.

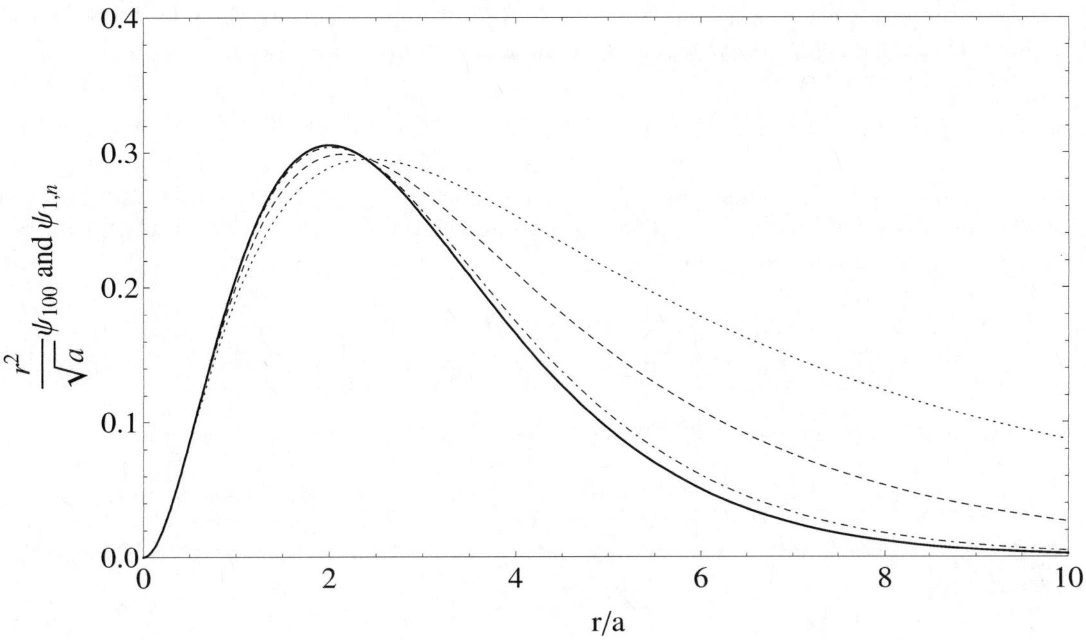

Figure 1. Plot of $\Psi_{1,n}(r)$ for values of $n = 5, 10, 50$. The solid line represents the analytical function $\frac{r^2}{\sqrt{a}}\Psi_{100}(r)$, while $\Psi_{1,5}(r)$, $\Psi_{1,10}(r)$ and $\Psi_{1,50}(r)$ are given by the dotted, dashed and dotted-dashed curves respectively.

5. The case of the first excited state

In this section we are going to calculate the energy eigenvalue and wave function of the first excited state and compare the result with the analytic value of the energy (15) for the case $N = 2$.

From the variational method we know that the trial function of the first excited state must be orthogonal to the trial function ground state Ψ_1 (1). Thus we are looking for a Ψ_2 such that

$$\langle \Psi_1, \Psi_2 \rangle \equiv \int \Psi_1^* \Psi_2 \, r^2 \sin\theta dr d\theta d\phi = 0. \tag{20}$$

This Ψ_2 which is in agreement with the condition (20) is given by

$$\Psi_2 = C_2 (1 + \frac{8 - 3n}{6A} r) \frac{1}{(A + r)^{n/2}}. \tag{21}$$

In order for the radial part of the integral (20) to be a convergent one, the value of the integer n must be such that $n > 2$.

By using the normalization condition for Ψ_2, we are going to determine the coefficient $C_2 > 0$ in terms of the parameter A. Then, from

$$\int |\Psi_2|^2 \, r^2 \sin\theta dr d\theta d\phi = 1, \tag{22}$$

we get

$$C_2^2 = \frac{3(n-5)(n-4)(n-3)(n-2)(n-1)A^{n-3}}{8\pi(3(n-2)n + 4)}. \tag{23}$$

IOP Conf. Series: Journal of Physics: Conf. Series **1143** (2018) 012031 doi:10.1088/1742-6596/1143/1/012031

Next, we compute the expectation value of the Hamiltonian (6) using the trial wave function (21). Since the function (21) only depends on the radial coordinate, we have

$$\langle H\rangle_{\Psi_2} = \int (\Psi_2^* H \Psi_2)\, r^2 \sin\theta dr d\theta d\phi = 4\pi \int_0^\infty r^2\, \Psi_2^* H_r \Psi_2\, dr. \tag{24}$$

Replacing the trial wave function (21) into equation (24), and performing computations similar to the ones given in the case of the ground state trial function (1) and using the result of the integral (3), we can show that

$$E_2(A) \equiv \langle H\rangle_{\Psi_2} = \frac{C_2^2 A^{1-n}\left(3Ae^2\mu\left(-3n^2+n+4\right) + 2\pi\hbar(n-4)(3n(3n-8)-8)\epsilon_0\right)}{18\mu(n-4)(n-3)(n-1)(n+1)\epsilon_0}. \tag{25}$$

Now, replacing equation (23) into equation (25), we get

$$E_2(A) = \frac{\hbar(n-5)(n-4)(n-2)(3n(3n-8)-8)}{24A^2\mu(n+1)(3(n-2)n+4)} - \frac{e^2(n-5)(n-2)(3n-4)}{16\pi A\left(3n^2-6n+4\right)\epsilon_0}. \tag{26}$$

Thus computing the minimum value by means of the following equation

$$\left.\frac{dE_2}{dA}\right|_{A=A_{min}} = 0, \tag{27}$$

we obtain

$$A_{min} = \frac{4\pi\hbar(n-4)(3n(3n-8)-8)\epsilon_0}{3e^2\mu(n+1)(3n-4)}. \tag{28}$$

Evaluating the energy (26) at the minimum value (28), we arrive to

$$E_2(A_{min}) = -\frac{3e^4\mu(4-3n)^2(n-5)(n-2)(n+1)}{128\pi^2\hbar^2(n-4)(3(n-2)n+4)(3n(3n-8)-8)\epsilon_0^2}. \tag{29}$$

Taking the limit $n\to\infty$, we obtain

$$\lim_{n\to\infty} E_2(A_{min}) = -\frac{e^4\mu}{128\pi^2\hbar\epsilon_0^2} = -\frac{e^4\mu}{2(4\pi\epsilon_0)^2\hbar}\frac{1}{2^2}. \tag{30}$$

Therefore, as in the case of the ground state, surprisingly, when we take the limit $n\to\infty$, we arrive to the correct analytical result (15) which corresponds to the value of the first exited state.

Now let us analyze the wave function. Using equations (23) and (28) into equation (21), we have

$$\begin{aligned}\Psi_{2,min}(r) &= C_{2,min}\left(1+\tfrac{8-3n}{6A_{min}}r\right)\tfrac{1}{(A_{min}+r)^{n/2}}\\ &= \frac{9\sqrt{\frac{(n-5)(n-3)(n-2)(n-1)(n+1)^3(3n-4)^3\left(1+\frac{3(n+1)(3n-4)r}{a(n-4)(3n(3n-8)-8)}\right)^{-n}}{a^3(n-4)^2(3(n-2)n+4)(3n(3n-8)-8)^3}}\left(1+\frac{(8-3n)(n+1)(3n-4)r}{2a(n-4)(3n(3n-8)-8)}\right)}{2\sqrt{2\pi}}.\end{aligned} \tag{31}$$

Performing the limit $n\to\infty$, we can show that

$$\lim_{n\to\infty}\Psi_{2,min}(r) = \frac{1}{2a^{3/2}\sqrt{2\pi}}\left(1-\frac{r}{2a}\right)e^{-\frac{r}{2a}}. \tag{32}$$

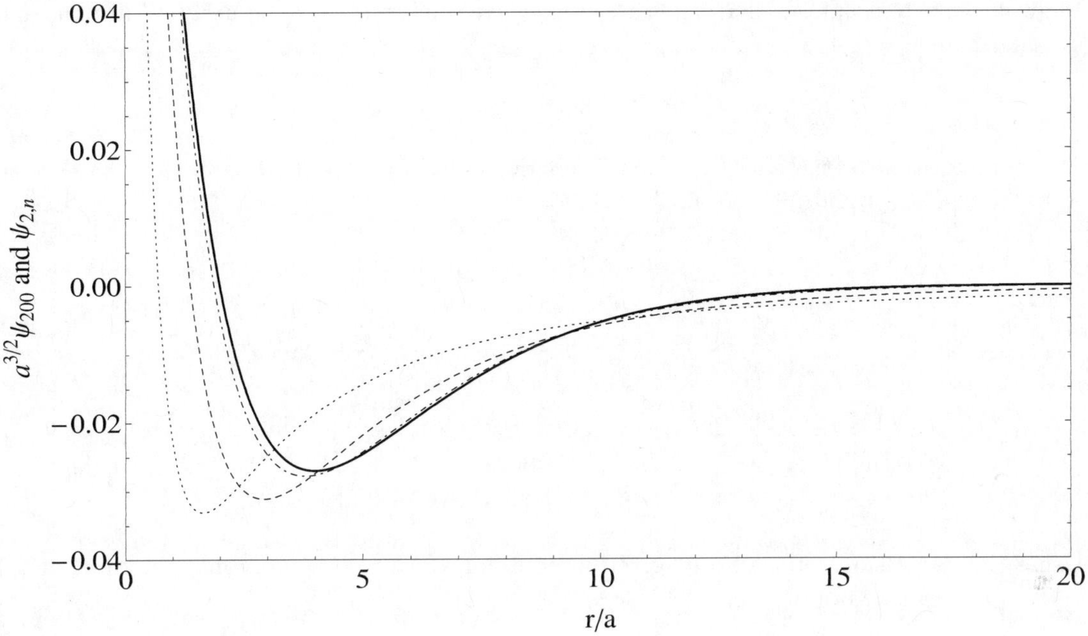

Figure 2. Plot of $a^{3/2}\Psi_{200}(r)$ and $\Psi_{2,n}(r) \equiv a^{3/2}\Psi_{2,min}(r)$ for values of $n = 6, 12, 60$. The continuous curve represents the analytical function $a^{3/2}\Psi_{200}(r)$, while $\Psi_{2,6}(r)$, $\Psi_{2,12}(r)$ and $\Psi_{2,60}(r)$ are given by the dotted, dashed and dotted-dashed curves respectively.

In standard textbooks, we can find the analytical expression for the wave function corresponding to the first exited stated (with $l = 0$)

$$\Psi_{200}(r) = \frac{1}{2a^{3/2}\sqrt{2\pi}}(1 - \frac{r}{2a})e^{-\frac{r}{2a}}. \quad (33)$$

Thus, as we can see, the limit $n \to \infty$ converges to the right answer.

In figure 2 we plot the functions $a^{3/2}\Psi_{200}(r)$ and $\Psi_{n,2}(r) \equiv a^{3/2}\Psi_{2,min}(r)$ for values of $n = 6, 12, 60$. We can see, as in the case of the ground state, when we increment the value of n the function $\Psi_{n,2}(r)$ is approaching to the analytical solution (continuous curve).

6. Results for higher level energies

Since using the variational method, the computations for higher level energies are almost the same as for the ground state and the first excited state, here we are only going to show the results.

For the second excited stated, we have used the trial wave function

$$\Psi_3 = C_3(1 + \frac{q_1}{A}r + \frac{q_2}{A^2}r^2)\frac{1}{(A+r)^{n/3}}, \quad (34)$$

where

$$q_1 = \frac{1440 - 2n(n(135n - 814) + 1152)}{9(5n(9n - 14) + 48)}, \qquad q_2 = \frac{(4n - 15)(5n - 36)(n(9n - 16) + 12)}{54(5n(9n - 14) + 48)}. \quad (35)$$

IOP Conf. Series: Journal of Physics: Conf. Series **1143** (2018) 012031 doi:10.1088/1742-6596/1143/1/012031

We would like to point out that, the three trial wave functions (1), (21) and (34) are mutually orthogonal, namely

$$\langle \Psi_1, \Psi_2 \rangle = \langle \Psi_1, \Psi_3 \rangle = \langle \Psi_2, \Psi_3 \rangle = 0. \tag{36}$$

The computation of the energy by means of the variational method associated to the trial wave function (34) provides the result

$$\begin{aligned} E_3 = & -(e^4(-21+2n)(3+2n) \\ & ((-9+n)(-15+2n)(2n^2+3n(-7+4q_1)+9(6+q_1(-8+3q_1)))+ \\ & 54(-9+n)(-15+2n+12q_1)q_2+2430q_2^2)^2\mu)/ \\ & (128\hbar^2(-9+n)(-6+n)\pi^2 \\ & ((-3+n)(-15+2n)(2n^2+9q_1(-3+2q_1)+n(-9+6q_1))+162(-15+2n)(-1+q_1)q_2+ \\ & 1458q_2^2)((-9+n)(-21+2n)(2n^2+9n(-3+2q_1)+9(10+3q_1(-5+2q_1)))+ \\ & 54(-21+2n)(-18+2n+15q_1)q_2+7290q_2^2)\epsilon_0^2). \end{aligned} \tag{37}$$

Using the expressions of q_1 and q_2 given in equation (35), we can compute the limit of (37) when $n \to \infty$, and we obtain

$$\lim_{n\to\infty} E_3 = -\frac{e^4\mu}{288\pi^2\hbar^2\epsilon_0^2} = -\frac{e^4\mu}{2(4\pi\epsilon_0)^2\hbar^2}\frac{1}{3^2}. \tag{38}$$

Therefore, as in the case of the first exited state, when we take the limit $n \to \infty$, we arrive to the correct analytical result (15) which corresponds to the value of the second excited state.

7. Summary and discussion

We have proposed a general trial function in the form of a sequence of functions to calculate the eigen-energies of the hydrogen atom. The most important advantage in comparison with previous trial functions [5, 6] is the inclusion of one integer variational parameter that defines the sequence of functions. The inclusion of this parameter has allowed us to show that when this parameter approaches to infinity we are able to reproduce the exact analytical solution of the hydrogen atom without the need to solve the differential equation and work with complicated special functions.

This type of trial functions are not only useful to find the energies of the ground state, these can also be employed to derive the energies and wave functions of excited states.

The relevance of this method lies in the possibility of including the analysis of multi-electronic atoms as well as systems like molecules. Therefore, the natural next step will be to apply this technique to more complex quantum mechanical systems.

References

[1] Mei W N 1996 *Int. J. Educ. Sci. Tech.* **27** 285
[2] Schwarzschild M 1997 *Physics Today* **50**(10) 19-21
[3] Friedmann T and Hagen C R 2015 *Journal of Mathematical Physics* **56** 112101.
[4] Khon W 1947 *Phys. Rev.* **71** 635
[5] Griffiths D J 2005 *Introduction to Quantum Mechanics* (New Jersey: Prentice Hall)
[6] Sommerfeld T 2011 *J.Chem. Educ.* **88** 1521-1524
[7] Bethe H A and Salpeter E E 1957 *Quantum Mechanics of One and Two Electron Atoms* (New York: Academic Press) **p. 150**.